CHROMATIC POLYNOMIALS
AND
CHROMATICITY OF GRAPHS

CHROMATIC POLYNOMIALS
A N D
CHROMATICITY OF GRAPHS

by

F M Dong
Nanyang Technological University, Singapore

K M Koh
National University of Singapore, Singapore

K L Teo
Massey University, New Zealand

World Scientific

NEW JERSEY • LONDON • SINGAPORE • BEIJING • SHANGHAI • HONG KONG • TAIPEI • CHENNAI

Published by

World Scientific Publishing Co. Pte. Ltd.

5 Toh Tuck Link, Singapore 596224

USA office: 27 Warren Street, Suite 401-402, Hackensack, NJ 07601

UK office: 57 Shelton Street, Covent Garden, London WC2H 9HE

British Library Cataloguing-in-Publication Data
A catalogue record for this book is available from the British Library.

CHROMATIC POLYNOMIALS AND CHROMATICITY OF GRAPHS

ISBN-13 978-981-256-317-0
ISBN-10 981-256-317-2
ISBN-13 978-981-256-383-5 (pbk)
ISBN-10 981-256-383-0 (pbk)

Printed in Singapore

Preface

For a century, one of the most famous problems in mathematics was to prove the four-colour theorem. This has spawned the development of many useful tools for solving graph colouring problems. In a paper in 1912, Birkhoff proposed a way of tackling the four-colour problem by introducing a function $P(M, \lambda)$, defined for all positive integers λ, to be the number of proper λ-colourings of a map M. It turns out that $P(M, \lambda)$ is a polynomial in λ, called the *chromatic polynomial* of M. If one could prove that $P(M, 4) > 0$ for all maps M, then this would give a positive answer to the four-colour problem. The polynomial $P(M, \lambda)$ is defined for all real and complex values of λ. It was hoped that many useful tools from algebra and analysis could be used to find or estimate the roots of the polynomial and hence lead to the resolution of the problem.

The notion of a chromatic polynomial was later generalized to that of an arbitrary graph by Whitney (1932), who established many fundamental results for it. In 1946, Birkhoff and Lewis obtained results concerning the distribution of real roots of chromatic polynomials of planar graphs and conjectured that these polynomials have no real roots greater than or equal to four. The conjecture remains open.

In 1968, Read aroused new interest in the study of chromatic polynomials with his well referenced introductory article on the subject. He asked if it is possible to find a set of necessary and sufficient algebraic conditions for a polynomial to be the chromatic polynomial of some graph. For example, it is true that the chromatic polynomial of a graph determines the numbers of vertices and edges and that its coefficients are integers which alternate in sign. Read observed that the absolute values of the coefficients appear to form a unimodal sequence.

Read asked: What is a necessary and sufficient condition for two graphs to be chromatically equivalent; that is, to have the same chromatic polynomial? In particular, Chao and Whitehead Jr. (1978) defined a graph to be chromatically unique if no other graphs share its chromatic polynomial. They found several families of such graphs. Since then many invariants under chromatic equivalence have been found and various families of and results on such graphs have been obtained successively. The question of chromatic equivalence and uniqueness is termed the chromaticity of graphs. This remains an active area of research.

Although Birkhoff's hope of using the chromatic polynomial to prove the four-colour theorem was not borne out, it has attracted a steady stream of attention through the years, especially concerning the location of its roots. More recently, Thomassen discovered a relation between hamiltonian paths and the roots of the chromatic polynomial. There has also been an influx of new ideas from statistical mechanics due to the recent discovery of a connection to the Potts Model in Physics.

This book is divided into three main parts, after providing a chapter on the basic concepts and terminology of graphs and a list of notation that are needed and used in the book. Part one covers the first three chapters. It is devoted in greater detail than the other two to the rudiment of chromatic polynomials; their basic properties are derived, and some practical methods for computing them are given. Furthermore, we provide several ways of constructing chromatically equivalent graphs; characterize chromatically unique graphs that are disconnected and those with connectivity 1. Further results on chromatic equivalence classes of families of graphs are mentioned.

Part two, which consists of eight chapters from Chapter 4 to Chapter 11, deals specifically with the chromaticity of multi-partite graphs, subdivisions of graphs, and members of those families whose colour classes have nice structures. By expanding a chromatic polynomial of a graph in terms of falling factorials, we construct a polynomial, called the adjoint polynomial of the graph. We study several invariants of this polynomial and roots of some particular ones. It was found that this polynomial was particularly useful in determining the chromaticity of graphs whose complements are of simpler structure. We also mention some related polynomials.

The last part of the book covers the last four chapters and is concerned with the distribution of roots of the chromatic polynomials both on the real line and in the complex plane. In particular, we study those chromatic polynomials that possess only integral roots. Furthermore, we study bounds and inequalities of the chromatic polynomials of families of graphs.

Acknowledgement

The authors would like to express their appreciation to several people and institutions for their assistance in writing this book. Special thanks to reviewers Gek-Ling Chia, Ru-Ying Liu, Yee-Hock Peng, Ronald Read, Chin-Ann Soh, Alan Sokal, Chung-Piaw Teo, Ioan Tomescu and Hai-Xing Zhao for making corrections and providing valuable suggestions for improvement. They thank Debbie Ormsby of Massey University, New Zealand, for her help in preparing part of the manuscript and drafting the figures. KMK wishes to thank the Mathematics Departments of Simon Fraser University, Canada, and Shanghai Jiao Tong University, China for their hospitality during his sabbatical leave in 2002. KLT is grateful for the financial support rendered by National University of Singapore towards this project, and valuable assistance given by his colleague Charles Little.

F.M. Dong, K.M. Koh and K.L. Teo
Singapore
December, 2004

Contents

Basic Concepts in Graph Theory

1. Graphs and adjacency

A (*simple*) *graph* $G = (V(G), E(G))$ consists of a non-empty finite set $V(G)$ together with a (possibly empty) set $E(G)$ of unordered pairs of distinct elements of $V(G)$. Each element in $V(G)$ is called a *vertex* of G, and each member $\{u, v\}$ in $E(G)$, where $u, v \in V(G)$, is called an *edge* of G, which is often denoted by uv or vu. The sets $V(G)$ and $E(G)$ are called the *vertex set* and *edge set* of G respectively. The symbols $v(G)$ and $e(G)$ are used to denote the number of vertices and the number of edges in G respectively. A graph G is said to be of *order* n if $v(G) = n$ and of *size* m if $e(G) = m$. A graph of order n and size m is also called an (n, m)-*graph*. A graph G is said to be *trivial* if $v(G) = 1$, and non-trivial otherwise.

For instance, the graph H in Figure 1 is a $(7, 11)$-graph with $V(H) = \{a, b, c, w, x, y, z\}$ and $E(H) = \{ac, ax, aw, bw, bx, by, bz, cy, wx, wy, xy\}$.

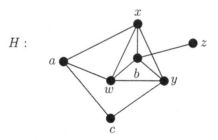

Figure 1

Remark By definition, any two distinct vertices of a (simple) graph are joined by at most one edge. If more than one edge is allowed to join some pairs of vertices, the resulting structure is called a *multigraph*. Unless otherwise stated, all graphs we consider in this book are simple graphs.

Let G be a graph. Two distinct vertices u and v in G are *adjacent* if uv is an edge of G. In this case, u is also called a *neighbour* of v, and vice versa. If $e = uv$ is an edge of G, we say that the edge e joins u and v, or the vertices u and v are the two *ends* of e. In this case, we also say that e is *incident* with u and v.

Let $A \subseteq V(G)$. Then A is called a *clique* in G if every two vertices in A are adjacent. On the other hand, A is called an *independent set* in G if no two vertices in A are adjacent in G. Thus for the graph H in Figure 1, the set $\{a, x, w\}$ is a clique while $\{a, b\}$ is an independent set. The *clique number* of G, denoted by $\omega(G)$, and the *independence number* of G, denoted by $\alpha(G)$, are defined as follows:

$$\omega(G) = \max\{|A| : A \text{ is a clique in } G\},$$
$$\alpha(G) = \max\{|A| : A \text{ is an independent set in } G\}.$$

Thus, in the graph H in Figure 1, we have $\omega(H) = 4$ (the largest clique being $\{b, w, x, y\}$) and $\alpha(H) = 3$ (a largest independent set being $\{c, w, z\}$).

Two edges of G are said to be *adjacent* if they have an end vertex in common. Thus, in Figure 1, ax and xy are adjacent while ac and xy are not. A set M of edges is called a *matching* in G if no two edges in M are adjacent. The matching number of G, denoted by $\alpha'(G)$, is defined as follows:

$$\alpha'(G) = \max\{|M| : M \text{ is a matching in } G\}.$$

Thus, in Figure 1, we have $\alpha'(H) = 3$ (a largest matching being $\{ac, xy, bz\}$).

2. Degrees

For $u \in V(G)$, the *neighbourhood* of u, denoted by $N_G(u)$ (or simply $N(u)$), is the set of neighbours of u; i.e., $N(u) = \{v \in V(G) : uv \in E(G)\}$.

Let G be a multigraph and $v \in V(G)$. The *degree* of v in G, denoted by $d_G(v)$ (or simply by $d(v)$), is the number of edges of G incident with v. Thus, if G is simple, then $d(v) = |N(v)|$. A vertex v is said to be *isolated* if $d(v) = 0$; and is called an *end-vertex* if $d(v) = 1$. Assume that $V(G) = \{v_1, v_2, \cdots, v_n\}$. Then the sequence $(d(v_1), d(v_2), \cdots, d(v_n))$ is called a *degree sequence* of G. A simple but useful relation between the degrees of the vertices and the size of a multigraph is given below.

Hand-shaking Lemma   *Let G be a multigraph of size m. Then*

$$\sum_{v \in V(G)} d(v) = 2m.$$

A vertex v in a multigraph is said to be *odd* (resp., *even*) if $d(v)$ is odd (resp., even). The following result follows readily from the above lemma.

Corollary *The number of odd vertices in any multigraph is always even.*

Given a graph G, the *maximum degree* $\Delta(G)$ and the *minimum degree* $\delta(G)$ of G are defined, respectively, as follows:

$$\Delta(G) = \max\{d(v) : v \in V(G)\},$$
$$\delta(G) = \min\{d(v) : v \in V(G)\}.$$

Thus, for the graph H in Figure 1, we have $\Delta(H) = 4$ and $\delta(H) = 1$.

A graph G is said to be *regular* if $\Delta(G) = \delta(G)$. More precisely, G is *k-regular* if $\Delta(G) = k = \delta(G)$; i.e., $d(v) = k$ for each $v \in V(G)$. A 3-regular graph is also called a *cubic* graph.

3. Isomorphic graphs and subgraphs

Two graphs G and H are said to be *isomorphic*, in notation: $G \cong H$, if there exists a bijection $\varphi : V(G) \to V(H)$ which preserves adjacency; i.e., $uv \in E(G)$ if and only if $\varphi(u)\varphi(v) \in E(H)$. Such a bijection φ is called an *isomorphism* of G onto H. Any isomorphism of G onto itself is called an *automorphism* of G. A graph G is said to be *vertex-transitive* (resp., *edge-transitive*) if for any two vertices u and v (resp., two edges $e_1 = u_1 v_1$ and $e_2 = u_2 v_2$), there exists an automorphism φ of G such that $\varphi(u) = v$ (resp, $\varphi\{u_1, v_1\} = \{u_2, v_2\}$).

Let H and G be two graphs. A graph H is called a *subgraph* of G if $V(H) \subseteq V(G)$ and $E(H) \subseteq E(G)$; and a *proper subgraph* of G if, in addition, $H \neq G$. A subgraph H of G is said to be *spanning* if $V(H) = V(G)$. Let $A \subseteq V(G)$ and $A \neq \emptyset$. The subgraph of G *induced* by A, denoted by $G[A]$ or $[A]$, is the graph having the vertex set A and its edge set consisting of all the edges of G joining vertices in A. We call $G[A]$ *the subgraph of G induced by A*. Let $F \subseteq E(G)$ and $F \neq \emptyset$. The subgraph of G *induced* by F, denoted by $G[F]$ or $[F]$, is the graph with $E([F]) = F$ and

$$V([F]) = \{v \in V(G) : v \text{ is an end of some edge in } F\}.$$

Thus, in Figure 1, if $A = \{a, w, x, y\}$ and $F = \{aw, bw, by, bz\}$, then $[A]$ and $[F]$ are shown in Figure 2.

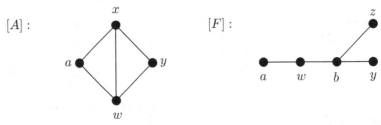

Figure 2

Let $A \subset V(G)$. The graph $G - A$ is the subgraph of G obtained by removing from G each vertex in A together with the edges incident with it. If $A = \{a\}$, we write $G - a$ for $G - \{a\}$. Evidently, $G - A = [V(G)\backslash A]$, which is an induced subgraph of G.

Let $F \subseteq E(G)$. The graph $G - F$ is the subgraph of G obtained by removing each edge in F from G. If $F = \{f\}$, we write $G - f$ for $G - \{f\}$. Clearly, $G - F$ is a spanning subgraph of G.

4. Connectedness

Let G be a graph. A *walk* in G is a finite sequence $v_0 v_1 v_2 \cdots v_k$ of vertices such that $v_i v_{i+1}$ is an edge for each $i = 0, 1, \cdots, k - 1$. The above walk is also called a $v_0 - v_k$ *walk*, and the *length* of the walk is k, which is the number of occurrences of its edges. Let $u, v \in V(G)$. A $u - v$ walk is said to be *closed* if $u = v$ and its length is at least two. A walk is a *trail* (resp., *path*) if no edge (resp., vertex) is repeated. Clearly, every path is a trail, but not conversely. A path is said to be *hamiltonian* (or *spanning*) if it contains all the vertices in G. A path $v_0 v_1 \cdots v_k$ is called a *chain* if $d_G(v_i) = 2$ for each $i = 1, 2, \cdots, k - 1$; and is a *maximal chain* if, in addition, $d_G(v_0) \geq 3$ and $d_G(v_k) \geq 3$. A *cycle* is a closed walk $v_1 v_2 \cdots v_k v_1$ in which the v_i's are distinct. Such a cycle of order k is also called a *k-cycle*. A 3-cycle is often called a *triangle*. The *girth* of G, denoted by $g(G)$, is the order of a smallest cycle in G. A k-cycle C in G, where $k \geq 4$, is said to be *pure* if $[V(C)] \cong C$; i.e., there is no edge in G joining two non-consecutive vertices along C (such an edge, if it exists, is called a *chord* of C). Thus, in Figure 1, $g(H) = 3$, and while $axyca$ is a pure cycle, the cycle $axywa$ is not pure.

A graph G is *connected* if any two vertices in G are joined by a path. It is *disconnected* otherwise. A *component* of G is a connected subgraph of G which is not a proper subgraph of any connected subgraph of G. We denote by $c(G)$ the number of components of G. For any two vertices u, v

in a connected graph G, the *distance* $d_G(u, v)$ (or simply $d(u, v)$) between u and v is the length of a shortest $u - v$ path in G. The *diameter* of G, denoted by *diam* G, is defined as *diam* $G = \max\{d(x, y) : x, y \in V(G)\}$. Thus, in Figure 1, we have *diam* $H = d(z, a) = d(c, z) = 3$.

5. Special graphs

Let $\mathbb{N} = \{1, 2, 3, \cdots\}$ be the set of positive integers, and let $n \in \mathbb{N}$.

A graph is an *empty graph* if it contains no edges. The empty graph of order n is denoted by O_n. In contrast with empty graphs, a graph is said to be *complete* if any two of its vertices are adjacent. The complete graph of order n is denoted by K_n. Clearly, $e(K_n) = \binom{n}{2}$. As a graph, any n-cycle is denoted by C_n. A cycle C_n is *even* (resp., *odd*) if n is even (resp., odd). A graph is said to be *unicyclic* if it contains exactly one cycle as a subgraph.

A *forest* is a graph containing no cycles; such a graph is also said to be *acyclic*. A *tree* is a connected forest. Thus every component of a forest is a tree. It is known that for every forest F, $v(F) = e(F) + c(F)$; and in particular, for any tree T, $v(T) = e(T) + 1$. Any path of order n as a graph is denoted by P_n.

Given $q \in \mathbb{N}$, the class of *q-trees* is defined recursively as follows: any complete graph K_q is a q-tree, and any q-tree of order $n + 1$ is a graph obtained from a q-tree G of order n, where $n \geq q$, by adding a new vertex and joining it to each vertex of a K_q in G. A graph G is called a *chordal* graph if G contains no pure cycle (i.e., every cycle in G contains a chord). It follows that every tree is a 1-tree, and every q-tree is a chordal graph, but not conversely (see Figure 3).

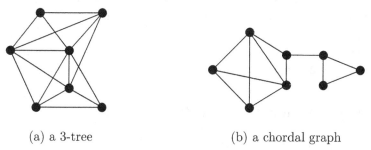

(a) a 3-tree   (b) a chordal graph

Figure 3

For $t \in \mathbb{N}$ and $t \geq 2$, a *t-partite* graph is a graph G whose $V(G)$ can be partitioned into t non-empty subsets $V_1, V_2, \cdots, V_t$ such that each edge of G joins a vertex in V_i to a vertex in V_j for some distinct i, j in $\{1, 2, \cdots, t\}$.

We call $(V_1, V_2, \cdots, V_t)$ a *t-partition* of G. A *complete* *t*-partite graph is a *t*-partite graph with a *t*-partition $(V_1, V_2, \cdots, V_t)$ such that every vertex in V_i is adjacent to every vertex in V_j for all distinct i, j in $\{1, 2, \cdots, t\}$. Such a complete *t*-partite graph is denoted by $K(p_1, p_2, \cdots, p_t)$ if $|V_i| = p_i$ for each $i = 1, 2, \cdots, t$.

A 2-partite graph is better known as a *bipartite* graph, and a 2-partition is also called a *bipartition*. Every tree is clearly bipartite and the tree $K(1, q)$ is also called a *star*. The following characterization of bipartite graphs, due to König (1916), is useful.

Theorem 1   *A graph is bipartite if and only if it contains no odd cycles.*

$\square$

A graph is said to be *planar* if it can be drawn in the plane in such a way that no two edges intersect except at a vertex. Any such drawing is a *plane drawing* of the graph. Let G be a planar graph. Then any plane drawing of G divides the set of points of the plane into regions, called *faces*; one face is unbounded and is the *infinite face*. The beautiful Euler's formula for plane graphs G which relates the order, size and the number of faces of G is stated below.

Theorem 2   *Let G be a connected plane graph of order n and size m that has f faces. Then*

$$n - m + f = 2.$$

$\square$

A planar graph G is said to be *outerplanar* if every vertex lies on the boundary of the same face in some plane drawing of G.

A connected multigraph G is said to be *eulerian* if there is a closed trail in G which contains each edge of G once. It is known that a connected multigraph G is eulerian if and only if every vertex of G is even.

6. New graphs formed from the given ones

Let G be a graph and $x, y \in V(G)$. We write $G \cdot xy$ to denote the graph obtained from G by contracting x and y and removing any loop and all but one of the multiple edges, if they arise. If $xy \notin E(G)$, we denote by $G + xy$ the graph obtained by adding a new edge xy to G. The *complement* of G, denoted by $\overline{G}$, is the graph with $V(\overline{G}) = V(G)$ such that two vertices are adjacent in $\overline{G}$ if and only if they are not adjacent in G. A graph G is said to be *self-complementary* if $\overline{G} \cong G$. An edge e in G is said to be *subdivided* if it is deleted and replaced by a chain connecting its ends, the internal vertices of this chain being new. A *subdivision* of G is a graph that can be obtained from G by a finite sequence of edge subdivisions. A subdivision of G is called a *G-homeomorph*.

For $k \geq 2$, let θ_k be the multigraph with 2 vertices and k edges. Any subdivision of θ_k is called a *multi-bridge graph*. For any $a_1, a_2, \cdots, a_k \in \mathbb{N}$, we denote by $\theta(a_1, a_2, \cdots, a_k)$ the graph obtained by replacing the edges of θ_k with paths of lengths $a_1, a_2, \cdots, a_k$ respectively. The graph $\theta(1, a_2, a_3)$ is called a *θ-graph* and the graph $\theta(a_1, a_2, a_3)$ is called a *generalized θ-graph*.

Let G and H be two graphs with two disjoint vertex sets. Their *disjoint union* is denoted by $G \cup H$, the disjoint union of k copies of G is denoted by kG and the disjoint union of a family of graphs $G_1, G_2, \cdots, G_k$ is denoted by $\cup_{i=1}^{k} G_i$. The *cartesian product* of G and H, denoted by $G \times H$, is the graph with vertex set $V(G) \times V(H)$ such that two vertices (x, y) and (x', y') are adjacent in $G \times H$ if and only if either (i) $x = x'$ and $yy' \in E(H)$ or (ii) $y = y'$ and $xx' \in E(G)$. The cartesian product of a family of graphs $G_1, G_2, \cdots, G_k$ is denoted by $\prod_{i=1}^{k} G_i$. The *join* of G and H, denoted by $G + H$, is the graph with $V(G + H) = V(G) \cup V(H)$ and
$$E(G + H) = E(G) \cup E(H) \cup \{xy : x \in V(G), y \in V(H)\}.$$
Observe that $K(p, q) = O_p + O_q$ and, more generally,

$$K(p_1, p_2, \cdots, p_t) = (\cdots ((O_{p_1} + O_{p_2}) + O_{p_3}) \cdots) + O_{p_t}.$$

The *wheel* of order n, denoted by W_n, is defined as $W_n = C_{n-1} + K_1$. Any edge joining K_1 and a vertex in C_{n-1} is called a *spoke* of W_n. For $1 \leq s \leq n - 2$, we denote by $W(n, s)$ the graph obtained from W_n by deleting all but s consecutive spokes, as shown in Figure 4. The *fan* of order n, denoted by F_n, is defined as $F_n = P_{n-1} + K_1$.

Figure 4 $W(8, 3)$

7. Graph parameters

Let G be a connected graph of order $n \geq 1$. A proper subset S of $V(G)$ is called a *cut* of G if $G - S$ is disconnected. A vertex w is called a *cut-vertex* of G if $\{w\}$ is a cut. The *vertex-connectivity* (the prefix 'vertex' can be omitted if there is no chance of confusion) of G, denoted by $\kappa(G)$, is defined by

$$\kappa(G) = \begin{cases} n - 1, & \text{if } G \cong K_n; \\ \min\{|S| : |S| \text{ is a cut of } G\}, & \text{if } G \not\cong K_n. \end{cases}$$

We define $\kappa(G) = 0$ if G is disconnected. We say G is *k-vertex-connected* (the prefix 'vertex' can be omitted if there is no chance of confusion), where $k \in \mathbb{N}$, if $\kappa(G) \geq k$. A cut S of G is called a *clique-cut* if S is a clique in G.

Let $u, v \in V(G)$. A family $\{Q_1, Q_2, \cdots, Q_k\}$ of $u - v$ paths in G is said to be *internally disjoint* if $V(Q_i) \cap V(Q_j) = \{u, v\}$ for all i, j with $1 \leq i < j \leq k$. The celebrated characterization of k-connected graphs, due to Whitney (1932) (see Bondy and Murty (1976)), is stated below.

Theorem 3   *A graph G is k-connected if and only if every two distinct vertices in G are joined by at least k internally disjoint paths.* $\square$

By definition, when $n \geq 3$, G is 2-connected if and only if G contains no cut-vertex. A *block* is a non-trivial connected graph which contains no cut-vertex. Thus G is a block if and only if $G \cong K_2$ when $n = 2$, and G is 2-connected when $n \geq 3$.

A *block* of G is a subgraph H of G such that H is itself a block, and is maximal with respect to this property. Let $b(G)$ denote the number of blocks in G. Thus, there are 9 blocks in the graph in Figure 5. An edge e in G is called a *bridge* if $G - e$ is disconnected; i.e., $\{e\}$ induces a block of G.

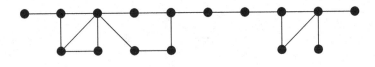

Figure 5

Let $k \in \mathbb{N}$. A *k-colouring* of G is a mapping

$$\varphi : V(G) \to \{1, 2, \cdots, k\}$$

(not necessarily onto) such that $\varphi(u) \neq \varphi(v)$ whenever $uv \in E(G)$. A graph G is said to be *k-colourable* if G admits a k-colouring. The *chromatic number* of G, denoted by $\chi(G)$, is the minimum value of k such that G is k-colourable. We say G is *k-chromatic* if $\chi(G) = k$.

Assume that $\chi(G) = k$ with a k-colouring θ. For each $i = 1, 2, \cdots, k$, let

$$V_i = \{v \in V(G) : \theta(v) = i\}.$$

We call V_i the *ith colour class* induced by θ. Clearly, each V_i is an independent set in G and $\{V_1, V_2, \cdots, V_k\}$ forms a partition of $V(G)$. Thus the chromatic number of G can, equivalently, be defined as the minimum number k such that $V(G)$ is partitioned into k independent sets. The graph G is said to be *uniquely k-colourable* if any two k-colourings of G induce the same partition of $V(G)$.

The *Turán graph* is a complete t-partite graph $K(p_1, p_2, \cdots, p_t)$ such that $|p_i - p_j| \leq 1$ for all $i, j = 1, 2, \cdots, t$. Such a Turán graph of order n is denoted by $K[t; n]$. Turán (1941) (see Bondy and Murty (1976)) proved the following result.

Theorem 4 *Let G be a graph of order n and $\omega(G) \leq k$, where $n, k \in \mathbb{N}$. Then $e(G) \leq e(K[k; n])$.* □

It follows that among the k-colourable graphs of order n, $K[k; n]$ has the maximum size.

A graph G is said to be *k-critical* if $\chi(G) = k$ and $\chi(H) < k$ for any proper subgraph H of G. Clearly, every k-chromatic graph contains a subgraph which is k-critical. Some useful properties of k-critical graphs are stated below.

Theorem 5 *Let G be a k-critical graph, where $k \in \mathbb{N}$. Then*

(i) G is connected;

(ii) $v(G) \neq k + 1$;

(iii) $k \leq 1 + \delta(G)$;

(iv) no cut of G is a clique;

(v) either $G = K_k$ or $2e(G) \geq 1 + (k-1)v(G)$. □

Notation

$$\mathbb{N} = \{1, 2, 3, \cdots\}$$
$$\mathbb{N}_0 = \{0, 1, 2, \cdots\}$$
$$\mathbb{Z} = \{\cdots, -2, -1, 0, 1, 2, \cdots\}$$
$$\mathbb{R} = \text{the set of real numbers}$$
$$\mathbb{C} = \text{the set of complex numbers}$$
$$a|b: \ a \text{ divides } b$$
$$a\nmid b: \ a \text{ does not divide } b$$
$$\lfloor x \rfloor = \text{the largest integer less than or equal to } x$$
$$\lceil x \rceil = \text{the smallest integer greater than or equal to } x$$
$$a \equiv b \ (\mathrm{mod} \ k): \ a \text{ is congruent to } b \text{ modulo } k; \text{ i.e., } k|(a-b)$$
$$|S| = \text{the number of elements in the finite set } S$$
$$s(r, n) = \text{(signed) Stirling number of the first kind}$$
$$S(r, n) = \text{Stirling number of the second kind}$$
$$\binom{n}{r} = \text{the number of } r\text{-element subsets of an } n\text{-element set} = \frac{n!}{r!(n-r)!}$$
$$(\lambda)_n = \lambda(\lambda - 1) \cdots (\lambda - n + 1)$$
$$A \times B = \{(a, b) : a \in A, b \in B\}$$
$$B \backslash A = \{x \in B : x \notin A\}$$

In what follows, G and H are graphs.

$$V(G) : \text{the vertex set of } G$$
$$E(G) : \text{the edge set of } G$$
$$v(G) : \text{the number of vertices in } G \text{ or the order of } G$$
$$e(G) : \text{the number of edges in } G \text{ or the size of } G$$
$$N(u) = N_G(u) : \text{the set of vertices } v \text{ such that } uv \in E(G)$$
$$\overline{G} : \text{the complement of } G$$

$[A]$: the subgraph of G induced by A, where $A \subseteq V(G)$

$[F]$: the subgraph of G induced by F, where $F \subseteq E(G)$

$G - v$: the subgraph of G obtained by removing v and all edges incident with v from G, where $v \in V(G)$

$G - e$: the subgraph of G obtained by removing e from G, where $e \in E(G)$

$G - F$: the subgraph of G obtained by removing all edges in F from G, where $F \subseteq E(G)$

$G - A$: the subgraph of G obtained by removing each vertex in A together with the edges incident with vertices in A from G, where $A \subseteq V(G)$

$G \cdot xy$: the graph obtained from G by contracting x and y and removing any loop and all but one of the multiple edges, if they arise, where $x, y \in V(G)$

$G + xy$: the graph obtained by adding a new edge xy to G, where $x, y \in V(G)$ and $xy \notin E(G)$

$\mathcal{G}[G_1 \cup_r G_2]$: the family of all K_r-gluings of G_1 and G_2

$d(v) = d_G(v)$: the degree of v in G, where $v \in V(G)$

$d(u, v) = d_G(u, v)$: the distance between u and v in G, where $u, v \in V(G)$

$diam\ G = \max\{d(x, y) : x, y \in V(G)\}$

$c(G)$: the number of components in G

$b(G)$: the number of blocks in G

$g(G)$: the girth of G

$r(G) = v(G) - c(G)$, the rank of G

$r_c(G) = e(G) - v(G) + c(G)$, the cyclomatic number (or nullity) of G

$t(G)$: the number of triangles in G

$\delta(G)$: the minimum degree of G

$\Delta(G)$: the maximum degree of G

$\alpha(G)$: the independence number of G

$\alpha'(G)$: the matching number of G

$\alpha(G, k)$: the number of partitions of $V(G)$ into k non-empty independent sets

$\kappa(G)$: the connectivity of G

$\omega(G)$: the clique number of G

$\chi(G)$: the chromatic number of G

$G + H$: the join of G and H

$G \cup H$: the disjoint union of G and H

kG : the disjoint union of k copies of G

$G \times H$: the cartesian product of G and H

$P(G, \lambda)$: the chromatic polynomial of G

$G \sim H$: G and H are χ-equivalent; i.e., $P(G, \lambda) = P(H, \lambda)$

$s_r(G) = P(G, r)/r!$

$n_G(H)$: the number of subgraphs in G which are isomorphic to H

$i_G(H)$: the number of induced subgraphs in G which are isomorphic to H

C_n : the cycle of order n

F_n : the fan of order n

K_n : the complete graph of order n

O_n : the empty graph of order n

P_n : the path of order n

W_n : the wheel of order n

$K(p, q)$: the complete bipartite graph with a bipartition (V_1, V_2) such that $|V_1| = p$ and $|V_2| = q$

$K(p_1, p_2, \cdots, p_t)$: the complete t-partite graph with a t-partition $(V_1, V_2, \cdots, V_t)$ such that $|V_i| = p_i$ for $i = 1, 2, \cdots, t$

$[G] = \{H : H \sim G\}$

$K[t; n]$: the Turán graph of order n and clique number t

$W(n, k)$: the subgraph of W_n obtained by deleting all but k consecutive spokes

$\theta(a_1, a_2, \cdots, a_k)$: the multi-bridge graph obtained by joining two distinct vertices by k internally-disjoint paths of lengths $a_1, a_2, \cdots, a_k$ respectively

$\theta_k(f)$: a special multi-bridge graph $\theta(a_1, a_2, \cdots, a_k)$, where $a_1 = a_2 = \cdots = a_k = f$

Chapter 1

The Number of λ-Colourings and Its Enumerations

1.1 Introduction

Let G be a graph and $\lambda \in \mathbb{N}$. A mapping $f : V(G) \longrightarrow \{1, 2, \cdots, \lambda\}$ is called a λ-*colouring* of G if $f(u) \neq f(v)$ whenever the vertices u and v are *adjacent* in G. Two λ-colourings f and g of G are regarded as *distinct* if $f(x) \neq g(x)$ for some vertex x in G. The number of distinct λ-colourings of G, which is the key notion of this monograph, is denoted by $P(G, \lambda)$.

By convention, $P(G, 0) = 0$. By definition, $P(G, \lambda) \geq 1$ if and only if G is λ-colourable. Thus, $P(G, \chi(G)) \geq 1$ and $P(G, r) = 0$ if $r \in \mathbb{N}_0$ and $r < \chi(G)$; that is,

$$\chi(G) = \min\{\lambda \in \mathbb{N} : P(G, \lambda) \geq 1\}. \tag{1.1}$$

There are six sections in this chapter. In Section 1.2, we shall compute $P(G, \lambda)$ by definition for some classes of graphs. Two more efficient ways of enumerating $P(G, \lambda)$ based on the Fundamental Reduction Theorem and Zykov's K_r-gluing of graphs will be presented in Section 1.3. In Section 1.4, we shall discuss the form in which $P(G, \lambda)$ is expressed in terms of $P(K_i, \lambda)$'s, i.e., $P(G, \lambda) = \sum_{i \geq 1} b_i P(K_i, \lambda)$, and give an interpretation for the coefficients b_i. These results will then be used in Section 1.5 for the introduction of the *umbral product* of polynomials, which enables us to

1

compute $P(G + H, \lambda)$, where $G + H$ is the join of the graphs G and H, in terms of $P(G, \lambda)$ and $P(H, \lambda)$. To end this chapter, we shall introduce in Section 1.6 the tree form of $P(G, \lambda)$, and discuss some features and state some results on $P(G, \lambda)$ in this form.

1.2   Examples

In this section, we shall enumerate $P(G, \lambda)$ for some special graphs G for illustration. We begin with two extremal classes of graphs.

Example 1.2.1   *For the empty graph O_n of order n, it is clear that*

$$P(O_n, \lambda) = \lambda^n. \tag{1.2}$$

More generally, if $G = \bigcup_{i=1}^{k} G_i$, then

$$P(G, \lambda) = \prod_{i=1}^{k} P(G_i, \lambda). \tag{1.3}$$

Example 1.2.2   *For the complete graph K_n of order n, we have*

$$P(K_n, \lambda) = \lambda(\lambda - 1) \cdots (\lambda - n + 1).$$

Observe that

$$
\begin{aligned}
P(K_1, \lambda) &= \lambda, \\
P(K_2, \lambda) &= \lambda(\lambda - 1) = \lambda^2 - \lambda, \\
P(K_3, \lambda) &= \lambda(\lambda - 1)(\lambda - 2) = \lambda^3 - 3\lambda^2 + 2\lambda, \\
P(K_4, \lambda) &= \lambda(\lambda - 1)(\lambda - 2)(\lambda - 3) = \lambda^4 - 6\lambda^3 + 11\lambda^2 - 6\lambda,
\end{aligned}
$$

etc.

In general, when $P(K_n, \lambda)$ is expressed in terms of the powers of λ, we have

$$P(K_n, \lambda) = \lambda(\lambda - 1) \cdots (\lambda - n + 1) = \sum_{k=1}^{n} s(n, k)\lambda^k, \tag{1.4}$$

where $s(n, k)$'s are the (signed) *Stirling numbers of the first kind*, which are defined recursively by

$$s(n, k) = s(n - 1, k - 1) - (n - 1)s(n - 1, k), \tag{1.5}$$

with the boundary conditions that

$$\begin{cases} s(r,0) = 0 & \text{for all } r \in \mathbb{N}, \\ s(r,r) = 1 & \text{for all } r \in \mathbb{N}_0. \end{cases}$$

Example 1.2.3 *Let H be a graph containing a K_r as a subgraph, and let G be the graph obtained from H by adding a new vertex w which is linked with each vertex in K_r (and no others) as shown in Figure 1.1. Then we have*

$$P(G, \lambda) = (\lambda - r)P(H, \lambda). \tag{1.6}$$

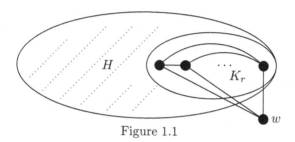

Figure 1.1

Remark It is known that every chordal graph can be constructed recursively using the above method starting with a collection of isolated vertices.

By applying (1.6) together with induction, the following results follow readily.

(i) If G is a chordal graph of order n, then

$$P(G, \lambda) = \lambda^{r_0}(\lambda - 1)^{r_1}(\lambda - 2)^{r_2} \cdots (\lambda - k)^{r_k}, \tag{1.7}$$

where $k \in \mathbb{N}$ and r_i's are in $\mathbb{N}$ such that $\sum_{i=0}^{k} r_i = n$. Note that $k = \chi(G) - 1$.

(ii) If G is a q-tree of order n, then

$$P(G, \lambda) = \lambda(\lambda - 1) \cdots (\lambda - q + 1)(\lambda - q)^{n-q}. \tag{1.8}$$

(iii) If G is a tree of order n, then

$$P(G, \lambda) = \lambda(\lambda - 1)^{n-1}. \tag{1.9}$$

Example 1.2.4 *Let us compute $P(C_4, \lambda)$, where the cycle C_4 is shown in Figure 1.2.*

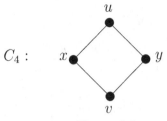

$$C_4:$$

Figure 1.2

Let f be a λ-colouring of C_4.

Case 1: $f(x) = f(y)$. There are $\lambda - 1$ ways to colour the vertices u and v independently, and thus the number of such λ-colourings f is $\lambda(\lambda - 1)^2$.

Case 2: $f(x) \neq f(y)$. There are $\lambda - 2$ ways to colour the vertices u and v independently, and thus the number of such λ-colourings f is $\lambda(\lambda - 1)(\lambda - 2)^2$.

We thus conclude that

$$
\begin{aligned}
P(C_4, \lambda) &= \lambda(\lambda - 1)^2 + \lambda(\lambda - 1)(\lambda - 2)^2 \\
&= \lambda^4 - 4\lambda^3 + 6\lambda^2 - 3\lambda \\
&= (\lambda - 1)^4 + (\lambda - 1). \quad\quad\quad (1.10)
\end{aligned}
$$

1.3 Basic results on enumeration of $P(G, \lambda)$

It is known that the problem of evaluating $\chi(G)$ is NP-complete and, by (1.1), the problem of evaluating $P(G, \lambda)$ is at least as hard as that of determining $\chi(G)$. In spite of this, there are results which are useful for evaluating $P(G, \lambda)$ more efficiently for some classes of graphs. Two of them will be introduced in this section.

The first result that we shall introduce provides us with a recursive way to compute $P(G, \lambda)$, and its proof is just an extension of the idea used in counting $P(C_4, \lambda)$ as shown in Example 1.2.4.

Theorem 1.3.1   *Let x and y be two non-adjacent vertices in a graph G. Then*

$$P(G, \lambda) = P(G + xy, \lambda) + P(G \cdot xy, \lambda). \quad\quad\quad (1.11)$$

Proof. Let f be a λ-colouring of G. We have either (i) $f(x) \neq f(y)$ or (ii) $f(x) = f(y)$. The number of λ-colourings f of G for which (i) holds equals $P(G + xy, \lambda)$ while the number of λ-colourings f of G for which (ii) holds equals $P(G \cdot xy, \lambda)$. The result thus follows. $\square$

Throughout this monograph, for convenience, we may sometimes use a drawing to denote its $P(G, \lambda)$.

Example 1.3.1 *Let G be the graph given below.*

By applying (1.11) repeatedly, we have

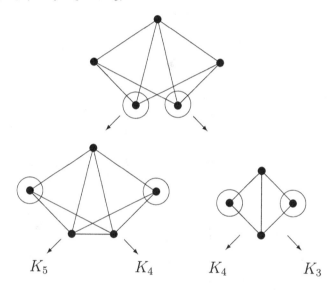

Thus

$$
\begin{aligned}
P(G, \lambda) &= P(K_5, \lambda) + 2P(K_4, \lambda) + P(K_3, \lambda) \\
&= \lambda(\lambda - 1)(\lambda - 2)(\lambda^2 - 5\lambda + 7) \\
&= \lambda^5 - 8\lambda^4 + 24\lambda^3 - 31\lambda^2 + 14\lambda.
\end{aligned}
$$

If we treat the graph $G + xy$ in Theorem 1.3.1 as a given graph H, then Theorem 1.3.1 can be equivalently restated as follows:

Let H be a graph and $e \in E(H)$. Then

$$
P(H, \lambda) = P(H - e, \lambda) - P(H \cdot e, \lambda). \tag{1.12}
$$

From now on, both (1.11) and (1.12) will be referred to as the *Fundamental Reduction Theorem* (or simply FRT). As was shown in Example 1.3.1, by applying (1.11) repeatedly, $P(G, \lambda)$ can eventually be expressed in terms of $P(K_r, \lambda)$'s. We now show an example where (1.12) is applied.

Example 1.3.2   *Let G be the graph given below.*

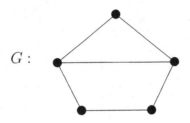

G :

By applying FRT (1.12) repeatedly, we have

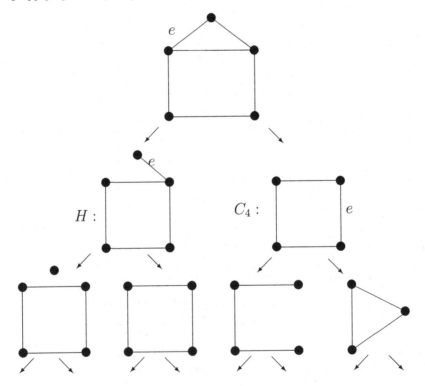

It is now obvious that if we proceed further, then $P(G, \lambda)$ could eventually be expressed in terms of $P(O_r, \lambda)$'s. However, to compute $P(G, \lambda)$ in this

case, we actually could stop at level two, and obtain by applying (1.6) and (1.10) that

$$
\begin{aligned}
P(G, \lambda) &= P(H, \lambda) - P(C_4, \lambda) \\
&= P(C_4, \lambda)(\lambda - 1) - P(C_4, \lambda) \\
&= \left((\lambda - 1)^4 + (\lambda - 1)\right)(\lambda - 2) \\
&= \lambda^5 - 6\lambda^4 + 14\lambda^3 - 15\lambda^2 + 6\lambda.
\end{aligned}
$$

Example 1.3.3   *We have computed $P(C_3, \lambda)$ $(= P(K_3, \lambda))$ and $P(C_4, \lambda)$. We shall now show by induction that, in general, for $n \geq 3$,*

$$
P(C_n, \lambda) = (\lambda - 1)^n + (-1)^n (\lambda - 1). \tag{1.13}
$$

It is easy to check that (1.13) holds when $n = 3$. Assume that $n \geq 4$ and that the result holds for C_{n-1}. By FRT,

$$
P(C_n, \lambda) = P(C_n - e, \lambda) - P(C_{n-1}, \lambda),
$$

where $e \in E(C_n)$. Now, by the induction hypothesis and by (1.9),

$$
\begin{aligned}
P(C_n, \lambda) &= \lambda(\lambda - 1)^{n-1} - (\lambda - 1)^{n-1} - (-1)^{n-1}(\lambda - 1) \\
&= (\lambda - 1)^n + (-1)^n (\lambda - 1),
\end{aligned}
$$

as required.

As shown in Example 1.3.1, if we apply (1.11) repeatedly beginning with a graph G, then we can eventually express $P(G, \lambda)$ in terms of $P(K_r, \lambda)$'s. This way of computing $P(G, \lambda)$ is particularly more efficient if G is dense (almost complete). On the other hand, if G is sparse (containing few edges), then we may apply (1.12) repeatedly, and eventually express $P(G, \lambda)$ in terms of $P(O_r, \lambda)$'s. However, whichever we use, as shown in Example 1.3.2, the procedure could be terminated so long as, for the resulting graphs H's, their $P(H, \lambda)$'s are known.

Let G_1 and G_2 be two graphs, and $r \in \mathbb{N}_0$ with $r \leq \min\{\omega(G_1), \omega(G_2)\}$, where $\omega(H)$ is the clique number of a graph H. Choose a K_r from each G_i, $i = 1, 2$, and form a new graph G from the union of G_1 and G_2 by identifying the two chosen K_r's in an arbitrary manner as shown in Figure 1.3. We call G a K_r-*gluing* of G_1 and G_2, and denote by $\mathcal{G}[G_1 \cup_r G_2]$ the family of all K_r-gluings of G_1 and G_2. When $r = 0$, G is just the disjoint union of G_1 and G_2; when $r = 1$, G is also called a *vertex-gluing* of G_1 and G_2; and when $r = 2$, G is also called an *edge-gluing* of G_1 and G_2.

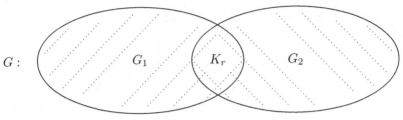

$$G:$$

Figure 1.3

The following result, due to Zykov (1949), provides a shortcut for evaluating $P(G, \lambda)$ if G is a K_r-gluing of some graphs.

Theorem 1.3.2   *Let G_1 and G_2 be two graphs and $G \in \mathcal{G}[G_1 \cup_r G_2]$. Then*

$$P(G, \lambda) = \frac{P(G_1, \lambda)P(G_2, \lambda)}{P(K_r, \lambda)}. \tag{1.14}$$

Proof.   For $i = 1, 2$, a given λ-colouring of K_r gives rise to $\frac{P(G_i, \lambda)}{P(K_r, \lambda)}$ λ-colourings of G_i. Thus

$$P(G, \lambda) = P(K_r, \lambda) \cdot \frac{P(G_1, \lambda)}{P(K_r, \lambda)} \cdot \frac{P(G_2, \lambda)}{P(K_r, \lambda)} = \frac{P(G_1, \lambda)P(G_2, \lambda)}{P(K_r, \lambda)}. \qquad \square$$

Corollary 1.3.1   *If G is a connected graph consisting of k blocks $B_1, B_2, \cdots, B_k$, then*

$$P(G, \lambda) = \frac{1}{\lambda^{k-1}} \prod_{i=1}^{k} P(B_i, \lambda). \qquad \square$$

Example 1.3.4   *The graph G in Example 1.3.2 is an edge-gluing of C_3 and C_4. Thus, by Theorem 1.3.2,*

$$\begin{aligned} P(G, \lambda) &= \frac{P(C_3, \lambda)P(C_4, \lambda)}{P(K_2, \lambda)} \\ &= \frac{\lambda(\lambda - 1)(\lambda - 2)\left((\lambda - 1)^4 + (\lambda - 1)\right)}{\lambda(\lambda - 1)} \\ &= \lambda^5 - 6\lambda^4 + 14\lambda^3 - 15\lambda^2 + 6\lambda. \end{aligned}$$

Example 1.3.5   *The graph G of Figure 1.1 is a K_r-gluing of H and K_{r+1}. Thus, by Theorem 1.3.2,*

$$P(G, \lambda) = \frac{P(H, \lambda)P(K_{r+1}, \lambda)}{P(K_r, \lambda)} = P(H, \lambda)(\lambda - r).$$

By applying FRT and Theorem 1.3.2, Read (1986) obtained the following useful result (see Exercise 1.9(b)).

Theorem 1.3.3   *Let H be a connected graph, and u and v two non-adjacent distinct vertices of H. Let G be the graph obtained from H by adding a $u - v$ path P of length r as shown in Figure 1.4. Then*

$$P(G, \lambda) = \frac{1}{\lambda(\lambda - 1)} P(C_{r+1}, \lambda) P(H, \lambda) + (-1)^r P(H \cdot uv, \lambda). \qquad (1.15)$$

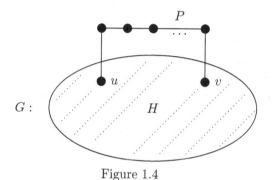

Figure 1.4

Example 1.3.6   *By applying Theorem 1.3.3, we can express $P(G, \lambda)$, where $G = \theta(r, s, t)$ is the generalized θ-graph and $r, s, t \geq 2$, in terms of $P(C_i, \lambda)$'s as shown below.*

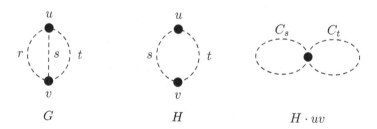

$$P(G, \lambda) = \frac{1}{\lambda(\lambda - 1)} P(C_{r+1}, \lambda) P(C_{s+t}, \lambda) + (-1)^r \frac{1}{\lambda} P(C_s, \lambda) P(C_t, \lambda).$$

1.4 $P(G, \lambda)$ **in factorial form**

Let G be a graph. We have seen that, by applying (1.12) repeatedly, we would have

$$P(G, \lambda) = \sum_{i \geq 1} a_i P(O_i, \lambda) = \sum_{i \geq 1} a_i \lambda^i, \qquad (1.16)$$

where a_i's are some constants. In this situation, we say that $P(G, \lambda)$ is expressed in *power* form.

On the other hand, if we apply (1.11) repeatedly, then we would arrive at

$$P(G, \lambda) = \sum_{i \geq 1} b_i P(K_i, \lambda) = \sum_{i \geq 1} b_i (\lambda)_i, \qquad (1.17)$$

where b_i's are some constants and

$$(\lambda)_i = \lambda(\lambda - 1) \cdots (\lambda - i + 1). \qquad (1.18)$$

In this situation, we say that $P(G, \lambda)$ is expressed in *factorial* form.

A natural question arises. What can be said about the constants a_i's and b_i's that appear in (1.16) and (1.17) respectively?

While this question for a_i's will be discussed in next chapter, we shall give in this section an interpretation for b_i's.

Consider the following G (see Example 1.3.1) again.

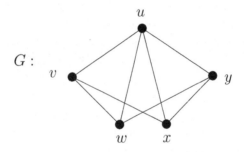

For $i = 1, 2, \cdots, 5$, let $\alpha(G, i)$ denote the number of ways of partitioning $V(G)$ into i independent sets. Observe that, in this case,

$$\{\{u\}, \{v\}, \{y\}, \{w\}, \{x\}\}$$

is the only partition with 5 independent sets;

$$\{\{u\}, \{v\}, \{y\}, \{w, x\}\} \quad \text{and} \quad \{\{u\}, \{w\}, \{x\}, \{v, y\}\}$$

are the two partitions with 4 independent sets; $\{\{u\}, \{v, y\}, \{w, x\}\}$ is the unique partition with 3 independent sets; and there are no partitions with i independent sets for $i = 1, 2$. Thus, we have:

$$\alpha(G, 1) = \alpha(G, 2) = 0, \ \alpha(G, 3) = 1, \ \alpha(G, 4) = 2 \text{ and } \alpha(G, 5) = 1.$$

On the other hand, as shown in Example 1.3.1, $P(G, \lambda) = (\lambda)_5 + 2(\lambda)_4 + (\lambda)_3$. The comparison of these two observations suggests that the constants b_i in (1.17) is equal to $\alpha(G, i)$. It is indeed the case as shown below (see Read (1968), for instance).

Theorem 1.4.1 *Let G be a graph of order n. Then*

$$P(G, \lambda) = \sum_{i=1}^{n} \alpha(G, i)(\lambda)_i, \qquad (1.19)$$

where $\alpha(G, i)$ is the number of ways of partitioning $V(G)$ into i independent sets.

Proof. There is a bijection between the family of colourings of G using *exactly* i colours from $\{1, 2, \cdots, \lambda\}$ and the family of partitions of $V(G)$ into i independent sets; and further, for any such partition, there are $(\lambda)_i$ λ-colourings of G. Thus the number of λ-colourings of G is given by $\sum_{i=1}^{n} \alpha(G, i)(\lambda)_i$. $\qquad\qquad\square$

If $\chi(G) = r$, then $\alpha(G, i) = 0$ when $i < r$, and $(r)_i = 0$ when $i > r$. Thus

$$P(G, r) = \sum_{i=1}^{n} \alpha(G, i)(r)_i = \alpha(G, r)r!.$$

Corollary 1.4.1 *If $\chi(G) = r$, then $\alpha(G, r) = P(G, r)/r!$.* $\qquad\square$

An independent set in G corresponds to a clique in $\overline{G}$, and vice versa. Thus, as pointed out in Frucht (1985) and R.Y Liu (1987),

$b_i(= \alpha(G, i))$ is the number of spanning subgraphs of $\overline{G}$ consisting of i components, each of which being a complete graph. (1.20)

For instance, if $G = P_3$, then

$\overline{G}$:

By (1.20), $b_2 = b_3 = 1$, $b_1 = 0$ and so, by Theorem 1.4.1, $P(G, \lambda) = (\lambda)_3 + (\lambda)_2$. If $H = C_4$, then

$\overline{H}$:

Thus, by (1.20), $b_4 = 1$, $b_3 = 2$, $b_2 = 1$, $b_1 = 0$ and so, by Theorem 1.4.1,

$$P(H, \lambda) = (\lambda)_4 + 2(\lambda)_3 + (\lambda)_2.$$

1.5   The join of graphs and the umbral product

By means of the expression of $P(G, \lambda)$ in factorial form (1.17), we shall introduce in this section a way of expressing $P(G + H, \lambda)$, where $G + H$ is the join of the graphs G and H, in terms of $P(G, \lambda)$ and $P(H, \lambda)$.

Let $G = P_3$ and $H = C_4$ as considered in Section 1.4, and we wish now to express $P(G + H, \lambda)$ in factorial form. Observe that

$$\overline{G + H} = \overline{G} \cup \overline{H}:$$

Thus, by (1.20), we have $P(G + H, \lambda) = (\lambda)_7 + 3(\lambda)_6 + 3(\lambda)_5 + (\lambda)_4$.

Assume, for any two graphs G and H, that $P(G, \lambda)$ and $P(H, \lambda)$ are expressed in factorial form. Following Read (1968), the *umbral product* of $P(G, \lambda)$ and $P(H, \lambda)$, denoted by $P(G, \lambda) \otimes P(H, \lambda)$, is an expression also in factorial form obtained by performing the usual polynomial multiplication where "factorials" are multiplied as "powers" (i.e., $(\lambda)_i \otimes (\lambda)_j = (\lambda)_{i+j}$).

Thus, for $G = P_3$ and $H = C_4$,

$$
\begin{aligned}
P(G, \lambda) \otimes P(H, \lambda) &= ((\lambda)_3 + (\lambda)_2) \otimes ((\lambda)_4 + 2(\lambda)_3 + (\lambda)_2) \\
&= (\lambda)_7 + 3(\lambda)_6 + 3(\lambda)_5 + (\lambda)_4 \\
&= P(G + H, \lambda).
\end{aligned}
$$

In general, we have (see Zykov (1949) and Exercise 1.16):

Theorem 1.5.1   *Let G_1 and G_2 be any two graphs with $P(G_i, \lambda)$ expressed in factorial form, $i = 1, 2$. Then*

$$P(G_1 + G_2, \lambda) = P(G_1, \lambda) \otimes P(G_2, \lambda). \qquad (1.21)$$

$\square$

Corollary 1.5.1   *For any graph* H,

$$P(H + K_1, \lambda) = \lambda P(H, \lambda - 1). \tag{1.22}$$

Proof. Assume that $P(H, \lambda) = \sum_{i \geq 1} b_i(\lambda)_i$. By Theorem 1.5.1,

$$
\begin{aligned}
P(H + K_1, \lambda) &= P(H, \lambda) \otimes (\lambda)_1 = \sum_{i \geq 1} b_i(\lambda)_{i+1} \\
&= \lambda \sum_{i \geq 1} b_i(\lambda - 1)_i = \lambda P(H, \lambda - 1). \qquad \square
\end{aligned}
$$

In particular, for the wheel W_n of order $n \geq 4$,

$$
\begin{aligned}
P(W_n, \lambda) &= P(C_{n-1} + K_1, \lambda) = \lambda P(C_{n-1}, \lambda - 1) \\
&= \lambda \left((\lambda - 2)^{n-1} + (-1)^{n-1}(\lambda - 2) \right). \tag{1.23}
\end{aligned}
$$

We shall next compute $P(K(p, q), \lambda)$. For this purpose, we first introduce a sequence of numbers $S(n, k)$, called the *Stirling numbers of the second kind*, which are defined recursively as

$$S(n, k) = S(n - 1, k - 1) + kS(n - 1, k), \tag{1.24}$$

where $n, k \in \mathbb{N}$, with the boundary conditions that

$$
\begin{cases}
S(r, 1) = 1 & \text{for all } r \in \mathbb{N}, \\
S(r, r) = 1 & \text{for all } r \in \mathbb{N}_0, \\
S(r, 0) = S(0, r) = 0 & \text{for all } r \in \mathbb{N}.
\end{cases} \tag{1.25}
$$

Combinatorially, $S(n, k)$ counts the number of ways of distributing n distinct objects into k identical boxes such that no box is empty. It is known that when the power λ^n is expressed in terms of the factorials $(\lambda)_k$'s, we have

$$\lambda^n = \sum_{k=1}^{n} S(n, k)(\lambda)_k. \tag{1.26}$$

Thus, by Theorem 1.5.1, we have (see Swenson (1973))

$$
\begin{aligned}
P(K(p, q), \lambda) &= P(O_p + O_q, \lambda) = P(O_p, \lambda) \otimes P(O_q, \lambda) = \lambda^p \otimes \lambda^q \\
&= \left(\sum_{r=1}^{p} S(p, r)(\lambda)_r \right) \otimes \left(\sum_{s=1}^{q} S(q, s)(\lambda)_s \right) \\
&= \sum_{r=1}^{p} \sum_{s=1}^{q} S(p, r)S(q, s)(\lambda)_{r+s}. \tag{1.27}
\end{aligned}
$$

More generally, for the complete m-partite graph $K(p_1, p_2, \cdots, p_m)$, we have (see Laskar and Hare (1975)):

$$P(K(p_1, p_2, \cdots, p_m), \lambda) = \sum_{r_m=1}^{p_m} \cdots \sum_{r_1=1}^{p_1} \prod_{i=1}^{m} S(p_i, r_i)(\lambda)_{r_1 + \cdots + r_m}. \quad (1.28)$$

1.6 $P(G, \lambda)$ in tree form

We have seen that for any graph G, $P(G, \lambda)$ can be expressed in either power form (1.16) or factorial form (1.17) (see also Theorem 1.4.1). In this final section, we shall introduce another form for $P(G, \lambda)$ which has its merit from a computational point of view.

Consider the following graph G:

By applying FRT (1.12) repeatedly beginning with G, we may have the following tree of computation:

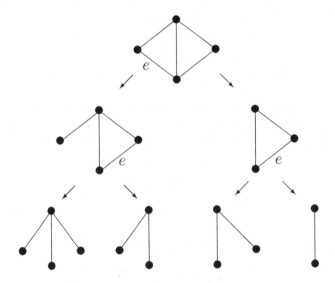

We note that in each step where (1.12) is applied on a current graph H, an edge e of H is chosen so that $H - e$ is connected, and we end up with four graphs which are trees. Since $P(T_n, \lambda) = \lambda(\lambda - 1)^{n-1}$ for any tree T_n

of order n, we arrive at

$$
\begin{aligned}
P(G, \lambda) &= \lambda(\lambda - 1)^3 - \lambda(\lambda - 1)^2 - \lambda(\lambda - 1)^2 + \lambda(\lambda - 1) \\
&= \lambda(\lambda - 1)^3 - 2\lambda(\lambda - 1)^2 + \lambda(\lambda - 1).
\end{aligned}
$$

In general, given any connected graph G of order n, we have

$$
P(G, \lambda) = \sum_{i=1}^{n-1} t_i \lambda(\lambda - 1)^i, \tag{1.29}
$$

where t_i's are some constants, and we say that $P(G, \lambda)$ is expressed in *tree form*.

The method of computing $P(G, \lambda)$ in tree form was suggested by Nijenhuis and Wilf (1978) (see also James and Riha (1975)). Given a connected graph G with a relatively small number of edges, comparing with the strategy by reducing G to empty graphs, it is clear that the strategy by reducing G to trees takes fewer number of steps. Thus, from a computational point of view, (1.29) has its own advantage.

It had been observed in Tutte (1973) and Nijenhuis and Wilf (1978) that $t_{n-1} = 1$ and the coefficients t_i's in (1.29) alternate in sign.

Frank and Shier (1986) showed that the absolute value of t_i, which is the number of trees of order $i + 1$ at the end of the computation tree, is an *invariant* regardless of any procedure where FRT (1.12) is repeatedly applied until the resulting graphs are trees.

We state below some results on $P(G, \lambda)$ in tree form for certain classes of graphs, which can be found in Frank and Shier (1986) or Adam and Broere (1993) (see Exercises 1.18 and 1.19).

(1) $G \cong C_n$, where $n \geq 3$, if and only if $|t_i| = 1$ for all $i = 1, 2, \cdots, n - 1$.

(2) G is bipartite if and only if $\sum_{i=1}^{n-1} t_i \neq 0$.

(3) $P(W_n, \lambda) = (-1)^{n-1} \lambda(\lambda - 1) + \sum_{i=0}^{n-2} (-1)^i \binom{n-1}{i} \lambda(\lambda - 1)^{n-i-1}$.

(4) $P(K_n, \lambda) = \sum_{i=1}^{n-1} s(n-1, i) \lambda(\lambda - 1)^i$, where $s(n-1, i)$'s are the Stirling numbers of the first kind as introduced in (1.5).

(5) $P(K(2, q), \lambda) = \lambda(\lambda - 1)^q + \sum_{i=1}^{q+1} (-1)^{q+1-i} \binom{q}{i-1} \lambda(\lambda - 1)^i$.

For more information on computing $P(G, \lambda)$ for more families of graphs, and on certain improved methods or tricks on computing $P(G, \lambda)$, the reader is referred to Loerine (1980), Read (1981, 1987a, 1988b) and Read and Tutte (1988).

Exercise 1

1.1 Let F be a forest of order n and with c components. Show that

$$P(F, \lambda) = \lambda^c (\lambda - 1)^{n-c}.$$

1.2 Prove that if G is a connected graph, then for any $\lambda \in \mathbb{N}$,

$$P(G, \lambda) \leq \lambda (\lambda - 1)^{v(G)-1}.$$

Show that the converse is not true.

(See Read (1968).)

1.3 Verify that

$$P(C_{n-2} + K_2, \lambda) = \lambda(\lambda - 1)\left((\lambda - 3)^{n-2} + (-1)^n (\lambda - 3) \right).$$

1.4 For the graph G shown below, verify that

$$P(G, \lambda) = \lambda(\lambda - 1)(\lambda - 2)(\lambda - 3)^3(\lambda - 4).$$

(See Read (1975).)

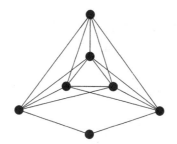

1.5 Let G be a connected chordal graph with $\chi(G) \leq 3$. Show that

$$P(G, \lambda) = \lambda(\lambda - 1)^{v(G)-t(G)-1}(\lambda - 2)^{t(G)},$$

where $t(G)$ is the number of triangles in G.

1.6 It is clear that there are exactly two non-isomorphic (n, m)-graphs with $m = \binom{n}{2} - 2$ and $n \geq 4$. Show that their respective chromatic polynomials are

$$\lambda(\lambda - 1) \cdots (\lambda - n + 4)(\lambda - n + 3)^2(\lambda - n + 2)$$

and

$$\lambda(\lambda - 1) \cdots (\lambda - n + 3)(\lambda^2 - (2n - 5)\lambda + n^2 + 5n + 7).$$

1.7 Show that

(i)
$$P(C_p + O_2, \lambda)$$
$$= \lambda \left((\lambda - 1)(\lambda - 3)^p + (\lambda - 2)^p + (-1)^p(\lambda^2 - 3\lambda + 1) \right);$$

(ii) and more generally,

$$P(C_p + O_q, \lambda) = \sum_{r=1}^{q} \sum_{t=2}^{p} (-1)^p m_{r,t}(\lambda)_{r+t},$$

where

$$m_{r,t} = S(q,r) \sum_{k=t}^{p} (-1)^k \binom{p}{k} S(k,t).$$

(See Koh and K.L. Teo (1990).)

1.8 Show that

$$P(C_n \times P_2, \lambda)$$
$$= (\lambda^2 - 3\lambda + 3)^n + (\lambda - 1)\left((3 - \lambda)^n + (1 - \lambda)^n\right) + \lambda^2 - 3\lambda + 1.$$

1.9 (i) For $n, k \in \mathbb{N}$ with $k \leq n - 2$, let $W(n, k)$ denote the graph obtained from the wheel W_n by deleting all but k consecutive spokes. Show that
$$P(W(n, k), \lambda) = (\lambda - 2)^{k-1}((\lambda - 1)^{n-k+1}$$
$$+ (-1)^{n-k}) + (-1)^{n-1}\lambda(\lambda - 2).$$

(ii) Let H be a connected graph, and u and v be two non-adjacent distinct vertices of H. Let G be the graph obtained from H by adding a $u - v$ path P of length r as shown in Figure 1.4. Show that

$$P(G, \lambda) = \frac{1}{\lambda(\lambda - 1)} P(C_{r+1}, \lambda) P(H, \lambda) + (-1)^r P(H \cdot uv, \lambda).$$

(See Read (1986).)

(iii) For any $n, k, m_1, \cdots, m_k \in \mathbb{N}$ with $m_1 + m_2 + \cdots + m_k = n - 1$, let $W_n(m_1, m_2, \cdots, m_k)$, called a *broken wheel of W_n*, be a graph obtained from W_n by deleting $n - 1 - k$ spokes (and so there are k spokes that remain) such that the number of rim edges between existing successive spokes are $m_1, m_2, \cdots, m_k$. For instance, $W_8(2, 1, 3, 1)$ denotes the graph below.

$W_8(2,1,3,1)$

Let $Q_r(\lambda) = \frac{1}{\lambda(\lambda-1)}P(C_{r+2},\lambda) = \frac{1}{\lambda}\left((\lambda-1)^{r+1}+(-1)^r\right)$. Show that

$$P(W_n(m_1,m_2,\cdots,m_k),\lambda) = \lambda\prod_{i=1}^{k}Q_{m_i}(\lambda) + (-1)^{n-1}\lambda(\lambda-2).$$

(See Read (1986).) Note that (a) is a special case of (c).

1.10 Let G be an outerplane graph of order n and size m having r_i interior regions of size i, $i = 3, 4, \cdots, k$. Show that

$$P(G,\lambda) = \frac{(\lambda-1)^m}{\lambda^{m-n}}\prod_{i=3}^{k}\left(1-(1-\lambda)^{k-i}\right)^{r_i}.$$

(See Li and Tian (1978) and also Wakelin and Woodall (1992).)

1.11 (i) Let G be a graph of order n with complement $\overline{G}$ such that $P(G,\lambda) = P(\overline{G},\lambda)$ but $G \not\cong \overline{G}$. Show that $n \geq 8$ and $n \equiv 0,1 \pmod 4$.

(ii) For each $n \geq 8$, $n \equiv 0,1 \pmod 4$, construct a graph G of order n such that $P(G,\lambda) = P(\overline{G},\lambda)$ but $G \not\cong \overline{G}$.

(See Koh and C.P. Teo (1991b), Liu, Zhou and Tan (1993), and also Xu and Liu (1995).)

1.12 Let G be a planar graph such that $P(G,\lambda) = P(\overline{G},\lambda)$. Show that $8 \leq v(G) \leq 9$. Construct all such graphs.

(See Liu, Zhou and Tan (1993).)

1.13 Show that

$$P(\overline{C}_n,\lambda) = \sum_{i=1}^{n}\frac{n}{i}\binom{i}{n-i}(\lambda)_i.$$

(See Frucht (1985); Ye (2001c) further characterized graphs G of order n such that their $P(G,\lambda)$ are given above.)

1.14   Let G be a graph of order n. By Theorem 1.4.1,

$$P(G, \lambda) = \sum_{i=1}^{n} \alpha(G, i)(\lambda)_i.$$

For any $r \in \mathbb{N}$, define

$$s_r(G) = \frac{P(G, r)}{r!}.$$

Thus $s_r(G) = 0$ if $r < \chi(G)$.

Show that

(i) if $\chi(G) \leq r \leq \chi(G) + 1$, then $s_r(G) \in \mathbb{N}$;

(ii) for $r \leq n + 1$, $s_r(G) = 1$ if and only if either $G \cong K_{r-1}$ or G is uniquely r-colourable.

1.15   Let G be a graph of order n and write

$$P(G, \lambda) = \sum_{i=1}^{n} b_i(\lambda)_i.$$

Show that if $(j + 2)^{j-1} \leq 2^n$, then $b_1 \leq b_2 \leq \cdots \leq b_j$.

(See Chvátal (1970).)

1.16   Let G_1 and G_2 be two graphs with their $P(G_i, \lambda)$ expressed in factorial form. Show that

$$P(G_1 + G_2, \lambda) = P(G_1, \lambda) \otimes P(G_2, \lambda).$$

1.17   Let G be a graph of order n. An *acyclic orientation* of G is a way of assigning a direction to each edge of G such that the resulting digraph contains no directed cycle. Show that $(-1)^n P(G, -1)$ is the number of acyclic orientations of G.

(See Stanley (1973).)

1.18   Let G be a connected graph of order n and write

$$P(G, \lambda) = \sum_{i=1}^{n-1} t_i \lambda (\lambda - 1)^i.$$

Show that

(i) $t_{n-1} = 1$ and the coefficients t_i's alternate in sign;

(ii) $G \cong C_n$ if and only if $|t_i| = 1$ for all $i = 1, 2, \cdots, n - 1$;

(iii) G is bipartite if and only if $\sum\limits_{i=1}^{n-1} t_i \neq 0$.

1.19 Verify that

(i) $P(W_n, \lambda) = (-1)^{n-1}\lambda(\lambda - 1) + \sum\limits_{i=0}^{n-2}(-1)^i \binom{n-1}{i}\lambda(\lambda - 1)^{n-i-1}$;

(ii) $P(K_n, \lambda) = \sum\limits_{i=1}^{n-1} s(n - 1, i)\lambda(\lambda - 1)^i$, where $s(n - 1, i)$'s are the Stirling numbers of the first kind as introduced in (1.5);

(iii) $P(K(2, q), \lambda) = \lambda(\lambda - 1)^q + \sum\limits_{i=1}^{q+1}(-1)^{q+1-i}\binom{q}{i-1}\lambda(\lambda - 1)^i$.

1.20 Show that

$$P\left(\bigcup_{i=1}^{n} K_{k_i} + O_2, \lambda\right) = \frac{\prod\limits_{i=1}^{n}(\lambda)_{k_i+1}}{\lambda^{n-1}(\lambda - 1)^{n-1}}\left(\prod_{i=1}^{n}(\lambda - k_i - 1) + (\lambda - 1)^{n-1}\right).$$

(See Dong and Koh (1998).)

Chapter 2

Chromatic Polynomials

2.1 Introduction

In Chapter 1, we define the notion $P(G, \lambda)$ as the number of λ-colourings of a given graph G, and introduce a number of ways to compute $P(G, \lambda)$. We also see that $P(G, \lambda)$ can be expressed in terms of λ in power, factorial or tree form. It turns out that $P(G, \lambda)$ is indeed a polynomial in λ, known as the *chromatic polynomial* of G.

Suppose that $P(G, \lambda)$ is expressed in power form $\sum a_i \lambda^i$. In this chapter, we shall present some useful results relating the coefficients a_i's and the numbers of certain subgraphs of G. We shall begin in Section 2.2 to establish the following result due to Whitney (1932b) that

$$a_i = \sum_{r=0}^{e(G)} (-1)^r N(i, r),$$

where $N(i, r)$ is the number of spanning subgraphs of G with i components and r edges. Whitney further observed that the above summation could be substantially simplified by introducing the notion of a *broken cycle*. This result, known as the Broken-cycle Theorem, together with various information on certain coefficients, will be presented in Section 2.3.

A real or complex number z is called a *root* of $P(G, \lambda)$ if $P(G, z) = 0$ (i.e., $(\lambda - z)|P(G, \lambda)$). The *multiplicity* of a root z of $P(G, \lambda)$ is the largest $k \in \mathbb{N}$ such that $(\lambda - z)^k|P(G, \lambda)$. It is obvious that the multiplicity of the root '0' of $P(G, \lambda)$ is the number of components of G (see Corollary 2.3.2). In Section 2.4, we shall present a result due to Woodall (1977) and, independently, Whitehead Jr. and Zhao (1984b) that the multiplicity of the root '1' of $P(G, \lambda)$ is the number of blocks of G.

Given $n, m \in \mathbb{N}$ such that $n - 1 \leq m \leq \binom{n}{2}$, what is the minimum value of $|a_i|$ for each $i = 1, 2, \cdots, n - 2$ when G runs through all (n, m)-graphs? In Section 2.5, we shall introduce the results of Sakaloglu and Satyanarayana (1997) and Rodriguez and Satyanarayana (1997) on this problem for connected graphs and connected planar graphs respectively.

Hong (1984) found two simple but interesting results which state that a_1 is divisible by $(\chi(G) - 1)!$, and that G is connected and bipartite if and only if a_1 is odd. These two results have recently been extended substantially by Dong, Koh and Soh (2004), and we shall present their results in Section 2.6.

To end this chapter, we shall discuss in Section 2.7 the Unimodal Conjecture and the Strong Logarithmic Concavity Conjecture on a_i's, and prove a related result that if G is a connected graph of order n, then

$$i|a_{i+1}| \leq (n - i)|a_i|$$

for $1 \leq i \leq n - 1$, where equality holds if and only if G is a tree.

2.2   An interpretation of the coefficients

In this section, we shall establish the following result which expresses the coefficients a_i's in $P(G, \lambda) = \sum_{i \geq 1} a_i \lambda^i$ in terms of the number of certain spanning subgraphs of a given graph G. This result for general graphs was given by Whitney (1932b), but its original idea was due to Birkhoff (1912) who applied it to evaluate the number of λ-colourings for planar maps.

Theorem 2.2.1   *Let G be an (n, m)-graph. Then*

$$P(G, \lambda) = \sum_{p=1}^{n} \left(\sum_{r=0}^{m} (-1)^r N(p, r) \right) \lambda^p, \tag{2.1}$$

where $N(p, r)$ denotes the number of spanning subgraphs of G with p components and r edges.

Proof.   For convenience, call a mapping $\theta : V(G) \rightarrow \{1, 2, \cdots, \lambda\}$ a λ-mapping of G. Let A be the set of all λ-mappings. Then $|A| = \lambda^n$. Let $E(G) = \{e_1, e_2, \cdots, e_m\}$. For each $i = 1, 2, \cdots, m$, let q_i denote the property that the two vertices incident with e_i have the same image under θ, and for any $\{i_1, i_2, \cdots, i_r\} \subseteq \{1, 2, \cdots, m\}$, let $N(q_{i_1} q_{i_2} \cdots q_{i_r})$ denote the number of mappings in A that satisfy each of the properties: $q_{i_1}, q_{i_2}, \cdots, q_{i_r}$. Thus $P(G, \lambda)$ counts the number of mappings in A that satisfy none of the q_i's. By the Principle of Inclusion and Exclusion, we have

$$P(G, \lambda) = \lambda^n - \sum_{i=1}^{m} N(q_i) + \sum_{1 \leq i < j \leq m} N(q_i q_j) + \cdots + (-1)^m N(q_1 q_2 \cdots q_m).$$

Consider a typical term $\sum_{1 \leq i_1 < i_2 < \cdots < i_r \leq m} N(q_{i_1} q_{i_2} \cdots q_{i_r})$. Let H be the spanning subgraph of G with r edges $e_{i_1}, e_{i_2}, \cdots, e_{i_r}$. Then a λ-mapping of G that satisfies properties $q_{i_1}, q_{i_2}, \cdots, q_{i_r}$ is indeed a λ-mapping of H that satisfies properties $q_{i_1}, q_{i_2}, \cdots, q_{i_r}$. Now, with respect to such a λ-mapping of H, all vertices in a component of H have the same image. Suppose there are p $(p = 1, 2, \cdots, n)$ components in H. Then the number of mappings in A satisfying properties $q_{i_1}, q_{i_2}, \cdots, q_{i_r}$ is λ^p. Since each spanning subgraph H with r edges and p components corresponds to some set of properties $q_{i_1}, q_{i_2}, \cdots, q_{i_r}$,

$$\sum_{1 \leq i_1 < i_2 < \cdots < i_r \leq m} N(q_{i_1} q_{i_2} \cdots q_{i_r}) = \sum_{p=1}^{n} N(p, r) \lambda^p.$$

Hence, noting that $N(n, 0) = 1$ and $N(p, 0) = 0$ for $1 \leq p \leq n - 1$, we have

$$
\begin{aligned}
P(G, \lambda) &= \lambda^n - \sum_{p=1}^{n} N(p, 1) \lambda^p + \sum_{p=1}^{n} N(p, 2) \lambda^p + \cdots \\
&\quad + (-1)^m \sum_{p=1}^{n} N(p, m) \lambda^p \\
&= \sum_{r=0}^{m} \sum_{p=1}^{n} (-1)^r N(p, r) \lambda^p = \sum_{p=1}^{n} \left(\sum_{r=0}^{m} (-1)^r N(p, r) \right) \lambda^p.
\end{aligned}
$$

$\square$

Remark For a short proof of Theorem 2.2.1, the reader may refer to Lin and Zhang (1989).

Example 2.2.1 *Consider the $(5, 6)$-graph in Figure 2.1.*

G :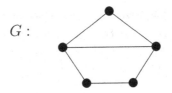

Figure 2.1

Let us determine the coefficient of λ^2 by Theorem 2.2.1. Clearly,

$$N(2,0) = N(2,1) = N(2,2) = N(2,5) = N(2,6) = 0.$$

On the other hand, the spanning subgraphs of G with 2 components and of size 3 and 4, respectively, are shown below; and we have $N(2,3) = 19$ and $N(2,4) = 4$. Thus, by Theorem 2.2.1, the coefficient of λ^2 is given by $-19 + 4 = -15$.

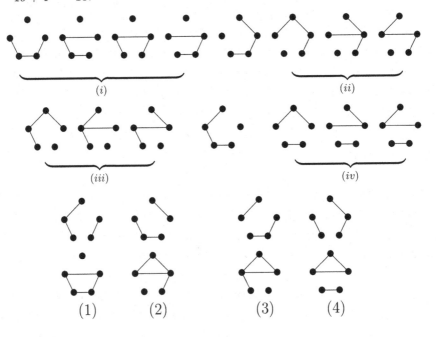

2.3 Broken-cycle Theorem

It is observed in Example 2.2.1 that the contribution of the spanning subgraph (1) (resp., (2), (3) and (4)) to λ^2 in $P(G, \lambda)$ cancels the contribution of one of the spanning subgraphs in group (i) (resp., (ii), (iii) and (iv)).

This suggests that it is not necessary to count all the spanning subgraphs of G in determining the coefficient of λ^i in $P(G, \lambda)$. This observation led Whitney to introduce the following notion of broken cycles to simplify the computation.

Thus, let G be an (n, m)-graph, and $\beta : E(G) \rightarrow \{1, 2, \cdots, m\}$ be a bijection. For any cycle C in G, let $e \in E(G)$ such that $\beta(e) > \beta(x)$ for any x in $E(C)\backslash\{e\}$. We call the path $C - e$ a *broken-cycle* in G with respect to (wrt) β. For instance, if we define an edge-labelling β of the graph G as shown in Figure 2.2 (i), then, as there are 3 cycles in G, there are 3 broken cycles as shown in Figure 2.2(ii).

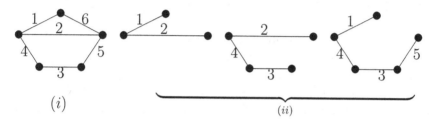

(i)   (ii)

Figure 2.2

We are now ready to establish the following celebrated result, known as the *Whitney's Broken-cycle Theorem*, by applying Theorem 2.2.1.

Theorem 2.3.1   *Let G be an (n, m)-graph, and let*

$$\beta : E(G) \rightarrow \{1, 2, \cdots, m\}$$

be any bijection. Then

$$P(G, \lambda) = \sum_{i=1}^{n}(-1)^{n-i}h_i(G)\lambda^i, \tag{2.2}$$

where $h_i(G)$ is the number of spanning subgraphs of G that have exactly $n - i$ edges and that contain no broken cycles wrt β.

Proof.   Assume that there are $q \geq 0$ broken cycles in G. For each broken cycle B, let $\beta(B) = \max\{\beta(f) : f \in E(B)\}$. Arrange the broken cycles as $B_1, B_2, \cdots, B_q$ such that $\beta(B_1) \leq \beta(B_2) \leq \cdots \leq \beta(B_q)$. Let $\mathcal{S}_1$ be the family of spanning subgraphs of G containing B_1 as a subgraph. For $2 \leq i \leq q$, let $\mathcal{S}_i$ be the family of spanning subgraphs of G containing B_i but not any of $B_1, B_2, \cdots, B_{i-1}$ as a subgraph. Let $\mathcal{S}_{q+1}$ be the family of spanning subgraphs of G that do not contain any B_i as a subgraph. Then

$\{S_1, S_2, \cdots, S_{q+1}\}$ forms a partition of the family of spanning subgraphs of G.

Consider S_1. Assume that $B_1 = C - f_1$ for some cycle C and $f_1 \in E(C)$. Observe that each member H in S_1 not containing f_1 gives rise to a unique member $H + f_1$ in S_1 containing f_1, and vice versa. Since $e(H + f_1) = e(H) + 1$ but $c(H + f_1) = c(H)$, by Theorem 2.2.1, the contributions of H and $H + f_1$ to $P(G, \lambda)$ are cancelled off.

Consider S_2. Assume that $B_2 = C - f_2$ for some cycle C and $f_2 \in E(C)$. It follows from the ordering of B's that $f_2 \notin E(B_1)$ (and so $f_2 \notin E(B_1) \cup E(B_2)$). Again, each member H in S_2 not containing f_2 gives rise to a unique member $H + f_2$ in S_2 containing f_2, and vice versa. Thus their contributions to $P(G, \lambda)$ are also cancelled off.

Likewise, the total contribution of the members in S_i to $P(G, \lambda)$ is zero, for each $i = 1, 2, \cdots, q$.

We are now left with the family S_{q+1}. Let $H \in S_{q+1}$ and assume that $e(H) = s$. As H is a forest, $0 \le s \le n-1$ and $c(H) = n - s$. Let $N'(n - s, s)$ denote the number of graphs in S_{q+1} with $n - s$ components and s edges. Thus, by Theorem 2.2.1,

$$
\begin{aligned}
P(G, \lambda) &= \sum_{p=1}^{n} \left(\sum_{r=0}^{m} (-1)^r N(p, r) \right) \lambda^p \\
&= \sum_{s=0}^{n-1} (-1)^s N'(n - s, s) \lambda^{n-s} \\
&= \sum_{s=0}^{n-1} (-1)^s h_{n-s}(G) \lambda^{n-s} \\
&= \sum_{i=1}^{n} (-1)^{n-i} h_i(G) \lambda^i,
\end{aligned}
$$

as required.   □

Notes (1) The conclusion of Theorem 2.3.1 is independent of the choice of any edge-labelling β.

(2) We may write h_i for $h_i(G)$ if there is no danger of confusion.

(3) The leading coefficient of $P(G, \lambda)$, i.e., $h_n(G)$, is equal to 1, by definition.

Example 2.3.1   *For the edge-labelled graph G in Figure 2.2(i), we have:*
$h_5 = 1$, $h_4 = 6$, $h_3 = \binom{6}{2} - 1 = 14$, $h_2 = \binom{6}{3} - 5 = 15$ *and* $h_1 = \binom{6}{4} - 9 = 6$.

Thus, by Theorem 2.3.1,

$$P(G, \lambda) = \lambda^5 - 6\lambda^4 + 14\lambda^3 - 15\lambda^2 + 6\lambda.$$

Example 2.3.2 *A tree T_n is an $(n, n-1)$-graph containing no broken cycles. Thus, by Theorem 2.3.1,*

$$P(T_n, \lambda) = \sum_{i=1}^{n} (-1)^{n-i} \binom{n-1}{n-i} \lambda^i = \sum_{i=1}^{n} (-1)^{n-i} \binom{n-1}{i-1} \lambda^i,$$

which agrees with (1.9).

Example 2.3.3 *For $n \geq 3$, the cycle C_n contains exactly one broken cycle with $n-1$ edges. Thus*

$$h_i = \begin{cases} n-1 & \text{if } i = 1, \\ \binom{n}{i} & \text{if } 2 \leq i \leq n, \end{cases}$$

and so

$$P(C_n, \lambda) = (-1)^{n-1}(n-1)\lambda + \sum_{i=2}^{n} (-1)^{n-i} \binom{n}{i} \lambda^i,$$

which agrees with (1.13).

Bari-Hall's Broken-cycle Formula Let G be an (n, m)-graph with a bijection $\beta : E(G) \to \{1, 2, \cdots, m\}$, and suppose that there are q broken cycles in G induced by β. For any r of these q broken cycles, where $1 \leq r \leq q$, any spanning subgraph of G whose edge set consists precisely of those edges in the given r broken cycles is called a *minimal* spanning subgraph of G *containing r broken cycles*. For instance, in the graph G in Figure 2.2(i), all the minimal spanning subgraphs of G containing 2 broken cycles are shown below:

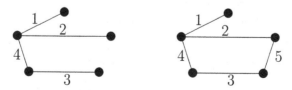

By applying the Principle of Inclusion and Exclusion, Bari and Hall (1977) gave the following expression for the coefficient h_{n-i} of λ^{n-i} in (2.2), which we shall refer to as *Bari-Hall's Broken-cycle formula*:

$$h_{n-i} = \binom{m}{i} + \sum_{j=2}^{i}\sum_{r=1}^{q}(-1)^r\binom{m-j}{i-j}n_{j,r}, \tag{2.3}$$

where $n_{j,r}$ is the number of minimal spanning subgraphs of G with j edges containing r broken cycles.

A broken cycle in G is said to be *proper* if it contains no other broken cycles. For instance, in the edge-labelled graph G of Figure 2.3 (i), out of the three broken cycles, only two, as shown in Figure 2.3 (ii), are proper.

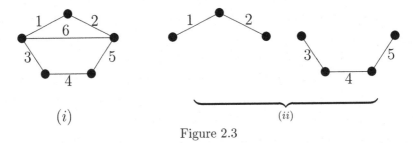

(i)   (ii)

Figure 2.3

Similar to what we have seen in the proof of Theorem 2.3.1, where cancellation occurs, Bari and Hall (1977) pointed out that, to apply (2.3), we may confine ourselves to spanning subgraphs of G containing *proper* broken cycles. That is, if there are p proper broken cycles, then we have:

$$h_{n-i} = \binom{m}{i} + \sum_{j=2}^{i}\sum_{r=1}^{p}(-1)^r\binom{m-j}{i-j}n_{j,r}^*, \tag{2.4}$$

where $n_{j,r}^*$ is the number of minimal spanning subgraphs of G with j edges containing r *proper* broken cycles.

Example 2.3.4   *Consider the edge-labelled graph G in Figure 2.3 (i). It can be checked that $n_{2,1}^* = 1$, $n_{2,2}^* = 0$, $n_{3,1}^* = 1$, $n_{3,2}^* = n_{4,1}^* = n_{4,2}^* = n_{5,1}^* = 0$ and $n_{5,2}^* = 1$. Thus, by (2.2) and (2.4),*

$$\begin{aligned}
P(G,\lambda) &= \lambda^5 - \binom{6}{1}\lambda^4 + \left(\binom{6}{2} - \binom{4}{0}\right)\lambda^3 \\
&\quad - \left(\binom{6}{3} - \binom{4}{1} - \binom{3}{0}\right)\lambda^2 + \left(\binom{6}{4} - \binom{4}{2} - \binom{3}{1}\right)\lambda \\
&= \lambda^5 - 6\lambda^4 + 14\lambda^3 - 15\lambda^2 + 6\lambda.
\end{aligned}$$

Let G and H be two graphs. We shall denote by $n_G(H)$ (resp., $i_G(H)$) the number of subgraphs (resp., induced subgraphs) of G which are isomorphic to H. If the graph G under consideration is clear from the context, then $n_G(H)$ and $i_G(H)$ will be replaced simply by $n(H)$ and $i(H)$ respectively. Thus, if G is the graph in Figure 2.4, then we have

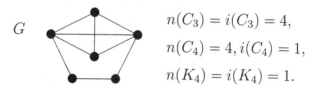

$$n(C_3) = i(C_3) = 4,$$
$$n(C_4) = 4, i(C_4) = 1,$$
$$n(K_4) = i(K_4) = 1.$$

Figure 2.4

Recall that $g = g(G)$ is the girth of G. As an immediate consequence of Theorem 2.3.1, we have:

Corollary 2.3.1   *Let G be an (n, m)-graph. Then*

$$P(G, \lambda) = \sum_{i=1}^{n} (-1)^{n-i} h_i \lambda^i$$

is a polynomial in λ such that

(i) the degree of $P(G, \lambda)$ is n and the leading coefficient is 1;

(ii) the coefficients are integers and alternate in sign;

(iii) the constant term is zero;

(iv) for $0 \le i \le g - 2$, $h_{n-i} = \binom{m}{i}$ (in particular, $h_{n-1} = m$);

(v) $h_{n-(g-1)} = \binom{m}{g-1} - n(C_g)$ (in particular, $h_{n-2} = \binom{m}{2} - n(C_3)$).   □

If G contains triangles (i.e., $g(G) = 3$), we get no information on h_{n-3} from the above corollary. In connection with this, the following two pieces of information on h_{n-3} and h_{n-4}, obtained by Farrell (1980b) (see Exercise 2.7), are useful.

Theorem 2.3.2   *Let G be an (n, m)-graph. Then*

(i) $h_{n-3} = \binom{m}{3} - (m-2)n(K_3) - i(C_4) + 2n(K_4);$

(ii)

$$
\begin{aligned}
h_{n-4} = \ & \binom{m}{4} - \binom{m-2}{2} n(K_3) + \binom{n(K_3)}{2} - (m-3)i(C_4) \\
& -(2m-9)n(K_4) - i(C_5) + i(K(2,3)) \\
& +2i(H) + 3i(W_5) - 6n(K_5),
\end{aligned} \tag{2.5}
$$

where H is the graph shown below.   □

$$ H : $$

Following Farrell's approach, Peng (1992a) managed to obtain expressions for h_{n-5} and h_{n-6} but confining to the class of bipartite graphs. While h_{n-6} is rather complicated to be stated here, we present in what follows his result on h_{n-5}.

Theorem 2.3.3   *Let G be an (n,m)-bipartite graph. Then*

$$
\begin{aligned}
h_{n-5} = \ & \binom{m}{5} - \binom{m-3}{2} n(C_4) - i(C_6) + (m-3)n(K(2,3)) \\
& -n(K(2,4)) + i(K(3,3) - e) + 4n(K(3,3)),
\end{aligned} \tag{2.6}
$$

where e is an edge in $K(3,3)$.   □

Farrell (1980b) (resp., Peng (1992a)) established Theorem 2.3.2 (resp., Theorem 2.3.3) by means of Theorem 2.2.1. (By applying (2.4) instead, Bielak, in an unpublished manuscript, offered not only a new proof of Theorem 2.3.2, but also an expression for h_{n-5} in terms of the numbers of 31 non-isomorphic induced subgraphs in G.)

Corollary 2.3.1 enables us to compute h_i's only for $i = n, n-1, \cdots, n - (g-1)$. When $g \geq 5$, Theorem 2.3.2 gives us no information on h_i if $i \leq n - g$. The following result, due to C.P. Teo and Koh (1994) and, independently, Peng (1994), provides us with an expression for computing h_{n-k}, where $g \leq k \leq \lceil \frac{3}{2}g \rceil - 3$, which is particularly useful if g is large.

Theorem 2.3.4   *Let G be an (n,m)-graph with $g \geq 4$. Then for each $k \in \mathbb{N}$ with $g \leq k \leq \lceil \frac{3}{2}g \rceil - 3$,*

$$
h_{n-k} = \binom{m}{k} - \sum_{r=g}^{k+1} n(C_r) \binom{m-r+1}{k-r+1}. \tag{2.7}
$$

To prove Theorem 2.3.4, we first give a lower bound for the number of edges of a subgraph induced by two cycles.

Lemma 2.3.1 *Let G be a graph with $g \geq 4$, and let C and C' be two distinct cycles in G. Then $e(H) \geq \lceil \frac{3}{2} g \rceil$, where $H = [E(C) \cup E(C')]$.*

Proof. Let $I = [E(C) \cap E(C')]$. Suppose on the contrary that $e(H) < \frac{3}{2} g$. Then

$$e(I) = e(C) + e(C') - e(H) > 2g - \frac{3}{2} g = \frac{g}{2},$$

and so

$$e(H) - e(I) < \frac{3}{2} g - \frac{g}{2} = g.$$

This, however, leads to a contradiction as $H - E(I)$ contains a cycle. $\qquad\square$

Proof of Theorem 2.3.4. We first claim that if $F \subseteq E(G)$ with $|F| \leq \lceil \frac{3}{2} g \rceil - 3$, then $[F]$ contains at most one broken cycle. Otherwise, let P and P' be two broken cycles contained in $[F]$ which are obtained from two cycles C and C' respectively. By Lemma 2.3.1,

$$e(E(C) \cup E(C')) \geq \lceil \frac{3}{2} g \rceil.$$

But then

$$|F| \geq |E(P) \cup E(P')| \geq \lceil \frac{3}{2} g \rceil - 2,$$

a contradiction. The claim thus follows.

As there is a bijection between the family of broken cycles of length $r - 1$ and the family of cycles of order r, it follows that the number of spanning subgraphs of G that have exactly k edges, where $g \leq k \leq \lceil \frac{3}{2} g \rceil - 3$, and that contain a broken cycle is given by

$$\sum_{r=g}^{k+1} n(C_r) \binom{m - (r-1)}{k - (r-1)}.$$

The result now follows by the definition of h_{n-k}. $\qquad\square$

Eisenberg (1971) and Meredith (1972) had, respectively, obtained lower and upper bounds for h_i for all i with $1 \leq i \leq n - 1$. Their results had been improved substantially by Li and Tian (1978) as shown below.

Theorem 2.3.5 *Let G be an (n, m)-graph with girth g. Then*

(i) $h_i \leq \binom{m}{n-i} - \binom{m-g+2}{n-i-g+2} + \binom{m-n(C_g)-g+2}{n-i-g+2}$ for $i = 1, 2, \cdots, n-1$;

(ii) $h_i \geq \sum_{j=0}^{l-1} \binom{m-n+j}{j}\binom{n-1-j}{i-1} - \sum_{k=3}^{l} \left\{ n(C_k) \sum_{j=0}^{l-k} \binom{m-n+j}{j}\binom{n-k-j}{i-1} \right\}$

for $i = 1, 2, \cdots, n-1$ and l with $3 \leq k \leq l \leq n$.　　　□

The coefficient h_1 of λ in $P(G, \lambda)$ is interesting in its own right. As shown below, four results pertaining to it are in order. The proofs of the first two results, noted respectively by Read (1968) and Eisenberg (1972), are left to the reader (see Exercise 2.4). We shall only prove the last two which were obtained by Hong (1984).

Theorem 2.3.6　　*Let G be a graph of order $n \geq 2$. Then*

(i) *G is connected if and only if $h_1 \geq 1$;*

(ii) *G is a tree if and only if $h_1 = 1$;*

(iii) *$(\chi(G) - 1)! | h_1$;*

(iv) *G is connected and bipartite if and only if h_1 is odd.*

Proof.　(iii) By Theorems 2.3.1 and 1.4.1,

$$\sum_{i=1}^{n} (-1)^{n-i} h_i \lambda^i = P(G, \lambda) = \sum_{i=1}^{n} \alpha(G, i)(\lambda)_i.$$

By comparing the coefficients of λ on both sides and noting that $\alpha(G, i) = 0$ for $i < \chi(G)$, we have

$$(-1)^{n-1} h_1 = \sum_{i=\chi(G)}^{n} (-1)^{i-1}(i-1)!\alpha(G, i), \qquad (2.8)$$

and the result follows.

(iv) Assume that G is connected and bipartite. Then G is uniquely 2-colourable, and by definition, $\alpha(G, 2) = 1$. Thus, by (2.8),

$$(-1)^{n-1} h_1 = -1 + \sum_{i=3}^{n} (-1)^{i-1}(i-1)!\alpha(G, i),$$

which implies that h_1 is odd.

Assume now that h_1 is odd. Then G is connected by (i). As $\chi(G) \geq 2$, $\alpha(G, 2)$ must be odd by (2.8). Thus $\alpha(G, 2) \neq 0$ and so G is bipartite. □

The following result follows immediately from Theorem 2.3.6(i) and (1.3).

Corollary 2.3.2 *The number $c(G)$ of components of a graph G is the multiplicity of the root '0' of $P(G, \lambda)$ (i.e., $c(G)$ is the smallest $r \in \mathbb{N}$ such that $h_r(G) > 0$).* □

2.4 The multiplicity of root '1'

Corollary 2.3.2 states that the multiplicity of the root '0' of $P(G, \lambda)$ is the number of components of a graph G. In this section, we shall give an interpretation of the multiplicity of the root '1' of $P(G, \lambda)$.

Let G be a connected graph with r blocks $B_1, B_2, \cdots, B_r$. By Corollary 1.3.1,

$$P(G, \lambda) = \frac{1}{\lambda^{r-1}} \prod_{i=1}^{r} P(B_i, \lambda). \tag{2.9}$$

As $P(B_i, 1) = 0$ for each $i = 1, 2, \cdots, r$, it follows that

$$(\lambda - 1)^r | P(G, \lambda).$$

Woodall (1977) and, independently, Whitehead Jr. and Zhao (1984b) showed that $(\lambda - 1)^k \nmid P(G, \lambda)$ if $k \in \mathbb{N}$ and $k \geq r + 1$. That is:

Theorem 2.4.1 *For any non-trivial connected graph G, the multiplicity of the root '1' of $P(G, \lambda)$ is the number of blocks in G.*

We shall prove Theorem 2.4.1 following the idea given in Whitehead Jr. and Zhao (1984b). Let $w = \lambda - 1$, and suppose that $P(G, \lambda)$ is expressed in powers of w. As $P(G, 1) = 0$, the constant term in this expression is zero. Let $a(G)$ be the coefficient of w in such an expression. To prove Theorem 2.4.1, by (2.9), it suffices to show that '1' is a *simple* root of $P(G, \lambda)$ if G is a block. But this follows immediately from the following result.

Lemma 2.4.1 *If G is a connected graph of order $n \geq 2$, then $a(G) = (-1)^n q$ for some $q \in \mathbb{N}_0$. Further, G is a block if and only if $q \geq 1$.*

Proof. If G contains more than one block, then $q = 0$ by (2.9). It remains to show that $a(G) = (-1)^n q$ for some $q \geq 1$ if G is a block. We shall prove this by induction on $n + e(G)$.

When $G = K_2$, we have $P(G, \lambda) = \lambda(\lambda - 1) = w^2 + w$, and so $a(G) = 1 = (-1)^2 \cdot 1$. Assume that G is a block of order $n \geq 3$. Let $v_1 v_2 \cdots v_k$ be a longest path in G. Observe that $G - v_1 v_2$ is connected while $G \cdot v_1 v_2$ is a block (see Exercise 2.9). By the induction hypothesis,

$$a(G - v_1 v_2) = (-1)^n q_1 \quad \text{and} \quad a(G \cdot v_1 v_2) = (-1)^{n-1} q_2,$$

where $q_1 \in \mathbb{N}_0$ and $q_2 \in \mathbb{N}$; and, by FRT, we have

$$a(G) = (-1)^n q_1 - (-1)^{n-1} q_2 = (-1)^n (q_1 + q_2),$$

where $q_1 + q_2 \in \mathbb{N}$. This completes the proof.   □

Corollary 2.4.1   *Let G be a graph of order $n \geq 3$. Then G is a block if and only if $(\lambda - 1)^2 \nmid P(G, \lambda)$.*   □

Remark   Divide the family $\mathcal{G}$ of connected graphs of order $n \geq 3$ into three disjoint classes as follows:

$$\mathcal{G}_1 = \{G \in \mathcal{G} : G \text{ is not a block}\},$$

$$\mathcal{G}_2 = \{G \in \mathcal{G} : G \text{ is a block and contains no } K_4 - \text{homeomorph as a}$$
$$\text{subgraph}\} \text{ and}$$

$$\mathcal{G}_3 = \mathcal{G} \backslash (\mathcal{G}_1 \cup \mathcal{G}_2).$$

Chao and Zhao (1983) used the coefficient $a(G)$ to determine whether a connected graph G is in $\mathcal{G}_i$, $i = 1, 2, 3$, as shown below:

(i) $G \in \mathcal{G}_1$ if and only if $a(G) = 0$;

(ii) $G \in \mathcal{G}_2$ if and only if $|a(G)| = 1$; and

(iii) $G \in \mathcal{G}_3$ if and only if $|a(G)| \geq 2$.

Problem 2.4.1   *Let G be a connected graph. Give an interpretation for the multiplicity of the root 2 of $P(G, \lambda)$.*

2.5 Least coefficients

For any $n, m \in \mathbb{N}$ with $m \leq \binom{n}{2}$, let $\mathcal{C}(n,m)$ denote the family of all connected (n,m)-graphs. By Theorem 2.3.1, for any graph $G \in \mathcal{C}(n,m)$, we have

$$P(G, \lambda) = \sum_{i=1}^{n} (-1)^{n-i} h_i(G) \lambda^i, \qquad (2.10)$$

where $h_i(G) \in \mathbb{N}$ for each $i = 1, 2, \cdots, n$. While $h_n(G) = 1$ and $h_{n-1}(G) = m$, in general, $h_i(G)$ is not completely determined by m and n for $1 \leq i \leq n-2$. In this section, we shall present some results, due to Sakaloglu and Satyanarayana (1997) and Rodriguez and Satyanarayana (1997), which determine

$$\min_{G \in \mathcal{C}(n,m)} h_i(G) \quad \text{and} \quad \min_{G \in \mathcal{C}_p(n,m)} h_i(G) \qquad (2.11)$$

for each $i = 1, 2, \cdots, n-2$, where $\mathcal{C}_p(n,m)$ is the family of planar graphs in $\mathcal{C}(n,m)$. Our approach here, however, is different from theirs.

We first introduce two families of graphs in $\mathcal{C}(n,m)$ which will play special roles in determining (2.11). Let $\mathcal{L}(n,m)$ be the family of graphs G in $\mathcal{C}(n,m)$ such that one block B of G has the clique number $\omega(B)$ at least $v(B) - 1$ and all other blocks are K_2's. Let $\mathcal{J}(n,m)$ be the family of graphs G in $\mathcal{C}(n,m)$ such that all blocks of G are cliques:

$$K_3, \ K_r, \ \underbrace{K_2, K_2, \cdots, K_2}_{b(G)-2},$$

where $r \geq 2$. The graphs (a) and (b) in Figure 2.5 respectively belong to $\mathcal{L}(8, 11)$ and $\mathcal{J}(8, 11)$.

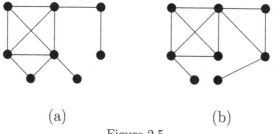

(a)   (b)

Figure 2.5

Note that $\mathcal{C}(n,m) \neq \emptyset$ when $n-1 \leq m \leq \binom{n}{2}$. It is not difficult to show that $\mathcal{L}(n,m) \neq \emptyset$ whenever $n-1 \leq m \leq \binom{n}{2}$; and $\mathcal{J}(n,m) \neq \emptyset$ if and only if $n \leq m \leq n + \binom{n-3}{2}$ and $m - n = \binom{q}{2}$ for some $q \in \mathbb{N}$ (see Exercise 2.10).

Lemma 2.5.1   *Let $G \in \mathcal{L}(n, m)$. Then*

(i) $\omega(G)$ is the minimum integer q such that $m - n + 2 \le \binom{q}{2}$; i.e.,

$$\omega(G) = \lceil (1 + \sqrt{8m - 8n + 17})/2 \rceil$$

and

(ii) $P(G, \lambda) = (\lambda)_{\omega(G)}(\lambda - 1)^{n - \omega(G) - 1}(\lambda - r)$, where $r = m - n + 2 - \binom{\omega(G) - 1}{2}$.

Proof.   (i) Let $q = \omega(G)$. By definition, we have

$$\binom{q}{2} + 1 + (n - q - 1) \le m \le \binom{q}{2} + q - 1 + (n - q - 1);$$

i.e.,

$$\binom{q - 1}{2} + 1 \le m - n + 2 \le \binom{q}{2}.$$

Hence q is the minimum integer such that $m - n + 2 \le \binom{q}{2}$.

(ii) If G is complete, then $\omega(G) = n$ and the result holds. Assume that G is not complete. If G has no end-vertices, then, by definition, G is 2-connected and $\omega(G) = n - 1$. Thus

$$P(G, \lambda) = (\lambda)_{n-1}(\lambda - r),$$

for some $r \in \mathbb{N}$. It is clear that $r = m - \binom{n-1}{2} = m - n + 2 - \binom{\omega(G) - 1}{2}$. Assume that G has an end-vertex x. Then

$$P(G, \lambda) = (\lambda - 1)P(G - x, \lambda),$$

where $G - x \in \mathcal{L}(n - 1, m - 1)$. The result then follows by induction. $\square$

We now state in what follows five results and leave their proofs to the reader (see Exercises 2.11-2.15).

Lemma 2.5.2   *(i) $\mathcal{J}(n, m) \ne \emptyset$ if and only if $n \le m \le n + \binom{n-3}{2}$ and $m - n = \binom{q}{2}$ for some $q \in \mathbb{N}$.*
(ii) If $n \le m \le n + \binom{n-3}{2}$ and $m - n = \binom{q}{2}$ for some $q \in \mathbb{N}$, then

$$P(G, \lambda) = (\lambda)_{q+1}(\lambda - 1)^{n-q-2}(\lambda - 2)$$

for every $G \in \mathcal{J}(n, m)$. $\square$

Lemma 2.5.3   *(i) For any two graphs $G, H \in \mathcal{L}(n, m) \cup \mathcal{J}(n, m)$, we have $P(G, \lambda) = P(H, \lambda)$.*

(ii) Let G be a graph of order n. If

$$P(G, \lambda) = (\lambda)_q (\lambda - 1)^{n-q-1} (\lambda - r)$$

for some $q, r \in \mathbb{N}$ with $2 \le q \le n - 1$ and $1 \le r \le q - 1$, then $G \in \mathcal{L}(n, m) \cup \mathcal{J}(n, m)$, where $m = \binom{q-1}{2} + n + r - 2$. □

Observe that Lemma 2.5.3 (i) follows directly from Lemmas 2.5.1 and 2.5.2. It follows from Lemma 2.5.3 (i) that, whenever we wish to compare the chromatic polynomial of a given G with that of H in $\mathcal{L}(n, m) \cup \mathcal{J}(n, m)$, any graph in $\mathcal{L}(n, m) \cup \mathcal{J}(n, m)$ can be considered as a candidate for H.

Lemma 2.5.4   *Let G be a chordal graph in $\mathcal{C}(n, m)$. Then, under any one of the following conditions, there exists a chordal graph $H \in \mathcal{C}(n, m)$ such that $h_i(G) > h_i(H)$ for every $i = 1, 2, \cdots, n - 2$:*

(i) $P(G, \lambda)$ has a root r, where $r \ge 2$, with multiplicity at least 3;

(ii) $P(G, \lambda)$ has two roots r_1, r_2, where $2 \le r_1 < r_2$, both with multiplicity at least 2. □

By Lemmas 2.5.3 and 2.5.4, the next result follows readily.

Lemma 2.5.5   *For any chordal graph G in $\mathcal{C}(n, m) \setminus (\mathcal{L}(n, m) \cup \mathcal{J}(n, m))$ and any $H \in \mathcal{L}(n, m) \cup \mathcal{J}(n, m)$,*

$$h_i(G) > h_i(H) \tag{2.12}$$

for every $i = 1, 2, \cdots, n - 2$. □

Lemma 2.5.6   *Let $G \in \mathcal{C}(n, m)$. Then G contains a vertex x such that $m - n + 1 \ge \binom{d(x)}{2}$.* □

The following result, due to Dong (2000), will be found useful. A vertex x in G is called *a simplicial vertex* if either $d(x) = 0$ or $[N(x)]$ is a complete subgraph of G.

Lemma 2.5.7   *Let G be a graph of order n, where $n \ge 3$, and $x \in V(G)$ such that $G - x$ is connected. Then there exists a family $\mathcal{S}$ of connected graphs of order $n - 2$ such that*

$$P(G, \lambda) = (\lambda - d(x)) P(G - x, \lambda) + \sum_{G' \in \mathcal{S}} P(G', \lambda), \tag{2.13}$$

where $\mathcal{S}$ is empty if and only if x is a simplicial vertex of G. □

We are now ready to establish the following result due to Rodriguez and Satyanarayana (1997).

Theorem 2.5.1   *For any* $G \in \mathcal{C}(n, m) \backslash (\mathcal{L}(n, m) \cup \mathcal{J}(n, m))$ *and any* $H \in$ $\mathcal{L}(n, m) \cup \mathcal{J}(n, m)$,

$$h_i(G) > h_i(H) \tag{2.14}$$

for every $i = 1, 2, \cdots, n - 2$.

Proof.   By Lemma 2.5.5, the result holds if G is chordal. Thus the result holds if $n \leq 3$. Assume that $n \geq 4$ and G is not chordal. It suffices to show that $h_i(G) > h_i(H)$ for some chordal graph $H \in \mathcal{C}(n, m)$ and for each $i = 1, 2, \cdots, n - 2$.

Case 1: G is 2-connected.

By Lemma 2.5.6, G has a vertex x such that $m - n + 1 \geq \binom{d(x)}{2}$. So $G - x \in \mathcal{C}(n - 1, m - d(x))$, and

$$m - d(x) - (n - 1) + 1 \geq \binom{d(x)}{2} - d(x) + 1 = \binom{d(x) - 1}{2}.$$

If $H_1 \in \mathcal{L}(n - 1, m - d(x))$, then $\omega(H_1) \geq d(x)$ by Lemma 2.5.1. Let H_2 be a graph obtained from H_1 by adding a new vertex w and adding new edges joining w to all the vertices in a $d(x)$-clique of H_1. Observe that H_2 is a chordal graph in $\mathcal{C}(n, m)$. By the induction hypothesis, $h_i(G - x) \geq h_i(H_1)$ for each $i = 1, 2, \cdots, n - 3$, where inequality is strict if $G - x$ is not chordal. Notice that $P(H_2, \lambda) = (\lambda - d(x))P(H_1, \lambda)$. Thus, if we write

$$(\lambda - d(x))P(G - x, \lambda) = \sum_{i=1}^{n} (-1)^{n-i} b_i \lambda^i,$$

then $b_i \geq h_i(H_2)$ for each $i = 1, 2, \cdots, n - 2$, where inequality is strict if $G - x$ is not chordal.

If x is a simplicial vertex of G, then $G - x$ is not chordal, and so

$$h_i(H_2) < b_i = h_i(G),$$

for each $i = 1, 2, \cdots, n - 2$.

Assume that x is not a simplicial vertex. By Lemma 2.5.7,

$$h_i(G) \geq h_i(H_2) + \sum_{G' \in \mathcal{S}} h_i(G') > h_i(H_2)$$

for each $i = 1, 2, \cdots, n - 2$, since $\mathcal{S}$ is non-empty and $h_i(G') > 0$.

Case 2: G is not 2-connected.

In this case, G has a block G_1 which is not chordal. Let $G_2, \cdots, G_s$ be the other blocks of G. Then

$$P(G, \lambda) = \frac{1}{\lambda^{s-1}} \prod_{j=1}^{s} P(G_j, \lambda).$$

Let $H_j \in \mathcal{L}(n_j, m_j)$, where $n_j = v(G_j)$ and $m_j = e(G_j)$ for each $j = 1, 2, \cdots, s$. By the induction hypothesis, $h_i(G_j) \geq h_i(H_j)$ for all $j = 1, 2, \cdots, s$ and $i = 1, 2, \cdots, n_j - 2$, where inequality is strict when $j = 1$, since G_1 is not chordal. Let H' be a graph in $\mathcal{G}[H_1 \cup_1 H_2 \cup_1 \cdots \cup_1 H_s]$. Observe that H' is a chordal graph in $\mathcal{C}(n, m)$ and

$$P(H', \lambda) = \frac{1}{\lambda^{s-1}} \prod_{j=1}^{s} P(H_j, \lambda).$$

Hence $h_i(G) > h_i(H')$ for each $i = 1, 2, \cdots, n - 2$. $\qquad\square$

Very recently, Pitteloud (2003) obtained a result on the least coefficients of the chromatic polynomials of disconnected graphs. For $c, n, m \in \mathbb{N}$, a graph G is called a (c, n, m)-*graph* if G is an (n, m)-graph with c components. It can be shown that there exists a (c, n, m)-graph if and only if $1 \leq c \leq n$ and $n - c \leq m \leq \binom{n-c+1}{2}$ (see Exercise 2.23).

Theorem 2.5.2   *Let $c, n, m \in \mathbb{N}$ such that $1 \leq c \leq n$ and $n - c \leq m \leq \binom{n-c+1}{2}$. If G and H are (c, n, m)-graphs such that*

$$P(H, \lambda) = \lambda^c (\lambda - 1)^z (\lambda - 2)(\lambda - 3) \cdots (\lambda - x + 1)(\lambda - x)(\lambda - y - 1)$$

and $P(G, \lambda) \neq P(H, \lambda)$, then $h_i(G) > h_i(H)$ for $i = c, c + 1, \cdots, n - 2$, where $x, y, z \in \mathbb{N}_0$ are defined by

$$\begin{cases} m - n + c = \binom{x}{2} + y, \ x > y \geq 0, \\ n - c = x + z. \end{cases} \qquad\square$$

Note that the integers x, y, z in Theorem 2.5.2 are uniquely determined by c, n, m. Since $0 \leq y < x$, we have

$$\binom{x}{2} + 1 \leq m - n + c + 1 \leq \binom{x+1}{2},$$

i.e., x is the minimum integer such that $m - n + c + 1 \leq \binom{x+1}{2}$. Thus x is uniquely determined by c, n, m and so are y and z. Once the values of

x, y, z are determined, it is easy to construct a chordal graph H which is a (c, n, m)-graph such that

$$P(H, \lambda) = \lambda^c (\lambda - 1)^z (\lambda - 2)(\lambda - 3) \cdots (\lambda - x)(\lambda - y - 1).$$

(See Exercise 2.24.)

We now turn our attention to the family $\mathcal{C}_p(n, m)$ of connected (n, m)-planar graphs. It is clear that this family is empty if $m > 3n - 6$ or $m < n - 1$. In the following, we shall present a result due to Sakaloglu and Satyanarayana (1997) on the least coefficients of $P(G, \lambda)$, where $G \in \mathcal{C}_p(n, m)$.

It is clear that if $m = n - 1$, then $\mathcal{C}_p(n, m)$ $(= \mathcal{C}(n, m))$ is the family of trees of order n. In what follows, we assume that $m \geq n$.

It can be shown that for a chordal graph $G \in \mathcal{C}_p(n, m)$,

$$P(G, \lambda) = \lambda(\lambda - 1)^{b(G)} (\lambda - 2)^{3n-m-2b(G)-3} (\lambda - 3)^{m-2n+b(G)+2} \quad (2.15)$$

(see Exercise 2.16), which indicates that $P(G, \lambda)$ depends only on n, m and $b(G)$. The following result follows directly from (2.15).

Lemma 2.5.8   *Let $G \in \mathcal{C}_p(n, m)$, where $m \geq n$. If G is chordal, then*

$$b(G) \leq \lfloor (3n - m - 4)/2 \rfloor. \qquad \square$$

Let $\mathcal{L}_p(n, m)$ be the family of chordal graphs G in $\mathcal{C}_p(n, m)$ such that $b(G) = \lfloor (3n - m - 4)/2 \rfloor$. It is easy to show that $\mathcal{L}_p(n, m) \neq \emptyset$ (see Exercise 2.17). By (2.15), all graphs in $\mathcal{L}_p(n, m)$ have the same chromatic polynomial.

Sakaloglu and Satyanarayana (1997) proved the following result.

Theorem 2.5.3   *If $G \in \mathcal{C}_p(n, m) \backslash \mathcal{L}_p(n, m)$ and $H \in \mathcal{L}_p(n, m)$, then $h_i(G) > h_i(H)$ for each $i = 1, 2, \cdots, n - 2$.* $\qquad \square$

2.6   Divisibility of the coefficients

In Theorem 2.3.6, we present Hong's results that (a) $(\chi(G) - 1)! | h_1$ and (b) G is connected and bipartite if and only if h_1 is odd. In this section, we shall establish two results, due to Dong, Koh and Soh (2004), on the divisibility of certain coefficients h_i in $P(G, \lambda)$, which extend Hong's results mentioned above.

We begin by introducing the following quantity '$f(r, k)$' and prove a result on the divisibility of $f(r, k)$. For $r \in \mathbb{N}_0$ and $k \in \mathbb{N}$, define

$$f(r, k) = \begin{cases} 1 & \text{if } r < 2k; \\ \lfloor \frac{r}{k} \rfloor! f(r - \lfloor \frac{r}{k} \rfloor, k) & \text{if } r \geq 2k. \end{cases} \qquad (2.16)$$

In particular, $f(r, 1) = r!$. The following table shows the values of $f(20, i)$ for $i = 1, 2, 3, 4$:

i	1	2	3	4
$f(20, i)$	$20!$	$10! \times 5! \times 2!$	$6! \times 4! \times 3! \times 2!$	$5! \times 3! \times 3! \times 2!$

Lemma 2.6.1 Let $n_1, n_2, \cdots, n_k \in \mathbb{N}_0$ and $k, r \in \mathbb{N}$. If $\sum_{i=1}^{k} n_i > r - k$, then $f(r, k) \mid \prod_{i=1}^{k} (n_i!)$.

Proof. When $k = 1$, $f(r, 1) = r!$. Thus, if $n_1 > r - 1$, we have $f(r, 1) | (n_1!)$, and the result follows.

Assume now that $k \geq 2$, and we shall prove the divisibility by induction on r.

When $r \leq 2k - 1$, we have $f(r, k) = 1$, and the result holds. Assume that $r \geq 2k$. As $\sum_{i=1}^{k} n_i > r - k$, $n_i \geq \lfloor \frac{r}{k} \rfloor$ for some i, say $n_1 \geq \lfloor \frac{r}{k} \rfloor$. Observe that

$$\binom{n_1}{\lfloor \frac{r}{k} \rfloor} = \frac{n_1!}{\lfloor \frac{r}{k} \rfloor! \, (n_1 - \lfloor \frac{r}{k} \rfloor)!}.$$

Denote $n_1 - \lfloor \frac{r}{k} \rfloor$ by n_1'. We have

$$\prod_{i=1}^{k} (n_i!) = n_1'! \left(\prod_{i=2}^{k} (n_i!) \right) \lfloor \frac{r}{k} \rfloor! \binom{n_1}{\lfloor \frac{r}{k} \rfloor}. \qquad (2.17)$$

As $n_1' + \sum_{i=2}^{k} n_i = \sum_{i=1}^{k} n_i - \lfloor \frac{r}{k} \rfloor > (r - \lfloor \frac{r}{k} \rfloor) - k$, by the induction hypothesis,

$$f(r - \lfloor \frac{r}{k} \rfloor, k) \mid n_1'! \prod_{i=2}^{k} (n_i!).$$

It follows by (2.17) that $\prod_{i=1}^{k} (n_i!)$ is divisible by $\lfloor \frac{r}{k} \rfloor! f(r - \lfloor \frac{r}{k} \rfloor, k)$, which is $f(r, k)$ by (2.16). This completes the proof. $\square$

We next present two results on the divisibility of some coefficients of $P(K_n, \lambda)$. Lemma 2.6.1 will be used to prove the first one.

Lemma 2.6.2   *Let $n, r \in \mathbb{N}$ with $n > r$. Then*
(i) $f(r, k) \mid h_k(K_n)$ *for each $k = 1, 2, \cdots, r$;*
(ii) *when $n = pq$, where p is prime, $p \nmid h_q(K_n)$.*

Proof.   (i) Since

$$P(K_n, \lambda) = \prod_{i=0}^{n-1} (\lambda - i),$$

if we let $A = \{1, 2, \cdots, n-1\}$, then

$$h_k(K_n) = \sum_{\substack{X \subseteq A \\ |X| = n-k}} \left(\prod_{x \in X} x \right). \tag{2.18}$$

As $X \subseteq A$, it is clear that

$$\prod_{x \in X} x = \frac{(n-1)!}{\prod_{z \in A \setminus X} z}.$$

Let $A \setminus X = \{z_1, z_2, \cdots, z_{k-1}\}$, where $z_1 < z_2 < \cdots < z_{k-1}$, and let $z_0 = 0$ and $z_k = n$. Observe that

$$\prod_{i=1}^{k} (z_i - z_{i-1} - 1)! \binom{z_i - 1}{z_i - z_{i-1} - 1} = \frac{(z_k - 1)!}{z_1 z_2 \cdots z_{k-1}} = \frac{(n-1)!}{\prod_{z \in A \setminus X} z} = \prod_{x \in X} x.$$

Thus, if we let $n_i = z_i - z_{i-1} - 1$, $i = 1, 2, \cdots, k$, then

$$\prod_{i=1}^{k} (n_i!) \ \Big| \ \prod_{x \in X} x. \tag{2.19}$$

On the other hand, as

$$\sum_{i=1}^{k} n_i = \sum_{i=1}^{k} (z_i - z_{i-1} - 1) = z_k - z_0 - k = n - k > r - k,$$

we have, by Lemma 2.6.1,

$$f(r, k) \ \Big| \ \prod_{i=1}^{k} (n_i!).$$

Finally, by (2.19) and (2.18), the result in (i) follows.

(ii) Let $X^* = A\backslash\{p, 2p, \cdots, (q-1)p\}$ and note that $|X^*| = n - q$. As p is prime, for $X \subseteq A$ and $|X| = n - q$,

$$p \,\Big|\, \prod_{x \in X} x \qquad \text{if and only if} \qquad X \neq X^*.$$

It follows by (2.18) that $p \nmid h_q(K_n)$. $\qquad\qquad \square$

We are now ready to present our two main results in this section.

Theorem 2.6.1 *For any graph G with chromatic number χ,*

$$f(\chi(G) - 1, k) \,\Big|\, h_k(G),$$

for each $k = 1, 2, \cdots, \chi - 1$.

Proof. By Theorems 2.3.1 and 1.4.1, we have

$$\sum_{i=1}^{n} (-1)^{n-i} h_i(G) \lambda^i = P(G, \lambda) = \sum_{i=\chi}^{n} \alpha(G, i) P(K_i, \lambda),$$

where $n = v(G)$. For $1 \le k \le \chi - 1$, by comparing the coefficients of λ^k on both sides, we have

$$(-1)^{n-k} h_k(G) = \sum_{i=\chi}^{n} \alpha(G, i)(-1)^{i-k} h_k(K_i),$$

and so

$$h_k(G) = \sum_{i=\chi}^{n} (-1)^{n-i} \alpha(G, i) h_k(K_i).$$

The result now follows from Lemma 2.6.2 (i). $\qquad\qquad \square$

Theorem 2.6.1 states that $f(\chi(G) - 1, i) \mid h_i(G)$ for each $i = 1, 2, \cdots,$ $\chi(G) - 1$. As no explicit expression for $f(k, i)$ has been found, it is useful to find a factor of $f(k, i)$ which is quite close to it and has an explicit expression. Since $f(\chi(G) - 1, 1) = (\chi(G) - 1)!$, we see that Theorem 2.3.6(iii) is a special case of Theorem 2.6.1 when $k = 1$. We leave it to the reader to deduce the following result from Theorem 2.6.1 (see Exercise 2.21).

Corollary 2.6.1   *For any graph G, the coefficient $h_k(G)$ is divisible by*

$$\prod_{s \geq 1} \left(\lfloor \frac{(k-1)^{s-1}(\chi(G)-1)}{k^s} \rfloor! \right)$$

for each $k = 2, 3, \cdots, \chi(G) - 1$. $\square$

Corollary 2.6.2   *Let G be a graph and $k, q \in \mathbb{N}$. If $\chi(G) \geq kq + 1$, then $q! \mid h_k(G)$.*

Proof.   We may assume that $q \geq 2$. Since $\chi(G) - 1 \geq kq$, by definition, $f(kq, k)$, and so $f(\chi(G) - 1, k)$, is divisible by $\lfloor \frac{kq}{k} \rfloor!$, which is $q!$. By Theorem 2.6.1, $q! \mid h_k(G)$. $\square$

Let us recall Theorem 2.3.6(iv). Suppose $h_1(G)$ is odd. Then $2! \nmid h_1(G)$, and so by Corollary 2.6.2, $\chi(G) \leq 2$; that is, G is bipartite. The converse of this follows from the following theorem if we note that every connected and bipartite graph is uniquely 2-colourable.

Theorem 2.6.2   *Let G be a uniquely r-colourable graph. If $r = kp$ for some prime p and $k \in \mathbb{N}$, then $p \nmid h_k(G)$.*

Proof.   Again, we have

$$h_k(G) = \sum_{i=r}^{n} (-1)^{n-i} \alpha(G, i) h_k(K_i), \qquad (2.20)$$

where $n = v(G)$. By Corollary 2.6.1,

$$\lfloor \frac{i-1}{k} \rfloor! \;\Big|\; h_k(K_i)$$

for each $i = 1, 2, \cdots, n$. Thus, when $i > r = kp$, we have $\frac{i-1}{k} \geq p$, and so

$$p! \;\Big|\; h_k(K_i). \qquad (2.21)$$

Since G is uniquely r-colourable, $\alpha(G, r) = 1$. Suppose on the contrary that $p \mid h_k(G)$. Then, by (2.20) and (2.21), we would have $p \mid h_k(K_r)$. But this contradicts Lemma 2.6.2(ii). Hence $p \nmid h_k(G)$, as was to be shown. $\square$

2.7 Unimodal Conjecture

Read (1968) observed that for any graph G, it appears that the coefficients of $P(G, \lambda)$ always increase in absolute value first, and decrease eventually. In other words, if $P(G, \lambda) = \sum_{i=1}^{n} (-1)^{n-i} h_i \lambda^i$, then there always seems to exist $k \in \mathbb{N}$ with $2 \leq k \leq n-1$ such that

$$h_1 \leq h_2 \leq \cdots \leq h_{k-1} \leq h_k \geq h_{k+1} \geq \cdots \geq h_n. \qquad (2.22)$$

An equivalent form of this observation is that there do not exist any $k, j \in \mathbb{N}$ with $2 \leq k \leq n - j - 1$ such that

$$h_k > h_{k+1} = \cdots = h_{k+j} < h_{k+j+1}.$$

For example,

$$
\begin{aligned}
P(T_8, \lambda) &= \lambda^8 - 7\lambda^7 + 21\lambda^6 - 35\lambda^5 + 35\lambda^4 - 21\lambda^3 + 7\lambda^2 - \lambda, \\
P(C_8, \lambda) &= \lambda^8 - 8\lambda^7 + 28\lambda^6 - 56\lambda^5 + 70\lambda^4 - 56\lambda^3 + 28\lambda^2 - 7\lambda, \\
P(W_8, \lambda) &= \lambda^8 - 14\lambda^7 + 84\lambda^6 - 280\lambda^5 + 560\lambda^4 - 672\lambda^3 + 447\lambda^2 - 126\lambda.
\end{aligned}
$$

As pointed out by Read and Tutte (1988), it was Nijenhuis and Wilf (1978) who first formulated this observation into a conjecture, which is now known as the **Unimodal Conjecture (UC)**. This conjecture has been verified for various families of graphs, but remains unsettled for general situation. In connection with this, we have the following result.

Theorem 2.7.1   *Let G be a connected graph of order n. Then for any $i \in \mathbb{N}$ with $i \leq n - 1$,*

$$h_{i+1} \leq \left(\frac{n}{i} - 1 \right) h_i,$$

where equality holds if and only if G is a tree.

Proof.   We shall prove this result by induction on the order and size (when the order is fixed) of G. The result is trivially true if $n = 2, 3$. If G is a tree, then $P(G, \lambda) = \lambda(\lambda - 1)^{n-1}$, and it follows that $h_{i+1} = (\frac{n}{i} - 1)h_i$ for $i = 1, 2, \cdots, n-1$. Assume that the result is true for $n < k$ or for $n = k$ and G is of size less than m, where $k, m \in \mathbb{N}$ with $k \geq 4$ and $n - 1 < m \leq \binom{n}{2}$.

Suppose that $n = k$ and that G is a connected (n, m)-graph containing a cycle. Let e be an edge in a cycle of G. Then $G \cdot e$ is a connected graph of order $n-1$ and $G-e$ is a connected $(n, m-1)$-graph. By the induction hypothesis, the result holds for $G \cdot e$ and $G - e$. Thus

$$h_{i+1}(G-e) \leq \left(\frac{n}{i} - 1\right) h_i(G-e), \qquad i = 1, 2, \cdots, n-1,$$

and

$$h_{i+1}(G \cdot e) \leq \left(\frac{n-1}{i} - 1\right) h_i(G \cdot e), \qquad i = 1, 2, \cdots, n-2.$$

By FRT, we have

$$h_i = h_i(G-e) + h_i(G \cdot e), \qquad i = 1, 2, \cdots, n-2.$$

Let $i \in \mathbb{N}$ with $i \leq n-2$. We have

$$\begin{aligned}
h_{i+1} &= h_{i+1}(G-e) + h_{i+1}(G \cdot e) \\
&\leq \left(\frac{n}{i} - 1\right) h_i(G-e) + \left(\frac{n-1}{i} - 1\right) h_i(G \cdot e).
\end{aligned}$$

As $h_i(G \cdot e) \geq 1$, it follows that

$$\begin{aligned}
h_{i+1} &< \left(\frac{n}{i} - 1\right) (h_i(G-e) + h_i(G \cdot e)) \\
&= \left(\frac{n}{i} - 1\right) h_i.
\end{aligned}$$

Consider now the case $i = n-1$. As $h_n(G \cdot e) = 0$ and $h_{n-1}(G \cdot e) = 1$, we have

$$\begin{aligned}
h_n &= h_n(G-e) \\
&\leq \left(\frac{n}{n-1} - 1\right) h_{n-1}(G-e) \\
&= \left(\frac{n}{n-1} - 1\right) (h_{n-1} - 1) \\
&< \left(\frac{n}{n-1} - 1\right) h_{n-1},
\end{aligned}$$

completing the proof.   □

As a corollary of Theorem 2.7.1, we have the following result which can be found in Lovász (1993).

Corollary 2.7.1   *Let G be a connected graph of order n.*

(i) If n is odd, then

$$h_n < h_{n-1} < \cdots < h_{\frac{n+1}{2}}.$$

(ii) If n is even, then

$$h_n < h_{n-1} < \cdots < h_{\frac{n}{2}+1} \leq h_{\frac{n}{2}},$$

where $h_{\frac{n}{2}+1} = h_{\frac{n}{2}}$ if and only if G is a tree.

Proof. If n is odd, then $\frac{n}{i} - 1 < 1$ for all i with $\frac{n+1}{2} \leq i \leq n$. If n is even, then $\frac{n}{i} - 1 < 1$ for all i with $\frac{n}{2} + 1 \leq i \leq n$, and $\frac{n}{i} - 1 = 1$ when $i = \frac{n}{2}$. Thus, by Theorem 2.7.1, when n is odd, we have

$$h_n < h_{n-1} < \cdots < h_{\frac{n+1}{2}};$$

and when n is even, we have

$$h_n < h_{n-1} < \cdots < h_{\frac{n}{2}+1} \leq h_{\frac{n}{2}}.$$

If n is even and G is a tree, then

$$h_{\frac{n}{2}+1} = h_{\frac{n}{2}} = \binom{n-1}{\frac{n}{2}-1}.$$

If n is even but G is not a tree, then, by Theorem 2.7.1,

$$h_{\frac{n}{2}+1} < h_{\frac{n}{2}}.$$

The proof is thus complete. $\qquad\qquad\qquad\qquad\qquad\qquad\qquad\square$

Hoggar (1974) proposed to consider, instead of UC, the following conjecture:

For all $k \in \mathbb{N}_0$ with $k \leq n - 3$,

$$h_k h_{k+2} < h_{k+1}^2. \tag{2.23}$$

This conjecture, known as the **Strong Logarithmic Concavity Conjecture (SLCC)**, clearly implies UC.

It is clear that the multiplication of two chromatic polynomials is also a chromatic polynomial (indeed, $P(G, \lambda)P(H, \lambda) = P(G \cup H, \lambda)$). Hoggar (1974) showed that the absolute coefficients of the multiplication of two polynomials whose absolute coefficients satisfy (2.23) also satisfy (2.23), but may not be so if (2.23) is replaced by (2.22). This observation prompted Read and Tutte (1988) to remark that SLCC may be more natural than UC.

Gernert (1984, 1985) gave reports on results found before 1985 about classes of graphs for which SLCC holds. These include chordal graphs (Hoggar (1974)), outerplanar graphs, generalized θ-graphs (Weickert (1983), Gernert (1984)), certain line graphs (Acharya (1982), Beineke (1971), Petkovšek (1984)) and broken wheels (Read (1986)). A cycle C in a connected graph G is called a *separating cycle* if $G - V(C)$ is disconnected. If G is a connected graph for which SLCC fails to hold, then G possesses at least one separating cycle or otherwise G belongs to one of the classes of exceptional graphs spelled out in Gernert (1984). It was reported recently in Lundow and Markström (2002) that SLCC holds for graphs G with $v(G) \leq 11$, and graphs G of order 12 with $e(G) < 20$ or $e(G) > 45$.

Exercise 2

2.1 Let G be a graph containing at least two triangles such that there is at least one vertex of degree 2 in one of the triangles. Show that $(\lambda - 2)^2 | P(G, \lambda)$.

2.2 Let G be a graph, and G' and G'' be two induced subgraphs of G such that $V(G') \cup V(G'') = V(G)$ and $V(G') \cap V(G'') = \{x, y\}$. Prove that if $\chi(G') \geq 3$ and $\chi(G'') \geq 3$, then $(\lambda - 2)^2 | P(G, \lambda)$.

2.3 For any given graph G, let $Z(G)$ be the set of all roots of $P(G, \lambda)$. Show that for any G and any $k \in \mathbb{N}$, there exists a graph H such that

$$Z(H) = \{0, 1, 2, \cdots, k - 1\} \cup \{z + k : z \in Z(G), z \neq 0\}.$$

(See Farrell (1980a).)

2.4 Let G be a graph of order $n \geq 2$. Prove that

(i) G is connected if and only if $h_1(G) \geq 1$;

(ii) G is a tree if and only if $h_1(G) = 1$;

(iii) for $1 \leq i \leq n - 1$, $h_i(G) > 0$ implies that $h_{i+1}(G) > 0$.

2.5 For $n, r \in \mathbb{N}$, let G be a graph of order n and U_r^n a connected unicyclic graph of order n, where the cycle is of length r. Prove that

(i) $h_1(G) = 2$ if and only if $G \cong U_3^n$;

(ii) $h_1(G) = 3$ if and only if $G \cong U_4^n$;

(iii) $h_1(G) = 4$ if and only if $G \cong U_5^n$ or G is a connected graph of size $n + 1$ which contains two C_3's with at most two vertices in common;

(iv) $h_1(G) = 5$ if and only if $G \cong U_6^n$.

(See Hong (1984).)

2.6 Let G be a connected (n, m)-graph containing a unique cycle of order $n - s + 1$. Show that

$$h_r(G) = \begin{cases} \binom{m}{n-r} & \text{if } r > s, \\ \binom{m}{n-r} - \binom{m-n+s}{s-r} & \text{if } r \leq s. \end{cases}$$

(See Meredith (1972).)

2.7 Let G be an (n, m)-graph. Prove that

(i) $h_{n-3} = \binom{m}{3} - (m-2)n(K_3) - i(C_4) + 2n(K_4);$

(ii)

$$
\begin{aligned}
h_{n-4} = {} & \binom{m}{4} - \binom{m-2}{2}n(K_3) + \binom{n(K_3)}{2} \\
& - (m-3)i(C_4) - (2m-9)n(K_4) - i(C_5) \\
& + i(K(2,3)) + 2i(H) + 3i(W_5) - 6n(K_5),
\end{aligned}
$$

where H is the graph shown below:

$H :$

(See Farrell (1980b).)

2.8 Prove the Bari-Hall's broken-cycle formula (2.3).

(See Bari and Hall (1977).)

2.9 Let G be a 2 connected graph of order at least 3. Let $v_1 v_2 \cdots v_k$ be a longest path in G (it is possible that $v_1 = v_k$). Prove that $G \cdot v_1 v_2$ is 2-connected.

2.10 Let $\mathcal{L}(n, m)$ and $\mathcal{J}(n, m)$ be the two families of connected (n, m)-graphs defined in Section 2.5. Show that

(i) $\mathcal{L}(n, m) \neq \emptyset$ whenever $n - 1 \le m \le \binom{n}{2};$

(ii) $\mathcal{J}(n, m) \neq \emptyset$ if and only if $n \le m \le n + \binom{n-3}{2}$ and $m - n = \binom{q}{2}$ for some $q \in \mathbb{N}.$

2.11 Let $n, m \in \mathbb{N}$ such that $n \le m \le n + \binom{n-3}{2}$ and $m - n = \binom{q}{2}$ for some $q \in \mathbb{N}$. Prove that for every $G \in \mathcal{J}(n, m),$

$$
P(G, \lambda) = (\lambda)_{q+1}(\lambda - 1)^{n-q-2}(\lambda - 2).
$$

2.12 Show that for any two graphs $G, H \in \mathcal{L}(n, m) \cup \mathcal{J}(n, m)$, $P(G, \lambda) = P(H, \lambda).$

2.13 Let G be a graph of order n. Assume that

$$P(G, \lambda) = (\lambda)_q (\lambda - 1)^{n-q-1}(\lambda - r)$$

for some $q, r \in \mathbb{N}$ with $2 \leq q \leq n - 1$ and $1 \leq r \leq q - 1$. Prove that $G \in \mathcal{L}(m, n) \cup \mathcal{J}(n, m)$, where $m = \binom{q-1}{2} + n + r - 2$.

2.14 Let G be a connected (n, m)-graph. Prove that G contains a vertex x such that $m - n + 1 \geq \binom{d(x)}{2}$.

2.15 Let G be a chordal graph in $\mathcal{C}(n, m)$. Prove that under any one of the following conditions, there exists a chordal graph $H \in \mathcal{C}(n, m)$ such that $h_i(G) > h_i(H)$ for $i = 1, 2, \cdots, n - 2$:

   (i) $P(G, \lambda)$ has a root r, where $r \geq 2$, with multiplicity at least 3;

   (ii) $P(G, \lambda)$ has two roots r_1, r_2, where $2 \leq r_1 < r_2$, both with multiplicity at least 2.

2.16 Prove that for a chordal graph $G \in \mathcal{C}_p(n, m)$,

$$P(G, \lambda) = \lambda(\lambda - 1)^{b(G)}(\lambda - 2)^{3n - m - 2b(G) - 3}(\lambda - 3)^{m - 2n + b(G) + 2}.$$

2.17 Let $\mathcal{L}_p(n, m)$, where $n \leq m \leq 3n - 6$, be the family of chordal graphs G in $\mathcal{C}_p(n, m)$ such that $b(G) = \lfloor (3n - m - 4)/2 \rfloor$. Show that $\mathcal{L}_p(n, m) \neq \emptyset$.

2.18 A graph G is said to be *cycle-connected* if G contains no cycle C such that $G - V(C)$ is disconnected. Let G be a cycle-connected graph which is not a block. Show that SLCC holds for G.

   (See Gernert (1984).)

2.19 Show that SLCC holds for the following graphs:

   (i) the generalized θ-graphs $\theta(d, e, f)$;

   (ii)

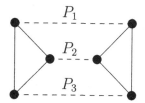

where $l(P_i) \geq 2$ for $i = 1, 2, 3$.

(See Gernert (1984).)

2.20 Show that SLCC holds for the general broken wheel $W_n(m_1, \cdots, m_k)$ as defined in Exercise 1.9(c).

(Note that this result was conjectured by Gernert (1984) and proved by Read (1986).)

2.21 Deduce from Theorem 2.6.1 the following result: For any graph G, $h_k(G)$ is divisible by

$$\prod_{s \geq 1} \left(\lfloor \frac{(k-1)^{s-1}(\chi(G) - 1)}{k^s} \rfloor ! \right)$$

for all $k = 2, 3, \cdots, \chi(G) - 1$.

2.22   (i) Construct a graph G such that $\chi(G)! | h_1(G)$.

(ii) Let G be a uniquely r-colourable graph, where $r \geq 2$. Show that $r! \nmid h_1(G)$.

(iii) Let G be a graph such that $(r-1)! | h_1(G)$ but $r! \nmid h_1(G)$. Show that $\chi(G) \leq r$ and either (1) $\chi(G) = r$ or (2) G is not uniquely k-colourable for any k.

(See Dong, Koh and Soh (2004).)

2.23 For $c, n, m \in \mathbb{N}$, prove that there exists a (c, n, m)-graph if and only if $1 \leq c \leq n$ and $n - c \leq m \leq \binom{n-c+1}{2}$.

2.24 Given $c, n, m \in \mathbb{N}$ with $1 \leq c \leq n$ and $n - c \leq m \leq \binom{n-c+1}{2}$, find $x, y, z \in \mathbb{N}_0$ in terms of c, n, m such that

$$\begin{cases} m - n + c = \binom{x}{2} + y, \ x > y \geq 0, \\ n - c = x + z. \end{cases}$$

Chapter 3

Chromatic Equivalence of Graphs

3.1 Introduction

Non-isomorphic graphs may possess the same chromatic polynomial. Two graphs G and H are said to be *chromatically equivalent* or, simply, χ-*equivalent*, written $G \sim H$, if $P(G, \lambda) = P(H, \lambda)$. It is evident that the relation '$\sim$' of being χ-equivalent is an equivalence relation on the family $\mathcal{G}$ of graphs, and thus $\mathcal{G}$ is partitioned into equivalence classes, called the χ-*equivalence classes*. Given $G \in \mathcal{G}$, let

$$[G] = \{H \in \mathcal{G} : H \sim G\}.$$

We call $[G]$ the χ-equivalence class determined by G. A family $\mathcal{F}$ of graphs is said to be χ-*closed* if $[G] \subseteq \mathcal{F}$ for each $G \in \mathcal{F}$. Thus a χ-closed family of graphs is a union of some χ-equivalence classes. A graph G is said to be *chromatically unique* or, simply, χ-*unique*, if $[G] = \{G\}$. The notion of χ-unique graphs was first introduced by Chao and Whitehead Jr. (1978a). The *chromaticity* of a graph G is the study of the following two basic problems on G:

(1) Is G χ-unique?

(2) If G is not χ-unique, what can be said about $[G]$?

In Section 3.2 below, we shall provide various examples of χ-equivalent graphs and introduce also some general methods of constructing χ-equivalent graphs. Those important properties which are shared by two χ-equivalent graphs will also be listed. After giving a number of typical χ-equivalence classes and χ-unique graphs, and a list of 'small' χ-unique graphs in Section 3.3, we shall proceed in Section 3.4 to present a characterization of disconnected χ-unique graphs due to Read (1987b) and Xu (1987) independently. The works by Chia (1986) and Read (1987b) on the study of χ-unique graphs G with $\kappa(G) = 1, 2$ (see also Xu (1987) for $\kappa(G) = 1$) will then be introduced in Sections 3.5 and 3.6 respectively. The latter will also include a result due to Koh and Goh (1990) that any edge-gluing of K_p and C_q is χ-unique. By applying some results in Section 3.5, we shall present in Section 3.7 a χ-equivalence class of graphs which extends that of trees. Some further results on the chromaticity of edge-gluings of graphs, K_3-gluings of graphs and 2-connected $(n, n + k)$-graphs with small k will be mentioned in the final section.

3.2   Chromatically equivalent graphs

Two graphs G and H are χ-equivalent if $P(G, \lambda) = P(H, \lambda)$. Thus by (1.9), any two trees of the same order are χ-equivalent. More generally, by Exercise 1.1, any two forests of the same order and having the same number of components are χ-equivalent; and by (1.8), any two q-trees of the same order are χ-equivalent. By Theorem 1.3.2, one can construct easily various 'trivial' χ-equivalent graphs using K_r-gluings as shown in Figure 3.1. In what follows, we shall introduce two general methods of construction to construct not so 'trivial' χ-equivalent graphs also by means of Theorem 1.3.2 (see Read (1987b) and Chia (1988a)).

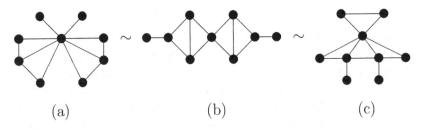

(a)   (b)   (c)

Figure 3.1

Construction 2.1   Let G_1 and G_2 be two connected graphs, and $r, s \in \mathbb{N}_0$ such that $r \leq s \leq \min\{\omega(G_1), \omega(G_2)\}$. Then any graph in $\mathcal{G}[G_1 \cup_r G_2]$ (see Figure 3.2(a)) is χ-equivalent to any graph in $\mathcal{G}[(G_1 \cup_s G_2) \cup_r K_s]$ (see Figure 3.2 (b)).

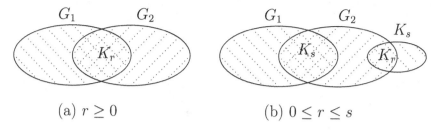

(a) $r \geq 0$ (b) $0 \leq r \leq s$

Figure 3.2

Indeed, if H_1 is a graph of Figure 3.2(a) and H_2 is a graph of Figure 3.2 (b), then, by Theorem 1.3.2,

$$P(H_2, \lambda) = \frac{P(G_1, \lambda)P(G_2, \lambda)}{(\lambda)_s} \cdot \frac{(\lambda)_s}{(\lambda)_r} = \frac{P(G_1, \lambda)P(G_2, \lambda)}{(\lambda)_r} = P(H_1, \lambda).$$

Example 3.2.1   *Let G_1 and G_2 be the graphs given below.*

By letting $r = 2$ and $s = 4$ in Construction 2.1, we have $H_1 \sim H_2$.

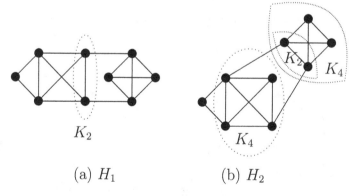

(a) H_1 (b) H_2

Construction 2.2   Let G_1 and G_2 be two connected graphs, and $r \in \mathbb{N}$ such that $r \leq \min\{\omega(G_1), \omega(G_2)\}$. Suppose that G is a K_r-gluing of G_1 and G_2 satisfying the condition:

(*)   for each $i = 1, 2$, there exists $u_i \in V(G_i) \backslash V(K_r)$ such that $V(K_r) \subset N(u_i)$.

Denote $A_i = N(u_i)\backslash V(K_r)$, and let G^* be the graph obtained from G by keeping the same vertex set but (i) deleting the $|A_1|$ edges joining u_1 and A_1 and (ii) adding $|A_1|$ new edges linking u_2 and A_1, as shown in Figure 3.3. Then $G^* \ncong G$ but $G^* \sim G$.

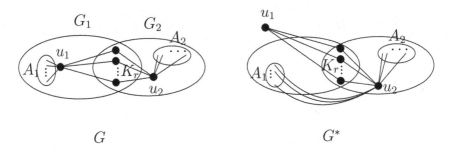

Figure 3.3

Indeed, as $A_i \neq \emptyset$ for each $i = 1, 2$, G and G^* have different degree sequences, and so $G \ncong G^*$. On the other hand, as $[(V(G_1)\backslash\{u_1\})\cup\{u_2\}] \cong G_1$ and $[V(K_r)\cup\{u_i\}] \cong K_{r+1}$ for each $i = 1, 2$, we have, by Theorem 1.3.2,

$$P(G^*, \lambda) = \frac{P(G_1, \lambda)P(G_2, \lambda)}{(\lambda)_{r+1}} \cdot \frac{(\lambda)_{r+1}}{(\lambda)_r} = \frac{P(G_1, \lambda)P(G_2, \lambda)}{(\lambda)_r} = P(G, \lambda).$$

Note that this construction is a special case of Construction 2.1.

Example 3.2.2 *Let G_1 and G_2 be the graphs given below.*

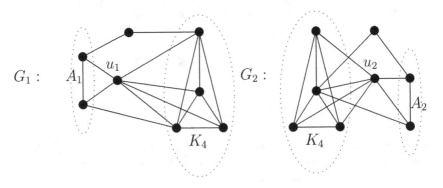

By taking $r = 4$ in Construction 2.2, we have the graphs G and G^* as shown below.

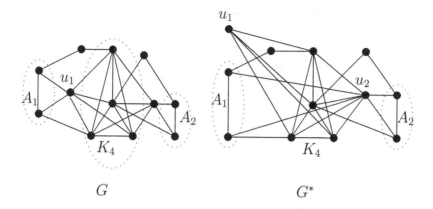

$$G \qquad\qquad G^*$$

Observe that $G^* \not\cong G$ but $G^* \sim G$.

We provide below some other 'non-trivial' pairs of χ-equivalent graphs.

Example 3.2.3 *(1) Bondy and Murty (1976).*

$$(a) \qquad\qquad (b)$$

(2) Chao and Whitehead Jr. (1979b).

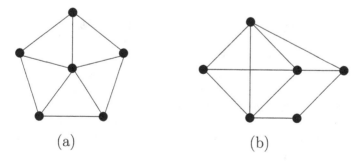

$$(a) \qquad\qquad (b)$$

(3) Chao and Whitehead Jr. (1979b).

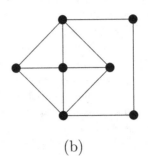

(a)   (b)

(4) Xu and Li (1984).

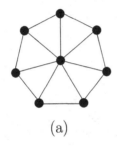
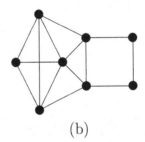

(a)   (b)

(5) Read (1987b).

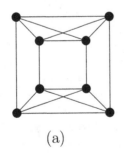
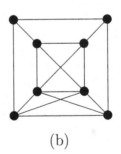

(a)   (b)

(6) Koh and Goh (1990).

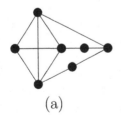
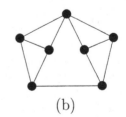

(a)   (b)

(7) Liu, Zhou and Tan (1993).

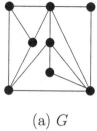

(a) G (b) $\overline{G}$

(8) Chee and Royle (1990)

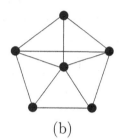

(a)   (b)

(9) Haggard and Mathies (2001)

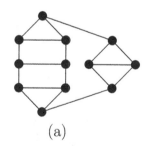

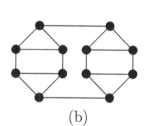

(a)   (b)

(10) Haggard and Mathies (2001)

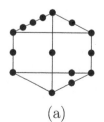
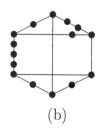
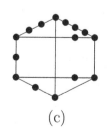

(a)   (b)   (c)   (d)

(11)

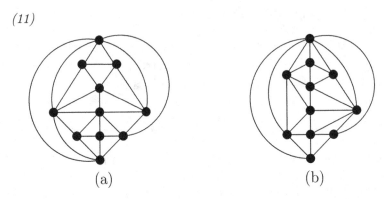

(a)   (b)

Note *This is just an example taken from the list of 25 pairs and 2 triples of χ-equivalent graphs given in Bari (1974).*

Pair (1) of χ-equivalent graphs is in fact an instance of the following more general construction due to Read (1987b).

Construction 2.3 Suppose that G is a connected graph having two non-adjacent vertices u and v, which form a cut such that $G - \{u, v\}$ has only two components, say G_1 and G_2. For each $i = 1, 2$, assume that there exists $x_i \in V(G_i) \cap N(u) \cap N(v)$. Let G^* be the graph obtained from G by identifying x_1 and x_2, and by adding a new vertex w adjacent only to both u and v, as shown in Figure 3.4. Then $G^* \sim G$. (See Exercise 3.2.)

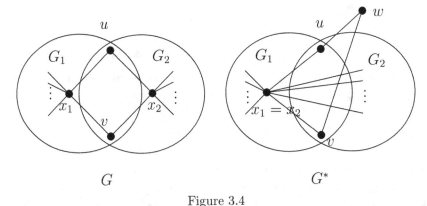

G G^*

Figure 3.4

Even though two χ-equivalent graphs need not be isomorphic, they do have certain properties in common. By definition and by Corollaries 2.3.1 and 2.3.2, Theorems 2.3.2, 2.3.4 and 2.4.1, we have the following results.

Theorem 3.2.1 *Let G and H be two χ-equivalent graphs. Then*

(i) $v(G) = v(H)$;

(ii) $e(G) = e(H)$;

(iii) $\chi(G) = \chi(H)$;

(iv) $n_G(C_3) = n_H(C_3)$;

(v) $i_G(C_4) - 2n_G(K_4) = i_H(C_4) - 2n_H(K_4)$;

(vi) $c(G) = c(H)$;

(vii) $b(G) = b(H)$;

(viii) G is connected if and only if H is connected;

(ix) G is 2-connected if and only if H is 2-connected;

(x) $g(G) = g(H)$;

(xi) $n_G(C_k) = n_H(C_k)$, where $g \le k \le \lceil 3g/2 \rceil - 2$;

(xii) G is bipartite if and only if H is bipartite.   □

3.3 Equivalence classes and χ- unique graphs

Recall that the relation '$\sim$' of being χ-equivalent is an equivalence relation on the family $\mathcal{G}$ of graphs, and thus $\mathcal{G}$ is partitioned into equivalence classes, called the χ-*equivalence classes*. For $G \in \mathcal{G}$, we write $[G] = \{H \in \mathcal{G} : H \sim G\}$; and G is said to be χ-*unique* if $[G] = \{G\}$; i.e., G does not share its chromatic polynomial with any other graph not isomorphic to it.

In what follows, we shall present some examples of χ-equivalence classes and χ-unique graphs.

Example 3.3.1   If T is a tree of order n, then by (1.9), $P(T, \lambda) = \lambda(\lambda - 1)^{n-1}$, and so any two trees of order n are χ-equivalent. If G is a graph such that $P(G, \lambda) = \lambda(\lambda - 1)^{n-1}$, then by Theorem 2.3.6(ii), $G \cong T$ for some tree T of order n. It follows that the family of trees of order n forms a χ-equivalence class. Indeed, more generally, given any $n, c \in \mathbb{N}$, the family of forests of order n and having c components forms a χ-equivalence class. We shall see in Chapter 6 that the family of q-trees of order n forms also a χ-equivalence class.

Example 3.3.2   It follows from Theorem 3.2.1 that any empty graph O_n, complete graph K_n and cycle C_m, where $n \ge 1$ and $m \ge 3$, are χ-unique.

Example 3.3.3   *The complete bipartite graph $K(p,p)$, where $p \geq 1$, is χ-unique.*

Indeed, let G be a graph such that $G \sim K(p,p)$. Then, by Theorem 3.2.1, $v(G) = 2p$, $e(G) = p^2$ and G is bipartite. Let V_1 and V_2 be the partite sets of G and assume that $|V_1| = p + k$, where $0 \leq k \leq p - 1$. Then $p^2 = e(G) \leq (p + k)(p - k) = p^2 - k^2$, and so $k = 0$. This shows that $G \cong K(p,p)$ and so $K(p,p)$ is χ-unique.   □

Example 3.3.4   *It follows from Lemma 2.5.3 and Theorem 2.5.1 that the family $\mathcal{L}(n,m) \cup \mathcal{J}(n,m)$ forms a χ-equivalence class, where $n \leq m \leq \binom{n}{2}$.*

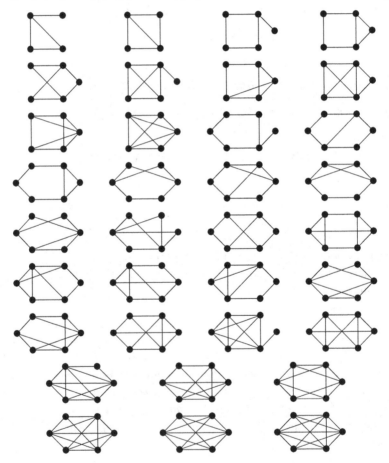

To end this section, we provide above the list of all connected χ-unique graphs of order n, $4 \leq n \leq 6$, excluding K_n, C_n and $K(p,p)$ (see Giudici and Vinke (1980)). For those of order n, $7 \leq n \leq 8$, the reader may refer to the list compiled by Li (1997).

3.4   Disconnected χ-unique graphs

There are disconnected graphs which are χ-unique. The empty graphs are such examples. Are there other disconnected χ-unique graphs? This problem was first studied by Giudici (1985). He showed that for a given graph H without cut-vertices and for any $k \in \mathbb{N}$, the graph $H \cup O_k$ is χ-unique if and only if H is χ-unique. Read (1987b) and Xu (1987) proved the following result which completely characterizes disconnected χ-unique graphs.

Theorem 3.4.1   *Let G be a disconnected graph. Then G is χ-unique if and only if $G \cong H \cup O_k$ for some $k \in \mathbb{N}$, where H is a connected χ-unique graph without cut-vertices (i.e., H is 2-connected or $H \cong K_i$, $i = 1, 2$).*

Proof.   Suppose that $G \cong H \cup O_k$, where H is a connected χ-unique graph without cut-vertices. We shall show that G is χ-unique. This is obvious if $H = K_1$. Thus, assume that $v(H) \geq 2$ and $b(H) = 1$. Let G' be a graph such that $G' \sim G$. Then $c(G') = k + 1$ and $b(G') = b(G) = 1$, and so $G' = H' \cup O_k$, where $b(H') = 1$. As

$$\lambda^k P(H', \lambda) = P(G', \lambda) = P(G, \lambda) = \lambda^k P(H, \lambda),$$

$H' \sim H$. Thus $H' \cong H$ since H is χ-unique. Hence $G' \cong G$, showing that G is χ-unique.

Conversely, suppose that G is a disconnected χ-unique graph. We first claim that G has at most one non-trivial component. Indeed, if G contains two non-trivial components, say G_1 and G_2, then by the result in Construction 2.1 ($r = 0$ and $s = 1$), the following two graphs are χ-equivalent.

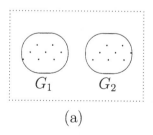
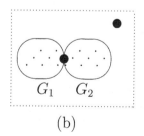

| (a) | (b) |

So G cannot be χ-unique, a contradiction.

Thus, we may write $G = H \cup O_k$ for some $k \in \mathbb{N}$, where H is a connected graph. As G is χ-unique, H must be χ-unique. It remains to prove that $b(H) = 1$. Indeed, if $b(H) \geq 2$, then by the result in Construction 2.1 ($r = 1$ and $s = 2$) again, the following two graphs are χ-equivalent,

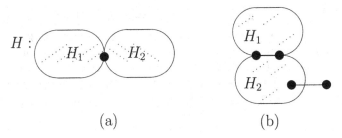

(a)   (b)

which shows that H is not χ-unique, a contradiction. The proof is thus complete.   □

3.5   One-connected χ-unique graphs

The study of disconnected χ-unique graphs reduces, by Theorem 3.4.1, to the study of 2-connected χ-unique graphs. What can be said about the χ-uniqueness of the class of graphs which are connected but not 2-connected? This problem had been solved by Chia (1986), Read (1987b) and Xu (1987) independently, and their solution is given below. For convenience, we shall denote by $G \cdot H$ a vertex-gluing of two connected graphs G and H.

Theorem 3.5.1   *Let G be a connected graph that is not 2-connected. Then G is χ-unique if and only if $G = H \cdot K_2$ (see Figure 3.5), where H is a vertex-transitive and χ-unique graph.*

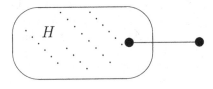

Figure 3.5

Note   The sufficiency of Theorem 3.5.1 was conjectured by Whitehead Jr. and Zhao (1984b). The proof that we shall give below requires the following result due to Watkins (1970).

Lemma 3.5.1 *Every vertex-transitive graph G with $\kappa(G) = 2$ is a cycle.*

□

Proof of Theorem 3.5.1. We shall assume that $v(H) \geq 2$.

[Sufficiency] Let $G = H \cdot K_2$, where H satisfies the assumption, and let G' be a graph such that $G' \sim G$. By Theorem 3.2.1, $b(G') = 2$. If $G' = G_1 \cdot G_2$, where G_i is a block with $v(G_i) \geq 3$ for each $i = 1, 2$, then by the result in Construction 2.1 ($r = 1$ and $s = 2$), the following two graphs are χ-equivalent, where H^* is an edge-gluing of G_1 and G_2.

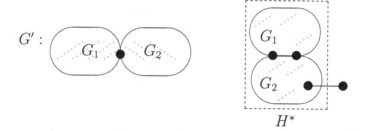

Thus

$$(\lambda - 1)P(H, \lambda) = P(G, \lambda) = P(G', \lambda) = (\lambda - 1)P(H^*, \lambda),$$

and so $H \sim H^*$. As H is χ-unique, $H^* \cong H$. Since $\kappa(H) = \kappa(H^*) = 2$ and H is vertex-transitive, by Lemma 3.5.1, $H^*(\cong H)$ is a cycle, which is not true. Thus $G' = H' \cdot K_2$, where H' is a block. But then, again, as $G' \sim G$, we have $H' \sim H$, and so $H' \cong H$ as H is χ-unique. This implies that $G' \cong G$, showing that G is χ-unique.

[Necessity] Suppose that G is a connected χ-unique graph, which is not 2-connected. Clearly, as G is χ-unique, G is a vertex-gluing of exactly 2 blocks, say B_1 and B_2. If $v(B_i) \geq 3$ for each $i = 1, 2$, then by the result in Construction 2.1, the following two non-isomorphic graphs are χ-equivalent:

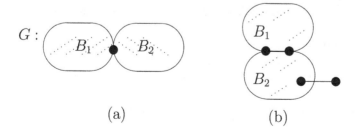

(a)   (b)

a contradiction. Thus $G = B \cdot K_2$, where B is a block. The assumption that G is χ-unique implies again that B has to be χ-unique and vertex transitive.   □

Remark   Chia (1986) asked if every vertex-transitive connected graph is χ-unique. Example 3.2.3 (5) given by Read (1987b) shows that the answer is 'no'. Farrell and Whitehead Jr. (1990) provided another example as shown in Figure 3.6. Chia (1995a) produced a big family of such graphs by generalizing that of Figure 3.6.

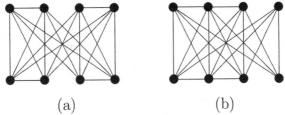

(a)   (b)

Figure 3.6   A vertex-transitive graph which is not χ-unique

3.6 χ-unique graphs with connectivity 2

It is now clear that the study of χ-unique graphs reduces to that of 2-connected χ-unique graphs. While the general situation of this task is too complicated, we shall confine ourselves mainly to the class of graphs of connectivity two.

Thus, let G be a 2-connected graph with a 2-element cut $\{u, v\}$. For the case when u and v are not adjacent, there is so far no known result on the chromaticity of G except the one that is contained in Construction 2.3 due to Read (1987b). In what follows, we shall assume that u and v are adjacent, and thus G is an edge-gluing of two graphs. The following result, which is a combination of work due to Read (1987b) and Chia (1988a), provides some necessary conditions for an edge-gluing of two blocks to be χ-unique. A cut S of G is called a k-*clique-cut* if S is a k-clique.

Theorem 3.6.1   *Let G be an edge-gluing of two blocks H_1 and H_2, where $e = uv$ is the gluing edge. If G is χ-unique, then*

(i) $\{u, v\}$ is the only 2-clique-cut of G;

(ii) both H_1 and H_2 are χ-unique;

(iii) either H_1 or H_2 satisfies the condition: for each $x \in V(H_i) \backslash \{u, v\}$, if $\{u, v\} \subseteq N(x)$, then $N(x) = \{u, v\}$;

(iv) either H_1 or H_2 is vertex-transitive;

(v) both H_1 and H_2 are edge-transitive;

(vi) if $H_i \not\cong C_3$ for each $i = 1, 2$, then either H_1 or H_2 contains no C_3. □

We leave it to the readers to prove the above result (see Exercise 3.6).

Chia (1997a) asked if the converse of Theorem 3.6.1 is true. While its general answer is not known, there are some results supporting its truth. One simple example is an edge-gluing of two cycles (i.e., a θ-graph (Chao and Whitehead Jr. (1978))). The following result, obtained by Koh and Goh (1990), is another example.

Theorem 3.6.2 *Any edge-gluing of a complete graph K_p and a cycle C_q, where $p, q \geq 3$, is χ-unique.*

To prove Theorem 3.6.2, let us first recall the notion of a critical graph with respect to vertex-colourings. A p-chromatic graph H, $p \geq 1$, is said to be p-*critical* if $\chi(F) < \chi(H)$ for any proper subgraph F of H. Every p-chromatic graph always contains a p-critical graph as a subgraph; and every critical graph, as expected, possesses special properties. Some of them which shall be applied in our proof are stated below.

Result 1 If H is a p-critical graph, then $\delta(H) \geq p - 1$.

Result 2 Let H be a p-critical graph. Then either $H \cong K_p$ or $v(H) \geq p + 2$.

It is known that $\chi(G) \leq \Delta(G) + 1$ for any graph G. The celebrated Brooks' theorem on vertex-colouring states that for a connected graph H that is neither an odd cycle nor a complete graph, $\chi(H) \leq \Delta(H)$. An equivalent result of Brooks' theorem is given below (see Bondy and Murty (1976), for instance).

Result 3 Let H be a p-critical graph that is not complete. Then

$$2e(H) \geq 1 + (p - 1)v(H).$$

Proof of Theorem 3.6.2.   Let $Q(p, q)$ denote the edge-gluing of K_p and C_q. We may assume that $p \geq 4$. Let G be a graph such that $G \sim Q(p, q)$. By Theorem 3.2.1, G is 2-connected, of order $n = p + q - 2$ and size $\binom{p}{2} + n - p + 1$, and $\chi(G) = p$.

Let H be a subgraph of G that is p-critical and let $h = v(H)$. Clearly, $h \geq p$ and, by Result 1, $\delta(H) \geq p - 1$.

Claim $h = p$.

Suppose that $h > p$. By Result 2, $h \geq p + 2$.

Case 1: $h \leq n - 1$. As G is 2-connected, we have

$$2\left(\binom{p}{2} + n - p + 1 \right) = \sum_{v \in V(G)} d(v) \geq h(p - 1) + 2(n - h) + 2,$$

which implies that $p \leq 3$, a contradiction.

Case 2: $h = n (\geq p + 2)$. Since H is p-critical and not complete, by Result 3, we have

$$2\left(\binom{p}{2} + n - p + 1 \right) = 2e(G) \geq 2e(H) \geq 1 + (p - 1)h = 1 + (p - 1)n.$$

This, however, implies that $2p \leq 7$, a contradiction.

Thus we have $h = p$, as claimed.

It follows from the claim that $H \cong K_p$. Let s be the number of vertices in H that have their neighbours in $G - V(H)$, and t the number of vertices in $G - V(H)$ that have their degrees at least 3 in G. As G is 2-connected, we have $s \geq 2$ and

$$2\left(\binom{p}{2} + n - p + 1 \right) \geq (p - s)(p - 1) + sp + 3t + 2(n - p - t).$$

The latter implies that $s + t \leq 2$ while the former forces that $s = 2$ and $t = 0$. We thus conclude that $G \cong Q(p, q)$, as required. $\square$

Tomescu (1994a) characterized all p-chromatic 2-connected graphs of order n, where $n \geq p \geq 4$, such that the total sum of the distances between the vertices is maximum, and deduced that if G is a p-chromatic 2-connected graph of order n such that $e(G)$ is minimum, then $G \cong Q(p, n - p + 2)$. The latter implies that the graph $Q(p, q)$ is χ-unique when $p \geq 4$.

The problems of studying the chromaticity of an edge-gluing of (A) a complete graph and a complete bipartite graph and (B) a complete bipartite graph and a cycle were proposed in Koh and Teo (1990). (Note that the complete bipartite graph $K(p,q)$, $q \geq p \geq 2$, is χ-unique, and this is discussed in Chapter 5.) Chia (1995b) obtained the following partial solution of (A).

Theorem 3.6.3 *Every edge-gluing of a complete graph K_m and a complete bipartite graph $K(p, p+r)$, where $0 \leq r \leq 1$, is χ-unique if (i) $m = 3$ and $p \geq 1$ or (ii) $m > p^2 - (2-r)p + 3 - r$ and $p \geq 3$.* $\square$

A partial solution of (B) was obtained independently by Xu, Liu and Peng (1994) and Chia and Ho (2001) as stated below.

Theorem 3.6.4 *For all $q \geq 1$ and $r \geq 3$, every edge-gluing of a complete bipartite graph $K(2,q)$ and a cycle C_r is χ-unique.* $\square$

Recently, Chia and Ho (2003, 2004) have also obtained the following results:

Theorem 3.6.5 *For each $r \geq 3$, the following graphs are χ-unique:*
(i) the edge-gluing of $K(2,2,2)$ and C_r;
(ii) the edge-gluing of $K(3,3)$ and C_r. $\square$

3.7　A chromatically equivalence class

In this section, we shall apply some results in Section 3.5 to produce a χ-equivalence class of graphs, which extends that of trees.

Let H be a connected graph. Any vertex-gluing of H and K_2 is called a *H-with a bridge*. For $k \geq 2$, if G is a H-with $k-1$ bridges, then any vertex-gluing of G and K_2 is called a *H-with k bridges*. It is obvious that if G is a H-with k bridges, then

$$P(G,\lambda) = (\lambda - 1)^k P(H,\lambda). \tag{3.1}$$

Thus every two H-with k bridges are χ-equivalent.

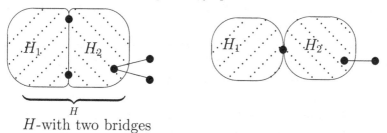

H-with two bridges

Figure 3.7　Two χ-equivalent graphs

On the other hand, it is not difficult to construct a graph G which is not a H-with k bridges but its $P(G, \lambda)$ has the form (3.1) as shown in Figure 3.7 (see Construction 2.1).

Whitehead Jr. and Zhao (1984b) observed that if H is a complete graph (resp., cycle) and G is a graph such that (3.1) holds, then G contains a subgraph isomorphic to a H-with k bridges. Generalizing these two observations, Zang and Yuan (1998) showed that if H is a χ-unique connected graph that does not contain any 2-clique-cut and G is a graph such that (3.1) holds, then G contains a subgraph isomorphic to a H-with k bridges. Their arguments, in fact, prove the following stronger result.

Theorem 3.7.1   *Let H be a χ-unique connected graph without any 2-clique-cut and $k \in \mathbb{N}_0$. Then the family of all H-with k bridges forms a χ-equivalence class.*

To prove Theorem 3.7.1, we first prove the following:

Lemma 3.7.1   *Let G be a vertex-transitive and χ-unique graph. Then G is a block containing no 2-clique-cut.*

Proof.   As G is vertex-transitive, G contains no cut-vertex, and so G is a block. If G contains a 2-clique-cut, then $\kappa(G) = 2$, and by Lemma 3.5.1, G is a cycle, which is not true.   □

Proof of Theorem 3.7.1.   It suffices to show that if G is a graph such that (3.1) holds, then G is a H-with k bridges.
Case 1: H is a block.

If $H = K_1$, then, by (3.1), $P(G, \lambda) = \lambda(\lambda - 1)^k$, and so G is a tree, which is a K_1-with k bridges.

If $k = 0$, then, by (3.1), $P(G, \lambda) = P(H, \lambda)$, and so $G \cong H$ as H is χ-unique.

We may assume now that $k \geq 1$ and $e(H) \geq 1$. As H is a block, by Theorem 2.4.1, the multiplicity of root 1 in $P(H, \lambda)$ is 1, and so the multiplicity of root 1 in $P(G, \lambda)$ is $k + 1$ by (3.1). By Theorem 2.4.1 again, let $B_1, \cdots, B_r$ be the r blocks with $v(B_i) \geq 3$, and the remaining $k + 1 - r$ blocks be K_2 in G. Suppose that $r \geq 2$. Form a new graph H' by edge-gluing the $B_1, \cdots, B_r$ together with a common edge (say) as shown in Figure 3.8.

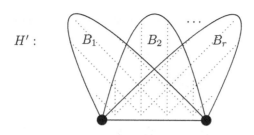

$H':$

Figure 3.8

Let G' be a H'-with k bridges. By applying the result in Construction 2.1 $(r = 1, s = 2)$ repeatedly, it follows that $G' \sim G$. Thus

$$(\lambda - 1)^k P(H', \lambda) = P(G', \lambda) = P(G, \lambda) = (\lambda - 1)^k P(H, \lambda),$$

and so $H' \sim H$. As H is χ-unique, we have $H' \cong H$, which is absurd as H' contains a 2-clique-cut while H does not. Thus $r = 0$ or $r = 1$. When $r = 0$, we have $\lambda(\lambda - 1)^{k+1} = P(G, \lambda) = (\lambda - 1)^k P(H, \lambda)$, and so $P(H, \lambda) = \lambda(\lambda - 1)$, i.e., $H = K_2$. It is clear in this case that G is a H-with k bridges. Assume now that $r = 1$. Then

$$(\lambda - 1)^k P(B_1, \lambda) = P(G, \lambda) = (\lambda - 1)^k P(H, \lambda),$$

and so $B_1 \sim H$. As H is χ-unique, $B_1 \cong H$. Thus G is a H-with k bridges.

Case 2: H contains a cut-vertex.

As H is χ-unique, by Theorem 3.5.1, $H = B \cdot K_2$, where B is a vertex-transitive and χ-unique graph. Thus

$$P(G, \lambda) = (\lambda - 1)^k P(H, \lambda) = (\lambda - 1)^{k+1} P(B, \lambda),$$

and it follows from Lemma 3.7.1 that B is a block containing no 2-clique-cut. By Case 1, G is a B-with $k + 1$ bridges, and so is a H-with k bridges. The proof is thus complete. $\square$

Corollary 3.7.1 *Let H be a vertex-transitive and χ-unique connected graph, and $k \in \mathbb{N}_0$. Then the family of all H-with k bridges forms a χ-equivalence class.* $\square$

3.8 Further results

3.8.1 A K_3-gluing of graphs

The technique used in proving Theorem 3.6.2 was further utilized by Koh and Goh (1990) to study the chromaticity of a K_3-gluing of K_p, $p \geq 4$, and a K_4-homeomorph which contains a C_3 (see Figure 3.9). Note that the χ-uniqueness of K_4-homeomorphs with a C_3 is presented in Chapter 4.

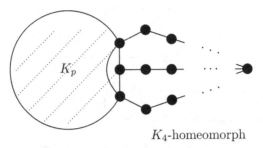

K_4-homeomorph

Figure 3.9

Their result is stated below.

Theorem 3.8.1   *Every K_3-gluing of K_p, $p \geq 4$, and a K_4-homeomorph containing a C_3 is χ-unique except G_1 and G_2 in Figure 3.10. Further, $[G_i] = \{G_i, R_i\}$ for each $i = 1, 2$, as shown in Figure 3.10.*   □

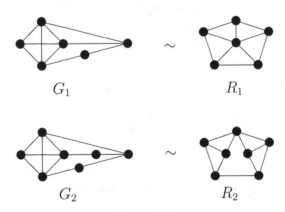

Figure 3.10

Let G be a graph of Figure 3.9. The proof of Theorem 3.8.1 given by Koh and Goh (1990) first made use of the notion of p-critical graph to show that if $G' \sim G$, then G' also contains K_p as a maximum clique. However, the rest of the proof to conclude that $G' \cong G$, where $G \not\cong G_i$, $i = 1, 2$, is, as expected, much more complicated than that of Theorem 3.6.2, and it makes an extensive use of Whitney's Broken-cycle Theorem before arriving at the conclusion.

3.8.2 Polygon-trees and related structures

A family of graphs, called *polygon-trees*, is defined recursively as follows. The smallest polygon-tree with a single polygon is a cycle. A polygon-tree with n polygons, $n \geq 2$, is an edge-gluing of a cycle and a polygon-tree with $n - 1$ polygons. Chao and Li (1985) studied the chromaticity of a subfamily of the above family, called k-*gon trees*, $k \geq 3$, where each polygon is a C_k (see Figure 3.11); and claimed that the following result is true.

Theorem 3.8.2   *Given any $r \geq 1$ and $k \geq 3$, the family of all k-gon trees with r k-gons forms a χ-equivalence class.*   □

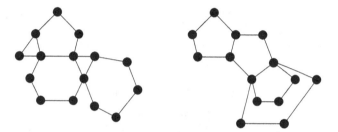

A polygon-tree   A 5-gon tree with four 5-gons

Figure 3.11

Wakelin and Woodall (1992), however, pointed out that the proof given in Chao and Li (1985) was incorrect, and a correct proof was provided therein. Indeed, Wakelin and Woodall (1992) established the following more general result.

Theorem 3.8.3   *A graph G is a polygon-tree with r_k k-gons, where $k = 3, 4, \cdots$, if and only if*

$$P(G, \lambda) = (-1)^n \lambda(\lambda - 1) \prod_{k \geq 3} \left(1 + (1 - \lambda) + (1 - \lambda)^2 + \cdots + (1 - \lambda)^{k-2}\right)^{r_k},$$

where $n = 2 + \sum_{k \geq 3} r_k(k - 2)$.   □

Corollary 3.8.1   *Given $r_k \in \mathbb{N}_0$ for $3 \leq k \leq s$, where $s \geq 3$, the family of polygon trees with r_k k-gons for $k = 3, 4, \cdots, s$ forms a χ-equivalence class.*   □

By a different approach, Theorem 3.8.3 was also proved by Xu (1994) independently in answering a problem asked by Whitehead Jr. (1988).

For each $k \geq 3$, denote by $\theta_k(k-1)$ the multi-bridge graph

$$\theta(\overbrace{k-1, k-1, \cdots, k-1}^{k}),$$

and for $r \geq 1$, let $\mathcal{B}(r, k)$ be the family of graphs obtained via edge-gluing r $\theta_k(k-1)$'s, similar to that in defining polygon-trees (see Figure 3.12). Koh and C.P. Teo (1996) obtained the following result, parallel to Theorem 3.8.2.

Theorem 3.8.4 *For each $k \geq 3$ and $r \geq 1$, the family $\mathcal{B}(r, k)$ forms a χ-equivalence class.* □

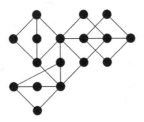

Figure 3.12   A graph in $\mathcal{B}(5, 3)$

3.8.3   2-connected $(n, n+k)$-graphs with small k

Evidently, every cycle is χ-unique. Note that the cycle C_n, where $n \geq 3$, is the only 2-connected (n, n)-graph. Chao and Whitehead Jr. (1978), who initiated the study of χ-unique graphs, showed that every θ-graph $\theta(1, b, c)$, where $b \geq c \geq 2$, is χ-unique. Thus result was extended by Loerinc (1978) to include all generalized θ-graphs. Note that generalized θ-graphs are the only 2-connected $(n, n+1)$-graphs. Chao and Zhao (1983) studied the family of connected $(n, n+2)$-graphs which contain no end-vertices via their chromatic polynomials. Motivated by this, K.L. Teo and Koh (1991) determined all equivalence classes of 2-connected $(n, n+2)$-graphs that have two triangles or an induced C_4. In particular, they found several families of χ-unique graphs.

Let $H_1(a), H_2(b, c), H_3(d, e, f), H_4(g, h)$ be respectively the graphs (1), (2), (3), (4) shown in Figure 3.13.

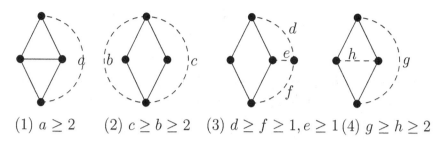

$$(1)\ a \geq 2 \qquad (2)\ c \geq b \geq 2 \qquad (3)\ d \geq f \geq 1, e \geq 1 \ (4)\ g \geq h \geq 2$$

Figure 3.13

Chao and Zhao expressed the chromatic polynomial of each 2-connected $(n, n+2)$-graph G as $P(G, w+1)$ and then multiplied it by $(w+1)^2$. The resulting polynomial $Q(G, w)$ can then be explicitly expressed in terms of the maximal chain lengths of the graph. K.L. Teo and Koh studied the coefficients of $Q(G, w)$, for all graphs G under investigation. They found the following families of χ-unique graphs.

Theorem 3.8.5 *(i) The graph $H_1(a)$ is χ-unique for all $a \geq 2$.*

(ii) The graph $H_2(b, c)$ is χ-unique if and only if $(b, c) \neq (3, 4)$.

(iii) The graph $H_3(d, e, f)$ is χ-unique for all $d \geq f \geq 1$ and $e \geq 1$.

(iv) The graph $H_4(g, h)$ is χ-unique for all $g \geq h \geq 2$. □

Note that (i) was also proved by Chao and Whitehead Jr. (1979b) and Chao and Zhao (1983).

Chen and Ouyang (1997b) determined the χ-equivalence classes of all 2-connected $(n, n+2)$-graphs of girth 5 that are not K_4-homeomorphs (see also Section 5.4). They found the following χ-unique graphs.

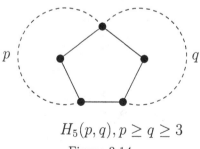

$$H_5(p, q), p \geq q \geq 3$$

Figure 3.14

Theorem 3.8.6 *(i) The graph $\theta(2, 3, a, b)$, where $a \geq b \geq 3$, is χ-unique if and only if $(a, b) \neq (4, 5)$.*

(ii) The graph $H_5(p,q)$ in Figure 3.14 is χ-unique for all $p \geq q \geq 3$. □

The chromaticity of subdivisions of $H_2(b,c)$, $H_3(d,e,f)$ and $H_5(p,q)$ will be studied in Chapter 5.

Koh and K.L. Teo (1994) further studied the chromaticity of 2-connected $(n, n+3)$-graphs with at least two triangles. Their paper contains an error, which was pointed out by Li and Feng (2002). The error was corrected by Dong, Teo and Koh (2002). We provide the χ-unique graphs obtained by these authors in the following theorem.

Let $M_1(a)$, $M_2(b)$, $M_3(c,d)$, $M_4(e,f,g)$ and $M_5(h,i,j)$ be respectively the graphs (1), (2), (3), (4), (5) shown in Figure 3.15.

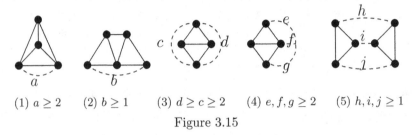

(1) $a \geq 2$ (2) $b \geq 1$ (3) $d \geq c \geq 2$ (4) $e, f, g \geq 2$ (5) $h, i, j \geq 1$

Figure 3.15

Theorem 3.8.7   *(i) $M_1(a)$ is χ-unique for all $a \geq 2$.*

(ii) $M_2(b)$ is χ-unique for all $b \in \mathbb{N}$.

(iii) $M_3(c,d)$ is χ-unique if and only if $d \neq c+1$.

(iv) $M_4(e,f,g)$ is χ-unique if and only if $f \geq 3$ and $(e,f,g) \neq (a, a+2, a)$ or $(a+1, a, a+1)$ for all $a \geq 2$.

(v) $M_5(h,i,j)$ is χ-unique for all $h, i, j \in \mathbb{N}$. □

Recently, Peng and Lau (2004) determined the χ-equivalence classes of all 2-connected $(n, n+4)$-graphs with at least four triangles. They found the following families of χ-unique graphs.

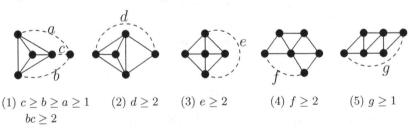

(1) $c \geq b \geq a \geq 1$ (2) $d \geq 2$ (3) $e \geq 2$ (4) $f \geq 2$ (5) $g \geq 1$
      $bc \geq 2$

Figure 3.16

Let $G_1(a, b, c)$, $G_2(d)$, $G_3(e)$, $G_4(f)$, and $G_5(g)$ be respectively the graphs (1), (2), (3), (4) and (5) shown in Figure 3.16.

Theorem 3.8.8   *(i) $G_1(a, b, c)$ is χ-unique if and only if $(a, b, c) \neq (1, 2, 2)$. (See Example 3.2.3(6).)*

(ii) $G_2(d)$ is χ-unique if and only if $d \neq 2$. (Note that $G_2(2) \sim W_6$; see Example 3.2.3(2).)

(iii) $G_3(e)$ is χ-unique if and only if $e \notin \{2, 3\}$. (See Examples 3.2.3 (1) and (3).)

(iv) $G_4(f)$ is χ-unique if and only if $f \neq 2$. (See (3).)

(v) $G_5(g)$ is χ-unique for all $g \geq 1$.   □

Note   Theorem 3.8.8 (i) is a special case of Theorem 3.8.1.

Exercise 3

3.1 Show that for each $i = 1, 2, \cdots, 9$, the two graphs of pair (i) in Example 3.2.3, are χ-equivalent.

3.2 Let G be a connected graph containing a cut $\{u, v\}$, where $uv \notin E(G)$. Let G^* be the graph obtained from G as described in Construction 2.3 of this chapter. Show that $G^* \sim G$.

(See Read (1987b).)

3.3 Let G be a vertex-transitive and χ-unique graph. Show that G contains no 2-clique-cut.

(See Xu (1987).)

3.4 Given the graph H below, determine $[H]$.

3.5 Let G be the graph as shown below, where $r \geq 2$.

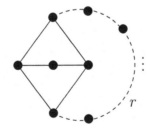

Show that G is χ-unique for all $r \geq 2$.

(See Chia (1995b).)

3.6 Let G be an edge-gluing of two blocks H_1 and H_2, where $e = uv$ is the gluing edge. Suppose that G is χ-unique. Prove that the six necessary conditions (i)-(vi) in Theorem 3.6.1 hold.

(See Read (1987b) and Chia (1988a).)

3.7 Prove that the χ-equivalence class of $W(7,5)$ consists of $W(7,5)$ and the graph (b) in Example 3.2.3 (3).

3.8 A connected graph is said to be *forest-like* if every pair of induced cycles have at most one edge in common. Prove that if G and H are forest-like graphs with the same order and size and the same number of induced cycles of each length, then $G \sim H$.

(See Chao and Whitehead Jr. (1978).)

3.9 Prove that any two graphs in $\mathcal{G}(G_0 \cup_{i_1} G_1 \cup_{i_2} \cdots \cup_{i_k} G_k)$ are χ-equivalent.

3.10 Show that the Petersen graph is χ-unique.

(See Chia (1988b).)

3.11 For any $n, r \in \mathbb{N}$ with $r \leq n - 3$, let $Q(n, r)$ denote the graph $K_n - E(H)$, where $H = K_2 \cup K(1, r)$.

(i) Verify that the graphs in each of the following pairs are χ-equivalent.

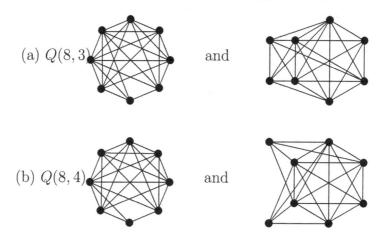

(a) $Q(8, 3)$ and

(b) $Q(8, 4)$ and

(ii) Show that $Q(n, r)$ is χ-unique if and only if $r \notin \{3, 4\}$.

(See Han (1986).)

3.12 (i) For $k \in \mathbb{N}$, let $\mathcal{U}_k$ denote the family of graphs G whose vertex set can be partitioned into k subsets $V_1, V_2, \cdots, V_k$ such that for any two vertices u, v in G, there is an automorphism θ of G with $\theta(u) = v$ if and only if $\{u, v\} \subseteq V_i$ for some $i = 1, 2, \cdots, k$. (Thus $\mathcal{U}_1$ consists of all vertex-transitive graphs.)

For $r = 0, 1$, consider a K_{r+1}-gluing $G_1 \cup_{r+1} G_2$ of two graphs G_1 and G_2. Let u be a vertex in this K_{r+1} such that $A_1 = N(u)\backslash V(G_2) \neq \emptyset$ and $A_2 = N(u)\backslash V(G_1) \neq \emptyset$. Let G be the graph obtained by joining a new vertex w to all the vertices in $K_{r+1} - u$.

Now, let α_r denote the operation:

> Remove from G all the edges joining u to A_1, but add $|A_1|$ new edges linking w to all vertices in A_1.

Let $G \in \mathcal{U}_k$ and for $r = 0, 1$, let H_r be a graph such that

$$P(H_r, \lambda) = (\lambda - r)P(G, \lambda).$$

Show that $[H_r]$ consists of the graphs which are obtained by applying the operations α_r and β_r on any graph in $[G]$ within a finite number of steps, where β_1 (resp., β_0) denotes the operation of linking a new vertex to a vertex in the graph (resp., adding a new isolated vertex to the graph).

(ii) Let G be the graph shown below ($P_3 \times P_2$):

(a) Verify that $G \in \mathcal{U}_2$.

(b) Verify that $P(G, \lambda) = \lambda(\lambda - 1)(\lambda^2 - 3\lambda + 3)^2$.

(c) Show that G is χ-unique.

(d) Let H_1 be a graph such that $P(H_1, \lambda) = \lambda(\lambda - 1)^2(\lambda^2 - 3\lambda + 3)^2$. Determine $[H_1]$.

(See Chia (1998).)

Chapter 4

Chromaticity of Multi-Partite Graphs

4.1 Introduction

All complete graphs K_n are χ-unique. A natural question is to what extent this result can be extended by deleting some edges from K_n. In particular, which complete t-partite graphs $K(p_1, p_2, \cdots, p_t)$, $p_i \in \mathbb{N}$, are χ-unique? Loerinc and Whitehead Jr. (1981) studied a class of graphs which are obtained from K_n by deleting a set of k ($0 \leq 2k \leq n$) independent edges. They showed that all these graphs are χ-unique. Clearly, each of them is isomorphic to some $K(p_1, p_2, \cdots, p_t)$, where $t \geq 2$ and $1 \leq p_i \leq 2$ for each $i = 1, 2, \cdots, t$. A more general result, which we state below, was obtained by Chao and Novacky Jr. (1982).

Theorem 4.1.1 *For each $t \geq 2$, $K(p_1, p_2, \cdots, p_t)$ is χ-unique if $|p_i - p_j| \leq 1$ for all $i, j = 1, 2, \cdots, t$.* $\qquad\square$

Indeed, the above result follows from Turan's Theorem (see Bondy and Murty (1976)) that the graph considered in the theorem is the only graph (up to isomorphism) of order $n = p_1 + p_2 + \cdots + p_t$, with chromatic number equal to t, having the maximum size.

The chromaticity of multi-partite graphs has attracted the attention of many researchers especially when the graphs are bipartite and tripartite. Much progress has been gained in the case of bipartite graphs and to a lesser extent tripartite graphs. In this chapter, we first deal with the chromaticity of complete multi-partite graphs and then complete bipartite graphs with some edges deleted. We finish this chapter by mentioning some results concerning complete tripartite and 4-partite graphs with some edges removed.

4.2 Complete bipartite graphs

As particular cases of Theorem 4.1.1, we see that for any $p \in \mathbb{N}$, the graphs $K(p, p)$ (see Example 3.3.3) and $K(p, p + 1)$ are χ-unique. When $p = 1$, we obtain the only two χ-unique stars. In 1978, Chao conjectured (see Salzberg, López and Giudici (1986)) that the graph $K(p, p + k)$ is χ-unique if $p \geq 2$ and $0 \leq k \leq 2$. This was later confirmed by Salzberg, López and Giudici (1986). In fact, they proved more generally that the graph $K(p, p + k)$ is χ-unique for all $p \geq 2$ and $0 \leq k \leq \max\{5, \sqrt{2p}\}$, and conjectured further that all complete bipartite graphs $K(p, q)$ are χ-unique when $p \geq q \geq 2$.

Through the study of some extremal properties of 3-colourings of certain bipartite graphs, Tomescu (1987) improved the above result slightly by showing that $K(p, q)$ is χ-unique if $p \geq 2$ and $0 \leq q - p < \sqrt{4(p + 1)}$. The conjecture was finally proved to be true by C.P. Teo and Koh (1990). A couple more alternative proofs appeared soon after (Dong (1993), C.P. Teo and Koh (1992)). We shall outline the proofs provided by Teo and Koh, and Dong.

Theorem 4.2.1 *The complete bipartite graph $K(p, q)$ is χ-unique for all p, q with $q \geq p \geq 2$.*

Let $G \sim K(p, q)$. Then, by Theorem 3.2.1(xii), G is bipartite. Write $G = (A, B; E)$, where A, B are the two partite sets of G and $E = E(G)$. Let $|A| = x$, $|B| = y$. Then $x + y = p + q$ and $xy \geq pq$. We may assume $x \leq y$. Hence

$$p \leq x \leq y \leq q.$$

Let us write $x = p + k$, $y = q - k$, where $0 \leq k \leq \frac{1}{2}(q - p)$. Thus G is obtained from $K(p + k, q - k)$ by deleting $s = (q - p)k - k^2$ edges. Suppose $A = \{a_1, a_2, \cdots, a_{p+k}\}$, $B = \{b_1, b_2, \cdots, b_{q-k}\}$. We define the *matching matrix* of G as the $(p + k) \times (q - k)$ matrix $M = (m_{ij})$ as follows:

$$m_{ij} = \begin{cases} 1 & \text{if } a_i b_j \in E(G); \\ 0 & \text{otherwise.} \end{cases}$$

Evidently, there is a 1-1 correspondence between the family of all C_4 : $\{a_i, a_j, b_u, b_v\}$ $(i < j, u < v)$ in G and the family of all 4-entry sets: $\{m_{iu}, m_{iv}, m_{ju}, m_{jv}\}$ with value 1 forming a rectangle in M as shown below:

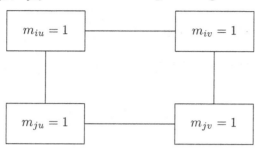

We call such a rectangle a *grid* in M. Thus $n_G(C_4)$ is equal to the number of such grids in M.

Since $e(G) = pq$, there are exactly pq entries in M with value 1. For each of such entry 1_i $(i = 1, 2, \cdots, pq)$, let n_i be the number of grids in M containing 1_i as an entry. We have

$$n_G(C_4) = \frac{1}{4} \sum_{i=1}^{pq} n_i.$$

C.P. Teo and Koh (1990) showed that for each $i = 1, 2, \cdots, pq$,

$$n_i \le (p-1)(q-1),$$

and for some i, the inequality is strict if $k \ne 0$. Thus

$$n_G(C_4) < \frac{1}{4} pq(p-1)(q-1) = \binom{p}{2}\binom{q}{2}.$$

It therefore follows from Theorem 3.2.1 that $K(p, q)$ is χ-unique. □

We next outline the proof given by Dong (1993) in establishing Theorem 4.2.1. Dong first proved the following:

Lemma 4.2.1 *Let $G = (A, B; E)$ be a bipartite graph.*

(i) *For any $e \in E$, the number of C_4 in G containing e is bounded above by $|E| - |A| - |B| + 1$.*

(ii) *If $x \in A, y \in B$ and $xy \notin E$, then for any edge e incident to x, the number of C_4 in G containing e is no more than $|E| - |A| - |B| - d_G(y) + 2$.* □

Dong then deduced from this the following:

Lemma 4.2.2 *Suppose that $G = (A, B; E)$ is a bipartite graph with $|A| \geq 2$ and $|B| \geq 2$. Then*

$$n_G(C_4) \leq |E|(|E| - |A| - |B| + 1)/4.$$

If $d_G(v) \geq 2$ for every $v \in V(G)$, then the equality holds if and only if G is a complete bipartite graph.   □

From Lemma 4.2.2, it follows easily that every complete bipartite graph $K(p, q)$, where $q \geq p \geq 2$, is χ-unique.

4.3   Complete tripartite graphs

While the chromaticity of all complete bipartite graphs has been settled, plenty of work is still needed to be done for tripartite graphs. The first serious attempt to classify complete tripartite graphs by their chromatic polynomials was carried out by Chia, Goh and Koh (1988). Their results have recently been extended by Zou (2000b) and Liu, Zhao and Ye (2002) and others by using a new tool.

Let $G = K(p, q, r)$ be a complete tripartite graph with $p \leq q \leq r$. Then $v(G) = p+q+r$, $e(G) = pq+qr+rp$, $\chi(G) = 3$, $n_G(C_3) = pqr$, $n_G(K_4) = 0$ and

$$n_G(C_4) = \binom{p}{2}\binom{q}{2} + \binom{q}{2}\binom{r}{2} + \binom{r}{2}\binom{p}{2}.$$

Now suppose $H \sim G$. Then $\chi(H) = \chi(G) = 3$. Hence H is a tripartite graph obtained from certain $K(u, v, w)$ by deleting s edges, where

$$s = (uv + vw + wu) - (pq + qr + rp).$$

If the difference $r - p$ is small, so is s. In this case, one can compare the number $n_H(C_i)$ and $n_G(C_i)$, for $i = 3, 4$, for each possible candidate H, and determine whether the given $K(p, q, r)$ is χ-unique. Using this idea, Chia, Goh and Koh (1988) established the χ-uniqueness of the following families of complete tripartite graphs:

  (I) $K(q, q, q + k)$, where $q \geq 2$ and $0 \leq k \leq 3$;

  (II) $K(q - k, q, q + k)$, where $q \geq 5$ and $0 \leq k \leq 2$;

(III) $K(q - k, q, q)$, where $q \geq k + 2$ and $0 \leq k \leq 3$.

Result (I) above does not cover the case when $p = 1$. In fact, it follows from a result on multi-partite graphs, which we shall mention later, that $K(1, q, r)$ is χ-unique if and only if $q, r \leq 2$.

Result (II) assumes $q \geq 5$. The remaining cases we have not discussed are $K(2, 3, 4)$, $K(3, 4, 5)$, and $K(2, 4, 6)$. The first two graphs were shown to be χ-unique by Chia, Goh and Koh (1988). Ten years later, Zou (1998b) proved that $K(2, 4, 6)$ is also χ-unique. We shall mention his approach later in a more general setting.

In connection with result (III), Chia, Goh and Koh (1988) conjectured that the graph $K(q - k, q, q)$ is χ-unique all $k \in \mathbb{N}_0$ and $q \geq k + 2$.

Given a graph G of order n and $k = \chi(G), \cdots, n$, recall (see Theorem 1.4.1) that $\alpha(G, k)$ is the number of ways of partitioning $V(G)$ into k independent sets. For convenience, we shall call $\alpha(G, k)$ the k-partition number of G.

In Section 4.2, we mentioned Tomescu's contribution in attempting to establish the χ-uniqueness of $K(p, q)$ for $p, q \geq 2$. An important tool he used is the 3-partition number of a bipartite graph. This has later been employed by several researchers to study the chromaticity of many families of bipartite graphs. We shall cover this in Section 4.5. As for tripartite graphs, Zou (2000b) likewise utilized the 4-partition numbers to prove that the above conjecture is true when $k = 4$. Zou and Shi (1999) later showed that it is also true provided that $q > k + \frac{k^2}{3}$.

By employing a method similar to that of Zou (2000b), Liu, Zhao and Ye (2002) eventually settled the conjecture affirmatively by proving the following stronger result.

Theorem 4.3.1 *Let* $G = K(p, q, r)$, *where* $2 \leq p \leq q \leq r$. *If* $H \sim G$, *then* $H = K(u, v, w) - S$, *where* $1 \leq u \leq v \leq w$, $q \leq w \leq r$ *and* $S \subseteq E(K(u, v, w))$. *In particular, if* $w = r$ *then* $H \cong G$. $\qquad \square$

We next describe formally the tool they used.

Let G be a graph of order n, and $k \in \mathbb{N}$. By Theorem 1.4.1,

$$P(G, \lambda) = \sum_{k=1}^{n} \alpha(G, k)(\lambda)_k.$$

Let us define

$$\xi(G, x) = \sum_{k=1}^{n} \alpha(G, k)x^k. \tag{4.1}$$

We shall call this the ξ-*polynomial* of G.

As $\{1, (\lambda)_1, (\lambda)_2, \cdots, (\lambda)_n\}$ is a basis for the polynomials of degree n, we have:

Lemma 4.3.1 *Two graphs G and H are χ-equivalent if and only if $\xi(G, x) = \xi(H, x)$; that is, $\alpha(G, k) = \alpha(H, k)$ for all k.* $\qquad\square$

Theorem 1.5.1 can now be rewritten as

$$\xi(G_1 + G_2, x) = \xi(G_1, x)\xi(G_2, x).$$

In particular,

$$\xi(K(p_1, p_2, \cdots, p_t), x) = \prod_{i=1}^{t} \xi(O_{p_i}, x).$$

Let us give a proof of Theorem 4.3.1, which is due to Liu, Zhao and Ye (2002).

Proof of Theorem 4.3.1. Suppose $H \sim G$. Then H is a tripartite graph obtained from some $F = K(u, v, w)$ by deleting s edges, say $e_1, e_2, \cdots, e_s$. We may assume $u \le v \le w$.

Let $t(e_i)$ be the number of triangles containing the edge e_i in F. Then $t(e_i) \le w$, and thus

$$n_H(C_3) \ge n_F(C_3) - sw,$$

and if the equality holds, then $t(e_i) = w$ for all $i = 1, 2, \cdots, s$.

Let $\Gamma = n_F(C_3) - n_G(C_3)$. Evidently, $\Gamma = uvw - pqr$.

By Theorem 3.2.1, $n_H(C_3) = n_G(C_3)$. Thus $\Gamma \le sw$. Let us write $f(w) = \Gamma - sw$. Thus $f(w) \le 0$. Since $e(H) = e(G)$ and $v(H) = v(G)$, we have

$$s = uv + vw + wu - pq - qr - rp$$

and

$$u + v + w = p + q + r.$$

Hence

$$
\begin{aligned}
f(w) &= uvw - pqr - [uv + (u + v)w - pq - qr - rp]w \\
&= uvw - pqr - [uv + (p + q + r - w)w - pq - qr - rp]w \\
&= (w - p)(w - q)(w - r) \qquad\qquad\qquad\qquad (4.2)
\end{aligned}
$$

From the fact that $u + v + w = p + q + r$ and $u \le r \le w$, we obtain

$$p \le \frac{p + q + r}{3} \le w.$$

Note that if $w = p$, then $p = q = r$. It follows from (4.2) that $f(w) \leq 0$ if and only if $q \leq w \leq r$.

Suppose now $w = r$. We consider three cases.

Case 1: $q < v \leq r$.

In this case, $u + v = q + p$ and $u < p \leq q$. Let $t = p - u = v - q$. Then $s = uv + ur + vr - rq - rp - qp = uv - qp = u(t+q) - q(t+u) = t(u-q) < 0$. This is a contradiction.

Case 2: $v = q$.

In this case, $u = p$ and $F = K(p, q, r)$. Thus $s = 0$ and so $H \cong G$.

Case 3: $u \leq v < q$.

Let the three partite sets of $F = K(u, v, w)$ be X_1, X_2, and X_3. Assume that $|X_1| = u$, $|X_2| = v$, and $|X_3| = w = r$. Note that $f(w) = f(r) = 0$. We have

$$n_G(C_3) = n_H(C_3) = n_F(C_3) - sr$$

and $t(e_i) = w = r$ for all $i = 1, 2, \cdots, s$. Since $u \leq v < q \leq r = w$, each e_i must join X_1 to X_2. So, $\overline{H}$ contains K_r as a component. Let us write $\overline{H} = \overline{H}_1 \cup K_r$. Thus $H = H_1 + O_r$. Now $\xi(H, x) = \xi(G, x)$, and so

$$\xi(H_1, x)\xi(O_r, x) = \xi(K(p, q), x)\xi(O_r, x).$$

Thus

$$\xi(H_1, x) = \xi(K(p, q), x),$$

which implies that

$$H_1 \sim K(p, q).$$

As $p \geq 2$, by Theorem 4.2.1, $K(p, q)$ is χ-unique. Hence $H_1 \cong K(p, q)$, which implies that $v = q$, a contradiction. $\qquad \square$

Corollary 4.3.1 *For any $k \in \mathbb{N}_0$, the graph $K(q - k, q, q)$ is χ-unique if $q \geq k + 2$.*

This result was proved independently by Chia and Ho (2005) by using a fairly similar method to that of Liu et al. Moreover, Chia and Ho also determined the chromatic class of $K(1, n, n)$; namely, the set of graphs of the form $T + O_n$, where T is a tree of order $n + 1$.

Employing a similar method, Liu, Zhao and Ye (2002) also established the following result.

Theorem 4.3.2 *For $q, k \in \mathbb{N}$ with $k \geq 2$ and $q \geq 2k$, the graph $K(q - k, q - 1, q)$ is χ-unique.* $\qquad \square$

Zou (2000b) (see also Zou and Shi (1999, 2000)) established the χ-uniqueness of general tripartite graphs under a rather complicated condition. Three consequences are of interest.

Theorem 4.3.3   (i) $K(p_1, p_2, p_3)$ is χ-unique if $|p_i - p_j| \leq 2$ for all i, j and $\min\{p_1, p_2, p_3\} \geq 3$.

(ii) The graph $K(q - k, q, q + k)$ is χ-unique if $q > \frac{2\sqrt{3}}{3}k + k^2$.

(iii) The graph $K(q, q, q + k)$ is χ-unique if $q > \frac{k+k^2}{3}$.   □

Part (i) of Theorem 4.3.3 has been extended by Chia and Ho (2005). They proved that the result is true when $|p_i - p_j| \leq 3$ and $\min\{p_1, p_2, p_3\} \geq 2$. Their proof was based mainly on the following result.

Lemma 4.3.2   If Y is χ-equivalent to $K(p_1, p_2, p_3)$, where $|p_i - p_j| \leq 3$, then Y is isomorphic to $H + O_t$ for some $t \in \{p_1, p_2, p_3\}$, and H is a bipartite graph.

Chia and Ho also found the following two chromatic classes:

$$[K(1, 2, 4)] = \{T + O_2 : T \text{ is a tree of order } 5\},$$

$$[K(1, 3, 4)] = \{T + O_3, S + O_4 : T \text{ is a tree of order } 5, S \text{ is a tree of order } 4\}.$$

They posed the following question.

Question 4.3.1   What is the chromatic equivalence class for the graph $K(1, m, n)$, where $2 \leq m < n$?

Problem 4.3.1 Find $p, q, r \geq 2$ such that $K(p, q, r)$ is not χ-unique and determine its χ-equivalence class.

4.4   Complete multi-partite graphs

In this section, we consider a complete multi-partite graph in general. Theorem 4.1.1 says that for $t \geq 2$ and $p_i \in \mathbb{N}$ ($i = 1, 2, \cdots, t$), the complete multi-partite graph $K(p_1, p_2, \cdots, p_t)$ is χ-unique if $|p_i - p_j| \leq 1$ for all $i, j = 1, 2, \cdots, t$. Koh and K.L. Teo (1990) put forward the following problem.

Problem 4.4.1 *For each $t \geq 2$, is the graph $K(p_1, p_2, \cdots, p_t)$ χ-unique if $|p_i - p_j| \leq 2$ for all $i, j = 1, 2, \cdots, t$ and if $\min\{p_1, p_2, \cdots, p_t\}$ is sufficiently large?*

In connection with this, Giudici and López showed (1985) that the complete t-partite graph $K(q - 1, q, \cdots, q, q + 1)$ is χ-unique if $t \geq 2$ and $q \geq 3$. Theorem 4.1.1 implies that $K(1, p_2, \cdots, p_t)$ is χ-unique if $\max\{p_2, p_3, \cdots, p_t\} \leq 2$. Chia (1998), and Li and Liu (1990) showed, independently, that the condition is also necessary. Assume

$$p_2 = \max\{p_2, \cdots, p_t\}.$$

Note that $K(1, p_2, \cdots, p_t)$ is isomorphic to the join $K(1, p_2) + O_{p_3} + \cdots + O_{p_t}$, and $K(1, p_2)$ is χ-unique if and only if $p_2 \leq 2$. Thus we have:

Theorem 4.4.1 *The complete t-partite graph $K(1, p_2, \cdots, p_t)$ is χ-unique if and only if $\max\{p_2, p_3, \cdots, p_t\} \leq 2$.* $\square$

The chromatic polynomial of $K(p_1, p_2, \cdots, p_t)$ was computed in Section 1.5. For the ξ-polynomial of $K(p_1, p_2, \cdots, p_t)$, we have:

Lemma 4.4.1 *For $p_1, p_2, \cdots, p_t \in \mathbb{N}$,*

$$\xi(K(p_1, p_2, \cdots, p_t), x) = \prod_{i=1}^{t} \xi(O_{p_i}, x),$$

and for $p \in \mathbb{N}$,

$$\xi(O_p, x) = \sum_{j=1}^{p} S(p, j) x^j,$$

where $S(p, j)$ is the Stirling number of the second kind. $\square$

Note that $S(p, 1) = 1$ and $S(p, 2) = 2^{p-1} - 1$. The following result follows from Lemma 4.4.1.

Lemma 4.4.2 *Let $G = K(p_1, p_2, \cdots, p_t)$. Then $\alpha(G, k) = 0$ for $k = 1, 2, \cdots, t-1$, $\alpha(G, t) = 1$ and*

$$\alpha(G, t + 1) = \sum_{i=1}^{t} 2^{p_i - 1} - t.$$ $\square$

Zhao, Li, Liu and Ye (2004) showed that the minimum real root of $\xi(O_p, x)$ is less than that of $\xi(O_q, x)$ if $p > q$. The following result now follows by induction on t.

Lemma 4.4.3 *Two complete t-partite graphs are χ-equivalent if and only if they are isomorphic.* $\square$

Of all the t-partite graphs of a given order n, the one mentioned in Theorem 4.1.1 is very special. It follows from Turan's Theorem that the graph is the unique (up to isomorphism) graph of order n with chromatic number t that has the maximum size. Denote this t-partite graph by $K[t; n]$. We have:

Lemma 4.4.4   *Let $G = K(p_1, p_2, \cdots, p_t)$, where $p_1 \leq p_2 \leq \cdots \leq p_t$, and let $n = v(G)$. Then*

(i) $e(G) \leq e(K[t; n])$, *and equality holds if and only if $G = K[t; n]$;*

(ii) $e(K[t; n]) - e(G) \geq p_t - p_1 - 1$. $\square$

Let H be a graph such that $H \sim K(p_1, p_2, \cdots, p_t)$. Then it follows from Lemma 4.4.2 that $H \cong F - S$, where $F = K(q_1, q_2, \cdots, q_t)$ for some $q_1, q_2, \cdots, q_t \in \mathbb{N}$ and $S \subseteq E(F)$. Evidently, $|S| = e(F) - e(G) \leq e(K[t; n]) - e(G)$, where $n = \sum_{i=1}^{t} p_i$.

Zou (2004) (see also Zhao, Li, Liu and Ye (2004)) proved the following:

Lemma 4.4.5 *Let $F = K(q_1, q_2, \cdots, q_t)$, where $q_i \in \mathbb{N}$ for all $i = 1, 2, \cdots, t$. Let $H = F - S$, where $S \subseteq E(F)$ and $0 \leq |S| < \min\{q_1, q_2, \cdots, q_t\}$. Then*

$$|S| \leq \alpha(H, t+1) - \alpha(F, t+1) \leq 2^{|S|} - 1.$$ $\square$

By applying Lemma 4.4.5, Zhao, Li, Liu and Ye (2004) proved the following result, which gives a positive answer to Problem 4.4.1.

Theorem 4.4.2   *Let $G = K(p_1, p_2, \cdots, p_t)$, where $p_1 \leq p_2 \leq \cdots \leq p_t$, and let $n = v(G)$. If $p_1 \geq e(K[t; n]) - e(G) + 1$, then G is χ-unique.*

We outline their proof in what follows.

Let $H \sim G$. Then $H \cong F - S$, where $F = K(q_1, q_2, \cdots, q_t)$ and $S \subseteq E(F)$. Furthermore,

$$|S| = e(F) - e(G) \leq e(K[t; n]) - e(G).$$

Let $\Gamma = \alpha(H, t+1) - \alpha(F, t+1)$. By Lemma 4.4.5, $|S| \leq \Gamma \leq 2^{|S|} - 1$.

Now by Lemma 4.4.2,

$$
\begin{aligned}
\alpha(G, t+1) - \alpha(H, t+1) &= \alpha(G, t+1) - \alpha(F, t+1) - \Gamma \\
&= \sum_{i=1}^{t} 2^{p_i-1} - \sum_{i=1}^{t} 2^{q_i-1} - \Gamma \\
&= 2^{p_1-1} \sum_{i=1}^{t} (2^{p_i-p_1} - 2^{q_i-p_1}) - \Gamma.
\end{aligned}
$$

Since $\alpha(G, t+1) = \alpha(H, t+1)$ by Lemma 4.3.1, the above expression is zero. As $\Gamma \geq 0$,

$$
M = \sum_{i=1}^{t} (2^{p_i-p_1} - 2^{q_i-p_1}) \geq 0.
$$

By making use of Lemma 4.4.5, Zhao, Li, Liu and Ye (2004) showed further that, under the assumption $p_1 \geq e(K[t; n]) - e(G) + 1$, M cannot be positive. Thus $M = 0$, and so $\Gamma = 0$, which, in turn, implies that $|S| = 0$. Hence $H \cong K(q_1, q_2, \cdots, q_t)$, and so $H \cong G$ by Lemma 4.4.3. $\qquad\square$

Several results on tripartite graphs follow from Theorem 4.4.2. For example, $K(q, q, q+2)$ is χ-unique for $q \geq 2$. In this case, $K[3; n] = K(q, q+1, q+1)$, where $n = 3q+2$. However, the χ-uniqueness of $K(q, q, q+3)$ does not follow from this theorem. In this case, $K[3; n] = K(q+1, q+1, q+1)$, where $n = 3q + 3$. Neither does the χ-uniqueness of $K(2, 4, 6)$. But can the lower bound for $\min\{p_1, p_2, \cdots, p_t\}$ in Theorem 4.4.2 be relaxed? The answer is no. For example, for $n = 5$ and $t = 3$, the graph $G = K(1, 1, 3)$ is not χ-unique. Note that in this case, the lower bound in Theorem 4.4.2 is

$$
e(K[3; 5]) - e(G) + 1 = e(K(1, 2, 2)) - e(K(1, 1, 3)) + 1 = 2.
$$

As a consequence of Theorem 4.4.2, Zhao, Li, Liu and Ye (2004) proved the following result which is more general than what Problem 4.4.1 is asking for.

Corollary 4.4.1 Let $p_1, p_2, \cdots, p_t, k \in \mathbb{N}$, where $p_1 \leq p_2 \leq \cdots \leq p_t$. Then the graph $K(p_1, p_2, \cdots, p_t)$ is χ-unique if $|p_i - p_j| \leq k$ for all i, j and $p_1 \geq tk^2/4 + 1$. $\qquad\square$

4.5 Complete bipartite graphs with some edges deleted

We have seen in Section 4.2 that the chromaticity of complete bipartite graphs has been completely determined. We now turn our attention to other bipartite graphs. Salzberg, López and Giudici (1986) proved that the graph obtained from $K(p,q)$ by removing an edge is χ-unique if $p \geq 3$ and $0 \leq q - p \leq 1$. Read (1988b) conjectured that all graphs obtained from $K(p,q)$, where $p, q \geq 3$, by removing an edge is χ-unique. This was later confirmed by C.P. Teo and Koh (1990). The chromaticity of graphs obtained from $K(p,q)$ by deleting $s \geq 2$ edges was later studied by several researchers, including Giudici and Lima De Sá (1990), Peng (1994), Borowiecki and Drgas-Burchardt (1993b), and S.J. Xu (1991a). Their results deal with cases when $|p - q|$ is "small".

For integers p, q and s with $q \geq p \geq 2$ and $s \geq 0$, let $\mathcal{K}^{-s}(p,q)$ (resp., $\mathcal{K}_2^{-s}(p,q)$) denote the family of connected (resp., 2-connected) bipartite graphs which are obtained from $K(p,q)$ by deleting a set of s edges. Chen (1998) showed that if $G \in \mathcal{K}^{-s}(p,q)$, where $3 \leq s \leq q - p$, p is "large" relative to s and $q - p$, and the set of s edges deleted forms a matching or a star (among other possibilities), then G is χ-unique.

Recently, Dong, Koh, Teo, Little and Hendy (2000a, 2000b, 2001) established many families of χ-unique bipartite graphs which include almost all of those mentioned earlier. In Dong, Koh, Teo, Little and Hendy (2001), they found a sharp upper bound for $\alpha(G,3)$ when $G \in \mathcal{K}^{-s}(p,q)$, where $0 \leq s \leq (p-1)(q-1)$; and another one when $G \in \mathcal{K}_2^{-s}(p,q)$, where $0 \leq s \leq p + q - 4$. These bounds are used to narrow down the search for graphs that may be χ-equivalent to a given one in $\mathcal{K}^{-s}(p,q)$ or $\mathcal{K}_2^{-s}(p,q)$.

In order to state their results more succinctly, we define, for each $x \in \mathbb{N}$ and $y \in \mathbb{N}_0$, the following function

$$g(x,y) = 2^{a+x} + 2^{a+d} - 2^x - 2^{a+1} + 1, \tag{4.3}$$

where $a, d \in \mathbb{N}_0$ and are determined by $y = ax + d$ and $d \leq x - 1$.

Theorem 4.5.1 *For $p, q, s \in \mathbb{N}_0$ with $2 \leq p \leq q$ and $s \leq (p-1)(q-1)$,*

$$\max_{G \in \mathcal{K}^{-s}(p,q)} \alpha(G,3) = (2^{p-1} + 2^{q-1} - 2) + g(q-1,s); \tag{4.4}$$

and the maximum is attained when $G = (A, B; E)$ with $|A| = p$ such that the following sequence is a degree sequence of vertices in A:

$$\underbrace{1, 1, \cdots, 1}_{a}, \; q - d, \; \underbrace{q, \cdots, q}_{p-a-1},$$

where a and d are integers determined by $s = a(q - 1) + d$, $a \geq 0$ and $0 \leq d \leq q - 2$. □

Theorem 4.5.2 *For $p, q, s \in \mathbb{N}$ with $3 \leq p \leq q$ and $q - 1 \leq s \leq p + q - 4$,*

$$\max_{G \in \mathcal{K}_2^{-s}(p,q)} \alpha(G, 3) = (2^{p-1} + 2^{q-1} - 2) + 2^{q-2} + 2^{s+3-q} - 3 - \lfloor 3/p \rfloor;$$

and the maximum is attained when $G = (A, B; E)$ with $A = \{x_1, \cdots, x_p\}$ and $B = \{y_1, \cdots, y_q\}$ such that either

(i) $p \geq 4$,

$$\begin{cases} N(x_1) = \{y_1, y_2\}, \\ N(x_2) = \{y_1, y_2, \cdots, y_{2q-s-2}\}, \\ N(x_i) = B \text{ for } i = 3, \cdots, p, \end{cases}$$

(ii) or $p = 3$, $s = q - 1$,

$$\begin{cases} N(x_1) = \{y_1, y_2\}, \\ N(x_2) = B - \{y_1\}, \\ N(x_3) = B. \end{cases}$$ □

As a consequence of the above two theorems, we have:

Corollary 4.5.1 *For $p, q, s \in \mathbb{N}_0$ with $3 \leq p \leq q$ and $s \leq p + q - 4$,*

$$\alpha(G, 3) \leq (2^{p-1} + 2^{q-1} - 2) + g(q - 2, s), \qquad (4.5)$$

for all $G \in \mathcal{K}_2^{-s}(p, q)$. □

Dong, Koh, Teo, Little and Hendy (2001) put forward the following conjecture.

Conjecture 4.5.1 *For $p, q, s \in \mathbb{N}_0$ with $3 \leq p \leq q$ and $s \leq (p - 2)(q - 2)$,*

$$\alpha(G, 3) \leq (2^{p-1} + 2^{q-1} - 2) + g(q - 2, s), \qquad (4.6)$$

for all $G \in \mathcal{K}_2^{-s}(p, q)$.

By using Theorems 4.5.1 and 4.5.2, Dong, Koh, Teo, Little and Hendy (2001) proved the following two results:

Theorem 4.5.3   *For $p, q, s \in \mathbb{N}_0$ with $2 \leq p \leq q$ and $s \leq (p-1)(q-1)$, and $G \in \mathcal{K}^{-s}(p, q)$, the χ-equivalence class determined by G is contained in the union of the families*

$$\mathcal{K}^{-s_i}(p - i, q + i).$$

where $\max\left\{-\frac{s}{p-1}, -\frac{q-p}{2}\right\} \leq i \leq \frac{s}{q-1}$ *and* $s_i = s - i(q - p + i)$ *for all i.* $\square$

Theorem 4.5.4   *For $p, q, s \in \mathbb{N}_0$ with $3 \leq p \leq q$, let $G \in \mathcal{K}_2^{-s}(p, q)$. Then the χ-equivalence class determined by G is a subfamily of $\mathcal{K}_2^{-s}(p, q)$ if either $s \leq p - 1$, or $s \leq 2p - 3$ and $q \geq p + 4$.* $\square$

Remark   Theorem 4.5.4 is part of a more general result; namely, Theorem 4.2 in Dong, Koh, Teo, Little and Hendy (2000b).

We now consider three types of bipartite graphs. For $p, q \in \mathbb{N}$ with $2 \leq p \leq q$, let $K(p, q) = (A, B; E)$ with $|A| = p$, $|B| = q$.

For $1 \leq s \leq p - 1$, let $H_1(p, q, s)$ denote the graph obtained from $K(p, q)$ by deleting a set of s edges that induces a star with centre in B; for $1 \leq s \leq q - 1$, let $H_2(p, q, s)$ denote the graph obtained from $K(p, q)$ by deleting a set of s edges that induces a star but with the centre in A; and for $1 \leq s \leq p - 1$, let $H_3(p, q, s)$ denote the graph obtained from $K(p, q)$ by deleting a set of s edges that forms a matching in $K(p, q)$. It is easy to see that

$$\begin{aligned}
\alpha(H_1(p, q, s), 3) &= \alpha(H_2(p, q, s), 3) = 2^{p-1} + 2^{q-1} - 2 + 2^s - 1, \\
\alpha(H_3(p, q, s), s) &= 2^{p-1} + 2^{q-1} - 2 + s.
\end{aligned}$$

Dong, Koh, Teo, Little and Hendy (2001) showed that the above are in fact, respectively, the upper and lower bound of $\alpha(G, 3)$, where $G \in \mathcal{K}^{-s}(p, q)$ (comparing this with Lemma 4.4.5). By using this, together with Theorems 4.5.3 and 4.5.4, they proved the following results, which include several existing results as special cases.

Theorem 4.5.5   *For $p, q, s \in \mathbb{N}_0$ with $3 \leq p \leq q$,*

(i) when $s \leq p - 2$, $H_i(p, q, s)$ is χ-unique for $i = 1, 2$;

(ii) when $p - 1 \leq s \leq q - 2$, $H_1(p, q, s)$ is χ-unique; and

(iii) when $s \leq p - 1$, $H_3(p, q, s)$ is χ-unique. $\square$

Let $G \in \mathcal{K}^{-s}(p,q)$, where $3 \le p \le q$ and $0 \le s \le p-1$. Then from what have been said before Theorem 4.5.5, we have

$$(2^{p-1} + 2^{q-1} - 2) + s \le \alpha(G,3) \le (2^{p-1} + 2^{q-1} - 2) + 2^s - 1.$$

Let us write $\alpha'(G,3) = \alpha(G,3) - (2^{p-1} + 2^{q-1} - 2)$. Then

$$s \le \alpha'(G,3) \le 2^s - 1.$$

Theorem 4.5.5 says that if $G \in \mathcal{K}_2^{-s}(p,q)$ and $\alpha'(G,3) = s$ or $2^s - 1$, then G is χ-unique. Dong, Koh, Teo, Little and Hendy (2000b) obtained further the following result.

Theorem 4.5.6 *Let $G \in \mathcal{K}_2^{-s}(p,q)$, where $p,q,s \in \mathbb{N}_0$ with $3 \le p \le q$ and $s \le p-1$. Then G is χ-unique if $s+1 \le \alpha'(G,3) \le s+4$.* $\square$

Let S be a set of edges in $K(p,q)$ whose removal results in a χ-unique graph $G \in \mathcal{K}_2^{-s}(p,q)$ in Theorem 4.5.6. Then $[S]$ is one of the following:

For $t = 1$: $P_3 \cup (s-2)K_2$.

For $t = 2$: $P_4 \cup (s-3)K_2$ or $2P_3 \cup (s-4)K_2$.

For $t = 3$: $P_5 \cup (s-4)K_2$, $P_4 \cup P_3 \cup (s-5)K_2$ or $3P_3 \cup (s-6)K_2$.

For $t = 4$:

$$P_6 \cup (s-5)K_2, P_5 \cup P_3 \cup (s-6)K_2, P_4 \cup 2P_3 \cup (s-7)K_2,$$
$$2P_4 \cup (s-6)K_2, K(1,3) \cup (s-3)K_2 \text{ or } 4P_3 \cup (s-8)K_2.$$

By applying Theorem 4.5.6, Dong, Koh, Teo, Little and Hendy (2000b) further establish the following theorem.

Theorem 4.5.7 *For any $G \in \mathcal{K}_2^{-s}(p,q)$ with $3 \le p \le q$ and $0 \le s \le \min\{4, p-1\}$, G is χ-unique.* $\square$

Figure 4.1

The graphs G in the above theorem include those obtained from $K(p,q)$ by removing the edges of a C_4 or the edges of the tree shown in Figure 4.1. Theorem 4.5.1 provides the following upper bound for $\alpha'(G,3)$:

$$\alpha'(G,3) \leq g(q-1,s),$$

where $G \in \mathcal{K}^{-s}(p,q)$ with $2 \leq p \leq q$ and $0 \leq s \leq (p-1)(q-1)$.

However, this bound is not good when the maximum degree of the subgraph of $K(p,q)$ induced by the edges deleted is small. For example, if G is obtained from $K(p,q)$ by deleting s independent edges, then $\alpha'(G,3) = s$ while $g(q-1,s) = 2^s - 1$. Dong, Koh, Teo, Little and Hendy (2000a) found an upper bound for $\alpha'(G,3)$ in terms of $\Delta([S])$, where S is the edge set removed. By this result, they proved the following result.

Theorem 4.5.8 *Let $p,q,s \in \mathbb{N}$ and $G \in \mathcal{K}_2^{-s}(p,q)$. Suppose $G = K(p,q) - S$, where $S \subseteq E(K(p,q))$ with $|S| = s$. Then G is χ-unique if either one of the following conditions holds:*

(i) $6 \leq s+1 \leq p \leq q$ and $\Delta([S]) = s - 1$;

(ii) $8 \leq s+1 \leq p \leq q$ and $\Delta([S]) = s - 2$. $\square$

Let S be a set of edges in $K(p,q)$ whose removal results in a graph $G \in \mathcal{K}_2^{-s}(p,q)$ in Theorem 4.5.8. Then $[S] = K(1,s-1) \cup K_2$, $K(1,s-2) \cup P_3$, $K(1,s-2) \cup 2K_2$ or one of the graphs shown below:

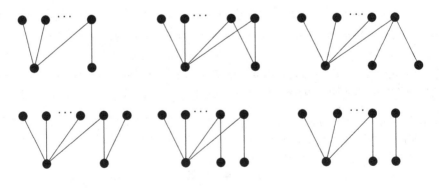

Figure 4.2

Remark Very recently, in his Ph.D. thesis supervised by YH Peng, Roslan (2004) has further extended some of the above results for bipartite graphs.

4.6 Further results

Apart from the bipartite case, there are not many known results on the chromaticity of general multi-partite graphs. Though it was not mentioned by the authors, the graphs appeared in Chia, Goh and Koh (1988) obtained from $K(p+1, p+1, p+1)$, $K(p, p+1, p+1)$ or $K(p+1, p+1, p+2)$ by removing one edge are χ-unique, for $p \geq 2$. Note that $K(2,2,2) - e$ is chromatically equivalent to $K(1,2,3)$. Further results have been obtained by Zhao, Liu and Ye (2003). The chromaticity of certain 4-partite graphs were recently studied by Zhao, Li, Liu and Wang (2002).

For positive integers $p_i, i = 1, 2, \cdots, t$, and $s \in \mathbb{N}_0$, let $\mathcal{K}^{-s}(p_1, p_2, \cdots, p_t)$ be the family of graphs obtained from $K(p_1, p_2, \cdots, p_t)$ by deleting s edges. The authors employed a method similar to the one used for proving Theorem 4.4.2 to prove the following result. The case $t = 3$ is due to Zhao, Liu and Ye (2003), and that for $t = 4$ is due to Zhao, Li, Liu and Wang (2002).

Lemma 4.6.1 *For $t = 2, 3$, let $G = K(p_1, p_2, \cdots, p_t)$, where $p_1 \leq p_2 \leq \cdots \leq p_t$. Suppose $n = p_1 + p_2 + \cdots + p_t$. Then the family $\mathcal{K}^{-s}(p_1, p_2, \cdots, p_t)$ is χ-closed in each of the following cases:*

(i) $G \cong K[t; n]$ and $p_1 \geq s + 2$,

(ii) $G \not\cong K[t; n]$ and $p_1 \geq e(K[t; n]) - e(G) + s + 1$. □

As direct consequences they obtained the following:

Corollary 4.6.1 *Let $j \geq k$ and $n \geq \max\{s + 2, \lfloor \frac{k^2}{3} \rfloor + s + 1\}$. Then $\mathcal{K}^{-s}(n, n + j, n + k)$ is χ-closed.* □

Corollary 4.6.2 *If $\max\{i, j, k\} = k \geq 2$ and $n \geq k^2/2 + s + 1$ or $\max\{i, j, k\} \leq 1$ and $n \geq s + 2$, then $\mathcal{K}^{-s}(n, n + i, n + j, n + k)$ is χ-closed.* □

By letting $s = 0$ in the above Corollaries, they obtained:

Theorem 4.6.1 *(i) If $j \geq k$ and $n \geq \max\{\lfloor k^2/3 \rfloor + 1, 2\}$, then $K(n, n + j, n + k)$ is χ-unique.*

(ii) If $\max\{i, j, k\} = k$ and $n \geq \max\{k^2/2 + 1, 2\}$, then $K(n, n + i, n + j, n + k)$ is χ-unique. □

Let $G = K(p_1, p_2, \cdots, p_t)$. Suppose the partite sets of G are A_i's with $|A_i| = p_i, i = 1, 2, \cdots, t$. For $1 \leq i, j \leq t$ and $i \neq j$, let

$$K_{i,j}^{-K(1,s)}(p_1, p_2, \cdots, p_t)$$

denote the graph obtained from G by removing $E(K(1, s))$ from the subgraph induced by $A_i \cup A_j$ with centre in A_i; and $K_{i,j}^{-sK_2}(p_1, p_2, \cdots, p_t)$ the graph obtained by deleting $E(sK_2)$ from the subgraph induced by $A_i \cup A_j$. Zhao, Liu and Ye (2003) used the analogous results for bipartite graphs and Lemma 4.6.1 to prove the following theorem.

Theorem 4.6.2 Let $G = K(p_1, p_2, p_3)$, where $p_1 \leq p_2 \leq p_3$. Let $n = v(G)$ and $p_1 \geq e(K[3; n]) - e(G) + s + 2$. Then

(i) $K_{ij}^{-K(1,s)}(p_1, p_2, p_3)$ is χ-unique for all $i, j = 1, 2, 3$ and $i \neq j$.

(ii) $K^{-sK_2}(p_1, p_2, p_3)$ is χ-unique.   □

Likewise, Zhao, Li, Liu and Wang (2002) proved the following theorem.

Theorem 4.6.3 Let $G = K(p_1, p_2, p_3, p_4)$ with $p_1 \leq p_2 \leq p_3 \leq p_4$. Let $n = v(G)$ and suppose $p_1 \geq e(K[4; n]) - e(G) + s + 2$. Then

(i) $K_{i,j}^{-K(1,s)}(p_1, p_2, p_3, p_4)$ is χ-unique for all $i, j = 1, 2, 3, 4$, $i \neq j$ and $p_2 + p_3 \neq p_1 + p_4$;

(ii) $K_{i,j}^{-K(1,s)}(p_1, p_2, p_3, p_4)$ is χ-unique for all $i, j = 1, 2, 3, 4$, $i \neq j$ and $p_1 = p_2$, $p_3 = p_4$;

(iii) $K_{i,j}^{-K(1,s)}(p_1, p_2, p_3, p_4)$ is χ-unique if (i, j) is a member of

$$\{(1, 2), (2, 1), (1, 3), (3, 1), (2, 4), (4, 2), (3, 4), (4, 3)\};$$

(iv) $K_{1,2}^{-sK_2}(p_1, p_2, p_3, p_4)$ is χ-unique.   □

Very recently, Zhao, Liu and Zhang (2004) have found several families of χ-unique 5-partite graphs. Let G be a 5-partite graph with $5n$ vertices. Define

$$\theta(G) = (\alpha(G, 6) - 2^{n+1} - 2^{n+1} + 5)/2^{n-2}.$$

The authors characterized all G with $\theta(G) = 0, 1, 2, 5/2, 7/2, 4$ and $17/4$. Consequently, they obtain many families of χ-unique graphs obtained by deleting the edges a star or matching from a complete 5-partite graph with $5n$ vertices.

We end this chapter by posing the following problem.

Problem 4.6.1 Study the chromaticity of the graphs obtained from any complete t-partite graph by removing the edges of a star or matching.

Exercise 4

4.1 Let $G = (A, B; E)$ be a connected bipartite graph with $|A| = p + r$, $|B| = q - r$ and $|E| = pq$, where $p, q, r \in \mathbb{N}$ such that $2 \leq p \leq q$ and $q - r \geq 2$. Show that if $r \neq q - p$ and $\delta(G) \geq 2$, then

$$n_G(C_4) < \binom{p}{2}\binom{q}{2}.$$

(See C.P. Teo and Koh (1990).)

4.2 Let $G = (A, B; E)$ be a connected bipartite graph with $|A| = p + r$, $|B| = q - r$ and $|E| = pq - 1$, where $p, q, r \in \mathbb{N}$ such that $3 \leq p \leq q$ and $q - r \geq 2$. Show that if $r \neq q - p$ and $\delta(G) \geq 2$, then

$$n_G(C_4) \neq \binom{p}{2}\binom{q}{2} - (p-1)(q-1).$$

Hence show that $K(p, q) - e$, where e is an edge of $K(p, q)$, is χ-unique. (See C.P. Teo and Koh (1990).)

4.3 Let $G = (A, B; E)$ be a bipartite graph with $|A|, |B| \geq 2$.

(i) Prove that

$$n_G(C_4) \leq |E|(|E| - |A| - |B| + 1)/4.$$

(ii) Show that if $d(v) \geq 2$ for all $v \in V(G)$, then the equality in (i) holds if and only if G is a complete bipartite graph.

(See Dong (1993).)

4.4 Let $G = K(p, q) - S$, where S is a matching of $K(p, q)$ and $2 \leq p \leq q$. Show that if $|S| \leq p$, then

$$n_G(C_4) = \binom{p}{2}\binom{q}{2} - |S|(p-1)(q-1) + \binom{|S|}{2}.$$

(See Giudici and Lima de Sá (1990).)

4.5 Let $G = K(p_1, p_2, p_3)$ with $|p_i - p_j| \leq 1$ and $p_i \geq 2$ for $i, j \in \{1, 2, 3\}$. Show that $G - e$ is χ-unique for all $e \in E(G)$.

(Chia, Goh and Koh (1988).)

4.6 Show that $K(2,4,6)$ is χ-unique.

(See Zou (1998b).)

4.7 Let G be a graph. Show that for each $k \in \mathbb{N}$,

$$\alpha(G, k) = \frac{1}{k!} \sum_{i=0}^{k} (-1)^i \binom{k}{i} P(G, k - i).$$

(See Tomescu (1985).)

4.8 Let $G = K(p_1, p_2, \cdots, p_t)$. Show that

$$\alpha(G, t + 1) = \sum_{i=1}^{t} 2^{p_i - 1} - t.$$

(Zou (2004).)

4.9 Let $G = K(p, q) - S$, where $2 \le p \le q$ and $S \subseteq E(K(p, q))$. Let uvw be a path in $[S]$ with $d_{[S]}(u) = 1$ and $d_{[S]}(v) = 2$. Show that for any $k \ge 2$,

$$\alpha(G, k) = \alpha(G + uv, k) + \alpha(G - \{u, v\}, k - 1) + \alpha(G - \{u, v, w\}, k - 1).$$

(See Dong, Koh, Teo, Little and Hendy (2000b).)

4.10 For $p_1, p_2, \cdots, p_t \ge 2$, let $G = K(p_1, p_2, \cdots, p_t)$ and $H = G - S$, where $S \subseteq E(G)$, and $0 \le |S| < \min\{p_1, p_2, \cdots, p_t\}$. Show that

$$|S| \le \alpha(H, t + 1) - \alpha(G, t + 1) \le 2^{|S|} - 1.$$

Give examples to show that the upper and lower bounds are sharp.

(See Zou (2004).)

4.11 Show that for a given $t \in \mathbb{N}$ with $t \ge 2$, two complete t-partite graphs are χ-equivalent if and only if they are isomorphic.

4.12 Let G be a non-empty bipartite graph. Show that

$$\alpha(G, 2) = 2^{c(G) - 1}.$$

(See Dong, Koh, Teo, Little and Hendy (2001).)

4.13 Let $G = K(p,q) - S$ with $p \geq q \geq 3$ and $|S| \leq p+q-4$, where $S \subseteq E(G)$. Show that G is 2-connected if and only if $\delta(G) \geq 2$.

(See Dong, Koh, Teo, Little and Hendy (2001).)

4.14 For $p,q \in \mathbb{N}$ with $2 \leq p \leq q$, show, by using Theorem 4.5.3, that $\mathcal{K}_2^{-s}(p,q)$ is χ-closed if $0 \leq s \leq p-1$.

(See Dong, Koh, Teo, Little and Hendy (2001).)

4.15 Show that for $3 \leq p \leq q$, $0 \leq s \leq p-1$ and $0 \leq t \leq 2^{p-1}-p-1$, the following family of graphs is χ-closed:

$$\{H \in \mathcal{K}^{-s}(p,q) : \alpha(H,3) - \alpha(K(p,q),3) = s+t\}.$$

(See Dong, Koh, Teo, Little and Hendy (2000b).)

4.16 Let $G = K(p,q) - S = (A,B;E)$, $S \subseteq E(K(p,q))$ and $|S| \leq p-1 \leq q-1$, and let

$$\Omega(G) = \{Q : Q \text{ is an independent set in } G \text{ with } Q \cap A \neq \emptyset, Q \cap B \neq \emptyset\}.$$

Show that $\alpha(G,3) - \alpha(K(p,q),3) = |\Omega(G)| \geq 2^{\Delta([S])} + |S| - 1 - \Delta([S])$.

(See Dong, Koh, Teo, Little and Hendy (2000b).)

4.17 For $p \leq q$, let $G = K(p,q) - S$, where $S \subseteq K(p,q)$. Let β_i be the number of vertices in $[S]$ of degree i. Show that for $|S| \leq p-1 \leq q-1$,

$$\alpha(G,3) - \alpha(K(p,q),3) \geq |S| + \sum_{i \geq 2} \beta_i(2^i - i - 1) + n_{[S]}(C_4),$$

and equality holds if and only if $|N_{[S]}(x) \cap N_{[S]}(y)| \leq 2$ for every $x,y \in A'$ or $x,y \in B'$, where A', B' are the two partite sets of $[S]$.

(See Dong, Koh, Teo, Little and Hendy (2000b).)

4.18 Let G be as in Exercise 4.16. Show that if $\alpha(G,3) - \alpha(K(p,q),3) = |S| + t$, where $1 \leq t \leq 4$, then either

(i) each component of $[S]$ is a path and $\beta_2 = t$, or

(ii) $[S] \cong K(1,3) \cup (|S|-3)K_2$.

(See Dong, Koh, Teo, Little and Hendy (2000b).)

4.19 Let G be as in Exercise 4.16. Suppose $G = (A, B; E)$ with $|A| = p, |B| = q$. Show that

$$\alpha(G,4)-\alpha(K(p,q),4) = \sum_{Q\in\Omega(G)} (2^{p-1-|Q\cap B|}+2^{q-1-|Q\cap A|}-2)+M(G),$$

where $\Omega(Q)$ is defined in Exercise 4.16, and

$$M(G) = |\{\{Q_1, Q_2\} : Q_1, Q_2 \in \Omega(G), Q_1 \cap Q_2 = \emptyset\}|.$$

(See Dong, Koh, Teo, Little and Hendy (2000b).)

4.20 Let G and $M(G)$ be as in Exercise 4.19. Show that if each component of $[S]$ is a path, then

$$M(G) = \binom{s+t}{2} - 3t - 3n_{[S]}(P_4) - n_{[S]}(P_5).$$

(See Dong, Koh, Teo, Little and Hendy (2000b).)

4.21 Let $G = K(n_1, n_2, \cdots, n_t)$ and $H = K(m_1, m_2, \cdots, m_t) - S$, where S is a set of edges of $K(m_1, m_2, \cdots, m_t)$ and $G[S]$ is of order n. Suppose $|S| + 1 \le n_1 \le n_2 \le \cdots \le n_t$.

   (i) Show that

$$\alpha(G, t+1) - \alpha(H, t+1) = 2^{n_1-1}M - \alpha,$$

   where $|S| \le \alpha \le 2^{|S|} - 1$, and

$$M = \sum_{i=1}^{t}(2^{n_i-n_1} - 2^{m_i-n_1}).$$

   (ii) Show that if $n_1 \ge e(K[t; n]) - e(G) + 1$ and $H \sim G$, then $M = 0$.

4.22 Let $G = K(n_1, n_2, \cdots, n_t)$. Show that

$$e(K[t; n]) - e(G) \le \sum_{1\le i<j\le t} \frac{(n_i - n_j)^2}{2t}.$$

4.23 Let $n_1 \le n_2 \le \cdots \le n_t$ be positive integers. Show that if $|n_i - n_j| \le k$ and $n_1 \ge \frac{tk^2}{4} + 1$, then $K(n_1, n_2, \cdots, n_t)$ is χ-unique.

4.24 Show that for $k \ge 2$ and $n \ge 2k$, $K(n - k, n - 1, n)$ is χ-unique.

4.25 Show that for $k \in \mathbb{N}$, $K(p, p, p + k)$ is χ-unique if $p > \frac{k(k+1)}{3}$.

Chapter 5

Chromaticity of Subdivisions of Graphs

5.1 Introduction

The operation of replacing an edge of a graph by a path is called *subdivision*. If H can be derived from G by a sequence of subdivisions, we say H is a *subdivision* of G. For each positive integer f, the graph $G(f)$ obtained from G by replacing each edge of G with a path of length f is called the *f-uniform subdivision* of G.

The reverse operation of subdivision is called *chain-contraction* (a chain is a path in which all its internal vertices are of degree 2). This operation replaces a $u - v$ chain by an edge uv. By contracting all maximal chains of a graph G, we arrive at a multigraph $M(G)$. Two graphs G and H are *homeomorphic* if $M(G) = M(H)$. If G is homeomorphic to H, we also say G is a *H-homeomorph*.

By using Theorem 1.3.3, Read and Whitehead Jr. (1999) expanded each $P(G, \lambda)$ in terms of the lengths of the maximal chains of G. An alternative expression of the chromatic polynomial of a general subdivision of a graph was derived by Brown and Hickman (2002b).

In this chapter, we explore the chromaticity of subdivisions of several families of graphs, including $K(p, q)$ and K_4. In particular, we study the chromaticity of the uniform subdivisions of some graphs.

In what follows, the label on each edge of $M(G)$ represents the corresponding maximal chain length in G.

5.2   Multi-bridge graphs

For each integer $k \geq 2$, let θ_k be the multigraph with 2 vertices and
k edges. Any subdivision of θ_k is called a *multi-bridge graph* (or, more
precisely, a *k-bridge graph*). For any $a_1, a_2, \cdots, a_k \in \mathbb{N}$, we denote by
$\theta(a_1, a_2, \cdots, a_k)$ the graph obtained by replacing the edges of θ_k with paths
of length $a_1, a_2, \cdots, a_k$ respectively (see Figure 5.1).

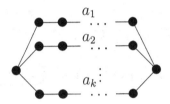

Figure 5.1

The f-uniform subdivision of θ_k will also be denoted by $\theta_k(f)$. If $a_i \geq 2$
for each i, then $\theta(a_1, a_2, \cdots, a_k)$ is a subdivision of $K(2, k)$.

A 2-bridge graph is simply a cycle, which is χ-unique. Chao and Whitehead Jr. (1979b), initiating the study of χ-unique graphs, showed that every 3-bridge graph $\theta(1, a_2, a_3)$ (called a *θ-graph*) is χ-unique. This result
was immediately extended by Loerinc (1978) to all 3-bridge graphs (called
generalized θ-graphs).

Note that if $a_1 = 1$, then $\theta(a_1, a_2, \cdots, a_k)$ is a graph obtained by gluing C_{a_2+1}, C_{a_3+1}, $\cdots$, C_{a_k+1} at a common edge. This is a polygon-tree.
It follows from Theorem 3.8.3 that the χ-equivalence class determined by
$\theta(1, a_2, a_3, \cdots, a_k)$ consists of all edge-gluings of $C_{a_2+1}, C_{a_3+1}, \cdots, C_{a_k+1}$.
From now on, we assume $k \geq 3$ and $a_i \geq 2$ for each $i = 1, 2, \cdots, k$.

5.2.1   Chromatic polynomials of multi-bridge graphs

Let $k, a_1, a_2, \cdots, a_k \in \mathbb{N}$, and $G = \theta(a_1, a_2, \cdots, a_k)$. Then, by Theorem 1.3.3, we obtain

$$P(G, \lambda) = \frac{1}{\lambda^{k-1}(\lambda - 1)^{k-1}} \prod_{i=1}^{k} ((\lambda - 1)^{a_i+1} + (-1)^{a_i+1}(\lambda - 1))$$

$$+ \frac{1}{\lambda^{k-1}} \prod_{i=1}^{k} ((\lambda - 1)^{a_i} + (-1)^{a_i}(\lambda - 1)). \tag{5.1}$$

Let us simplify the above expression by letting $\lambda = 1 - x$. Thus

$$
\begin{aligned}
P(G, 1-x) &= \frac{(-1)^{a_1+a_2+\cdots+a_k+1}}{(1-x)^{k-1}}\left(x\prod_{i=1}^{k}(x^{a_i}-1)-\prod_{i=1}^{k}(x^{a_i}-x)\right) \\
&= \frac{(-1)^{e(G)+1}}{(1-x)^{e(G)-v(G)+1}}\left(x\prod_{i=1}^{k}(x^{a_i}-1)-\prod_{i=1}^{k}(x^{a_i}-x)\right),
\end{aligned}
$$

as $e(G) = \sum_{i=1}^{k} a_i$ and $v(G) = \sum_{i=1}^{k} a_i - k + 2$.

In general, for any graph G and real number x, define

$$
\mathcal{Q}(G, x) = (-1)^{e(G)+1}(1-x)^{e(G)-v(G)+1}P(G, 1-x).
$$

We have:

Theorem 5.2.1   *For any* $k, a_1, a_2, \cdots, a_k \in \mathbb{N}$,

$$
\mathcal{Q}(\theta(a_1, a_2, \cdots, a_k), x) = x\prod_{i=1}^{k}(x^{a_i}-1)-\prod_{i=1}^{k}(x^{a_i}-x). \tag{5.2}
$$

In particular, $\mathcal{Q}(C_n, x) = x(x-1)(x^{n-1}-1)$.   □

It is clear that if $H \sim G$, then $\mathcal{Q}(H, x) = \mathcal{Q}(G, x)$. The converse of this statement is not true. For example, for all $m \in \mathbb{N}$,

$$
\mathcal{Q}(G, x) = \mathcal{Q}(G \cup mK_1, x).
$$

Note that if $\mathcal{Q}(H, x) = \mathcal{Q}(G, x)$, then

$$
\deg(\mathcal{Q}(H, x)) = e(H) + 1 \quad \text{and} \quad \deg(\mathcal{Q}(G, x)) = e(G) + 1,
$$

implying that $e(H) = e(G)$. Hence if, in addition to $\mathcal{Q}(H, x) = \mathcal{Q}(G, x)$, we also have $v(H) = v(G)$, then $H \sim G$.

Theorem 5.2.2   *For any graphs* G *and* H,

(i) *if* $H \sim G$, *then* $\mathcal{Q}(H, x) = \mathcal{Q}(G, x)$;

(ii) *if* $\mathcal{Q}(H, x) = \mathcal{Q}(G, x)$ *and* $v(H) = v(G)$, *then* $H \sim G$.   □

In view of the above theorem, we have

$$
[G] = \{H : \mathcal{Q}(H, x) = \mathcal{Q}(G, x), \ v(H) = v(G)\}.
$$

By Theorem 5.2.1, we are able to show that $\theta(a_1, a_2, \cdots, a_k)$ is χ-equivalent to $\theta(b_1, b_2, \cdots, b_k)$ if and only if they are isomorphic.

Lemma 5.2.1 *Suppose that*

$$\theta(a_1, a_2, \cdots, a_k) \sim \theta(b_1, b_2, \cdots, b_k),$$

where $k \geq 3$, $2 \leq a_1 \leq a_2 \leq \cdots \leq a_k$ and $2 \leq b_1 \leq b_2 \leq \cdots \leq b_k$. Then $a_i = b_i$ for all $i = 1, 2, \cdots, k$.

Proof. By Theorem 5.2.1, we have

$$x \prod_{i=1}^{k}(x^{a_i} - 1) - \prod_{i=1}^{k}(x^{a_i} - x) = x \prod_{i=1}^{k}(x^{b_i} - 1) - \prod_{i=1}^{k}(x^{b_i} - x). \quad (5.3)$$

After we cancel $(-1)^k x + (-1)^{k+1} x^k$ from both sides, the terms with lowest power on the left and right sides have powers $1 + a_1$ and $1 + b_1$ respectively. Hence $a_1 = b_1$. Suppose that $a_i = b_i$ for $i = 1, \cdots, m$ but $a_{m+1} \neq b_{m+1}$ for some integer m with $1 \leq m \leq k - 1$. It is clear that $b_1 + b_2 + \cdots + b_k = a_1 + a_2 + \cdots + a_k$. Let $S = a_1 + a_2 + \cdots + a_k$. Observe that

$$x \prod_{i=1}^{k}(x^{a_i} - 1) - \prod_{i=1}^{k}(x^{a_i} - x)$$

$$= (x-1)x^S + \sum_{j=2}^{k}(-1)^j(x - x^j) \sum_{1 \leq i_1 < \cdots < i_j \leq k} x^{S - a_{i_1} - \cdots - a_{i_j}}$$

and

$$x \prod_{i=1}^{k}(x^{b_i} - 1) - \prod_{i=1}^{k}(x^{b_i} - x)$$

$$= (x-1)x^S + \sum_{j=2}^{k}(-1)^j(x - x^j) \sum_{1 \leq i_1 < \cdots < i_j \leq k} x^{S - b_{i_1} - \cdots - b_{i_j}}.$$

Since $a_i = b_i$ for $i = 1, 2, \cdots, m$, by (5.1), we have

$$\sum_{j=2}^{k}(-1)^j(x - x^j) \sum_{\substack{1 \leq i_1 < \cdots < i_j \leq k \\ i_j > m}} x^{S - a_{i_1} - \cdots - a_{i_j}}$$

$$= \sum_{j=2}^{k}(-1)^j(x - x^j) \sum_{\substack{1 \leq i_1 < \cdots < i_j \leq k \\ i_j > m}} x^{S - b_{i_1} - \cdots - b_{i_j}}. \quad (5.4)$$

The terms with highest power on the left and right sides of (5.2) have powers $S - a_1 - a_{m+1} + 2$ and $S - b_1 - b_{m+1} + 2$ respectively. Hence $a_{m+1} = b_{m+1}$, a contradiction. Therefore $b_i = a_i$ for $i = 1, 2, \cdots, k$. $\qquad\square$

As an illustration, let us establish the χ-uniqueness of every generalized θ-graph (see Loerinc (1978)).

Theorem 5.2.3 *Every generalized θ-graph is χ-unique.*

Proof. Consider $G = \theta(a_1, a_2, a_3)$ with $a_1 \leq a_2 \leq a_3$. Let H be a graph such that $H \sim G$. Then H is 2-connected, of order $n = a_1 + a_2 + a_3 - 1$ and of size $m = n + 1$, by Theorem 3.2.1. Since $d(v) \geq 2$ for all $v \in V(H)$ and

$$\sum_{v \in V(H)} d(v) = 2m = 2(n+1) = 2n + 2,$$

we have

Case (i): $d(z) = 4$ for some $z \in V(H)$ and $d(x) = 2$ for all $x \in V(H) - \{z\}$;

Case (ii): $d(u) = d(v) = 3$ for some distinct $u, v \in V(H)$ and $d(x) = 2$ for all other $x \in V(H)$.

Case (i) implies that H is a vertex-gluing of two cycles, contradicting the fact that H is 2-connected.

Case (ii) implies that $H \cong \theta(b_1, b_2, b_3)$ for some $b_1 \leq b_2 \leq b_3$.

By Lemma 5.2.1, $H \cong G$. $\qquad\square$

Not all k-bridge graphs are χ-unique. For example, for all $b \geq 2$, $\theta(2, b, b+1, b+2)$ is χ-equivalent but not isomorphic to the edge-gluing of $\theta(3, b, b+1)$ and C_{b+2} (see Chen, Bao and Ouyang (1992)). Both graphs belong to a family of graphs known as generalized polygon-trees, which we shall discuss next.

5.2.2 Generalized polygon-trees

Recall that a polygon-tree is a graph obtained by edge-gluing a collection of cycles. A *generalized polygon-tree* (*g.p. tree* in short) is a subdivision of some polygon-tree. Dirac (1952) and Duffin (1965) characterized a g.p. tree as a 2-connected graph that does not contain any subdivision of K_4. By using this result, Chao and Zhao (1983) established the following result.

Lemma 5.2.2   *The family of g.p. trees is χ-closed.*   □

To partition the family of g.p. trees into smaller χ-closed subfamilies, S.J. Xu (1994) introduced a chromatic invariant for g.p. trees. A pair $\{x, y\}$ of non-adjacent vertices of a graph G is called a *communication pair* if there are at least three internally disjoint x-y paths in G. Let $\zeta(G)$ denote the number of communication pairs in G. For each $r \in \mathbb{N}$, let $\mathcal{GP}_r$ be the collection of all g.p. trees G with $\zeta(G) = r$. S.J. Xu (1994) established the following result.

Theorem 5.2.4   *For each $r \in \mathbb{N}$, the family $\mathcal{GP}_r$ is χ-closed.*   □

Let G be a k-bridge graph with $k \geq 3$. Then $G \in \mathcal{GP}_1$. If $H \sim G$, then by Theorem 5.2.4, either H is a k-bridge graph, which by Lemma 5.2.1 is isomorphic to G, or is an edge-gluing of some t-bridge graph, $3 \leq t \leq k-1$, and $k - t$ cycles.

In what follows, we shall denote by $\mathcal{G}_e(G_1, G_2, \cdots, G_k)$ the collection of all edge-gluings of all $G_1, G_2, \cdots, G_k$, where $k \geq 2$ and $e(G_i) \geq 1$ for all i. Thus we have:

Lemma 5.2.3   *Let $H \sim \theta(a_1, a_2, \cdots, a_k)$, where $k \geq 3$ and $a_i \geq 2$ for all i. Then one of the following is true:*

(i) $H \cong \theta(a_1, a_2, \cdots, a_k)$;

(ii) $H \in \mathcal{G}_e\left(\theta(b_1, b_2, \cdots, b_t), C_{b_{t+1}+1}, \cdots, C_{b_k+1}\right)$, *where* $3 \leq t \leq k-1$ *and* $b_i \geq 2$ *for all* $i = 1, 2, \cdots, k$.   □

We have narrowed down the search for graphs that are χ-equivalent but not isomorphic to $\theta(a_1, a_2, \cdots, a_k)$ to those edge-gluings of a multi-bridge graph and cycles. We shall find an expression for $\mathcal{Q}(H, x)$ for such a graph H.

Lemma 5.2.4   *Let G and H be non-empty graphs, and $M \in \mathcal{G}_e(G, H)$. Then*

$$\mathcal{Q}(M, x) = \frac{\mathcal{Q}(G, x)\mathcal{Q}(H, x)}{x(x-1)}.$$

Proof. Since $v(M) = v(G) + v(H) - 2$, $e(M) = e(G) + e(H) - 1$ and

$$P(M, \lambda) = \frac{P(G, \lambda)P(H, \lambda)}{\lambda(\lambda - 1)},$$

the result follows from the definition of the function $\mathcal{Q}$. $\quad\square$

By Theorem 5.2.1 and Lemma 5.2.4, we obtain:

Theorem 5.2.5 *Let* $k, t, b_1, b_2, \cdots, b_k \in \mathbb{N}$ *with* $3 \le t \le k-1$ *and* $b_i \ge 2$ *for all* $i = 1, 2, \cdots, k$. *If* $H \in \mathcal{G}_e\left(\theta(b_1, b_2, \cdots, b_t), C_{b_{t+1}+1}, \cdots, C_{b_k+1}\right)$, *then*

$$\mathcal{Q}(H, x) = x \prod_{i=1}^{k} (x^{b_i} - 1) - \prod_{i=1}^{t} (x^{b_i} - x) \prod_{i=t+1}^{k} (x^{b_i} - 1). \quad\square$$

5.2.3 Chromaticity of k-bridge graphs with $k = 4, 5, 6$

In this section, we provide known results on the chromaticity of the graph $\theta(a_1, a_2, \cdots, a_k)$ for $k = 4, 5, 6$. As an illustration, let us determine all χ-equivalence classes of 4-bridge graphs. The result, as shown below, was first proved by Chen, Bao and Ouyang (1992).

Theorem 5.2.6 *Let* $c, a_1, a_2, a_3, a_4 \in \mathbb{N}$ *with* $c \ge 2$ *and* $2 \le a_1 \le a_2 \le a_3 \le a_4$.

(i) $\theta(a_1, a_2, a_3, a_4)$ *is* χ-*unique if and only if for any* $c \ge 2$,

$$(a_1, a_2, a_3, a_4) \ne (2, c, c+1, c+2).$$

(ii) If $H \sim \theta(2, c, c+1, c+2)$ *and* $H \not\cong \theta(2, c, c+1, c+2)$, *then*

$$H \in \mathcal{G}_e\left(\theta(3, c, c+1), C_{c+2}\right).$$

Proof. Let $H \sim \theta(a_1, a_2, a_3, a_4)$ and suppose $H \not\cong \theta(a_1, a_2, a_3, a_4)$. Then by Lemma 5.2.3,

$$H \in \mathcal{G}_e\left(\theta(b_1, b_2, b_3), C_{b_4+1}\right).$$

By Theorems 5.2.1, 5.2.2 and 5.2.5, we have

$$x \prod_{i=1}^{4} (a^{a_i} - 1) - \prod_{i=1}^{4} (x^{a_i} - x) = x \prod_{i=1}^{4} (x^{b_i} - 1) - (x^{b_4} - 1) \prod_{i=1}^{3} (x^{b_i} - x). \quad (5.5)$$

Note that $g(H) = g(G) = a_1 + a_2$ by Theorem 3.2.1(x). Hence $b_4 + 1 \ge a_1 + a_2$ and $b_i + b_j \ge a_1 + a_2$, for $1 \le i < j \le 3$.

By studying the terms with power less than $a_1 + a_2 + 1$ in the expansion of (5.3), it is not hard to see that $a_1 = 2$, $b_1 = 3$, $a_2 = b_2$, $a_3 = b_3 = b_4 = a_2 + 1$ and $a_4 = a_2 + 2$. The proof is complete. $\quad\square$

By using a method similar to the one we used above, Bao and Chen (1994) showed that every 5-bridge graph is χ-unique if its shortest maximal chain is of length greater than 3. This is a special case of a general result (Theorem 5.2.9) due to Xu, Liu and Peng (1994). C.F. Ye (2001a) claimed that $\theta(1, 2, 2, a, b)$, where $a, b \geq 3$, is χ-unique. This is clearly false, as the graph is one of many edge gluings of C_3, C_3, C_{a+1} and C_{b+1}.

X.F. Li and Wei (2001) established the following result.

Theorem 5.2.7   *For $a, b \in \mathbb{N}$, where $b \geq a \geq 3$, the 5-bridge graph $\theta(2, 2, 2, a, b)$ is χ-unique if and only if $(a, b) \neq (3, 4)$. Moreover, the χ-equivalence class of $\theta(2, 2, 2, 3, 4)$ is*

$$\{\theta(2, 2, 2, 3, 4)\} \cup \mathcal{G}_e\left(\theta(2, 2, 3), C_4, C_4\right).$$ □

Li and Wei (2001) stated without proof that $\theta(a, a, a, b, c)$, where $c \geq b \geq a + 1 \geq 3$, is χ-unique if and only if $(b, c) \neq (a + 1, a + 2)$.

In a very similar fashion, C.F. Ye (2001b) showed that $\theta(2, 2, 2, 2, a, b)$, where $3 \geq a \geq b$, is χ-unique. He later (see C.F. Ye (2002)) extended this result to any k-bridge graph $\theta(2, 2, \cdots, a, b)$ with $b \geq a \geq 3$ and $k \geq 5$, and claimed that the graph $\theta(a, a, \cdots, a, b, c)$, where $c \geq b \geq a + 1 \geq 3$, is χ-unique. However, as we have seen in Theorem 5.2.7, this is not the case for $\theta(2, 2, 2, 3, 4)$.

Problem 5.2.1   *Study the chromaticity of $\theta(a_1, a_2, \cdots, a_5)$, where $2 \leq a_1 \leq a_2 \leq \cdots \leq a_5 \leq 3$.*

5.2.4   Chromaticity of general multi-bridge graphs

Although many results concerning the chromaticity of multi-bridge graphs have been produced, the problem is still far from completely solved. However, it is true that any f-uniform subdivision of θ_k is χ-unique. This result was proved independently by Dong (1993), Koh and C.P. Teo (1990b), and Xu, Liu and Peng (1994) by using different methods. We present here the proof given in Xu, Liu and Peng (1994).

Theorem 5.2.8   *For $k \geq 2$, the graph $\theta_k(f)$ is χ-unique.*

Proof.   The result is true for $k = 2$. Let us assume $k \geq 3$. Let $G = \theta_k(f)$, where $f \geq 2$. Let H be a graph such that $H \sim G$. Then, by Theorem 3.2.1,

$$v(H) = v(G) = (f - 1)k + 2,$$
$$g(H) = g(G) = 2f, \text{ and}$$
$$n_H(C_g) = n_G(C_g) = \binom{k}{2}.$$

In view of Lemma 5.2.3, we let H be an edge-gluing of $H' = \theta(b_1, b_2, \cdots, b_t)$, where $3 \leq t \leq k$ and $2 \leq b_1 \leq b_2 \leq \cdots \leq b_t$, and $k - t$ cycles. We need to show that $t = k$.

Since $g(H) = 2f$, if $b_1 < f$, then $b_j \geq 2f - b_1$ for all $j = 2, 3, \cdots, t$. It follows that $v(H') \geq (f - 1)t + 2$ and

$$n_{H'}(C_g) \leq \binom{t}{2}.$$

Thus

$$\binom{k}{2} = n_H(C_g) \leq \binom{t}{2} + \frac{k - t}{2},$$

and hence $k = t$, as desired. $\qquad\square$

Let us list some known results that have not been mentioned earlier.

In their study of g.p. trees, Xu, Liu and Peng (1994) discovered the following χ-unique multi-bridge graphs.

Theorem 5.2.9   For $k \geq 4$, $\theta(a_1, a_2, \cdots, a_k)$ is χ-unique if $k - 1 \leq a_1 \leq a_2 \leq \cdots \leq a_k$. $\qquad\square$

As a continuation of this work, Peng (1994) expressed some coefficients of the chromatic polynomial of $\theta(a_1, a_2, \cdots, a_k)$ explicitly in terms of certain subgraphs of $\theta(a_1, a_2, \cdots, a_k)$ and proved the following result.

Theorem 5.2.10   Let $h, s \in \mathbb{N}$ such that $h \geq s + 1 \geq 2$ or $s = h + 1$. Then for $k \geq 4$, $\theta(a_1, a_2, \cdots, a_k)$ is χ-unique if $a_2 - 1 = a_1 = h$, $a_j = h + s$ $(j = 3, \cdots, k - 1)$, $a_k \geq h + s$ and $a_k \notin \{2h, 2h + s, 2h + s - 1\}$. $\qquad\square$

Assume that $2 \leq a_1 \leq a_2 \leq \cdots \leq a_k$. What is the maximum value of M such that if $a_k - a_1 \leq M$, then $\theta(a_1, a_2, \cdots, a_k)$ is guaranteed to be χ-unique? Dong, Teo, Little, Hendy and Koh (2004) showed that $M = a_2 - 1$. The proof is similar to that for Theorem 5.2.6 and assume the truth of Theorem 5.2.9.

Theorem 5.2.11   If $2 \leq a_1 \leq a_2 \leq \cdots \leq a_k < a_1 + a_2$, where $k \geq 3$, then the graph $\theta(a_1, a_2, \cdots, a_k)$ is χ-unique. $\qquad\square$

Dong, Teo, Little, Hendy and Koh (2004) showed that Theorem 5.2.11 does not hold if the bound in the theorem is relaxed. Let $\epsilon = (a_1 + a_2) - a_k$. They showed that for each non-negative integer ϵ, there is a k-bridge graph that is not χ-unique.

Theorem 5.2.12 *(i) $\theta(2,2,2,3,4) \sim H$ for every H in the family $\mathcal{G}_e(\theta(2,2,3), C_4, C_4)$.*

(ii) For $k \geq 4$ and $a \geq 2$, $\theta(k-2, a, a+1, \cdots, a+k-2) \sim H$ for every H in the family $\mathcal{G}_e(\theta(k-1, a, a+1, \cdots, a+k-3), C_{a+k-2})$.

(iii) For $k \geq 5$, $\theta(2, 3, \cdots, k-1, k, k-3) \sim H$ for every graph H in the family $\mathcal{G}_e(\theta(2, 3, \cdots, k-1), C_{k-1}, C_k)$. $\qquad\qquad\square$

5.3   Chromaticity of generalized polygon-trees

Multi-bridge graphs are generalized polygon-trees (g.p. trees). In this section, we study the chromaticity of the g.p. trees obtained by subdividing $K_2 \times \theta_2$ and $K_1 + P_4$. These graphs can be represented as follows:

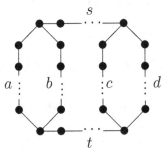

Figure 5.2

For completeness, we shall allow $s = 0$ and $t = 0$, in which case we obtain $\theta(a, b, c, d)$. We shall denote the graph in Figure 5.2 by $G_t^s(a, b; c, d)$. Let

$$\mathcal{C}_r(a, b; c, d) = \{G_t^s(a, b; c, d) : r = s, t\}.$$

By symmetry, we may assume $a \leq b$ and $a \leq c \leq d$.

The chromatic polynomial of each graph G in $\mathcal{C}_r(a, b; c, d)$ can be easily computed by using Theorem 1.3.3. Thus

$$P(G, 1-x) = \frac{(-1)^n x}{(1-x)^2} R(G, x),$$

where $n = a + b + c + d + r - 2$, and

$$\begin{aligned}
R(G, x) &= (x^{n+1} - x^{a+b+r} - x^{c+d+r} + x^{r+1} - x) - (1 + x + x^2) \\
&\quad + (x-1)(x^a + x^b + x^c + x^d) - (x^{a+c} + x^{a+d} + x^{b+c} + x^{b+d}).
\end{aligned}$$

From this, it follows that all graphs in $\mathcal{C}_r(a, b; c, d)$ are χ-equivalent (see also Chao and Zhao (1983)). Note that these graphs are $(n, n + 2)$-graphs, and if $min\{a, b, c, d\} \geq 2$, then the union of all $\mathcal{C}_r(a, b; c, d)$ is the family $\mathcal{GP}_2$, which is χ-closed (Theorem 5.2.4). A natural question is:

For which r, a, b, c, d, is $\mathcal{C}_r(a, b; c, d)$ a χ-equivalence class?

This question has attracted the attention of several researchers (Chao and Zhao (1983), Xu, Liu and Peng (1994), Dong, Teo, Little, Hendy and Koh (2004), Peng, Little, Teo and Wang (1997), Omoomi and Peng (2001), Omoomi and Peng (2003a) and Omoomi and Peng (2003b)) and is now completely settled. Here is a summary provided by Omoomi and Peng (2003b).

Theorem 5.3.1 *The family of graphs $\mathcal{C}_r(a, b; c, d)$ is a χ-equivalence class if and only if it is not one of the following families of graphs:*

(i) $\mathcal{C}_r(1, b; c, d)$ *for any* $r \geq 0$, $c \neq r + 1$ *or* $d \neq r + 1$;

(ii) $\mathcal{C}_0(2, b; b + 1, b + 2)$ *for any* $b \geq 2$;

(iii) $\mathcal{C}_1(2, 3; 3, 5)$;

(iv) $\mathcal{C}_1(3, 5; 5, 8)$;

(v) $\mathcal{C}_1(3, b; b + 1, b + 3)$ *for any* $b \geq 3$;

(vi) $\mathcal{C}_1(3, c + 3; c, c + 1)$ *for any* $c \geq 3$;

(vii) $\mathcal{C}_1(3, 3; c, c + 2)$ *for any* $c \geq 3$;

(viii) $\mathcal{C}_1(3, b; 3, b + 2)$ *for any* $b \geq 3$;

(ix) $\mathcal{C}_3(2, 3; 2, 4)$;

(x) $\mathcal{C}_5(2, 6; 4, 5)$;

(xi) $\mathcal{C}_3(2, c + 1; c, c + 2)$ *for any* $c \geq 3$;

(xii) $\mathcal{C}_3(2, c + 2; c, c + 1)$ *for any* $c \geq 3$;

(xiii) $\mathcal{C}_r(2, 4; 3, r + 2)$ *for any* $r \geq 2$;

(xiv) $\mathcal{C}_r(2, r + 2; 3, 4)$ *for any* $r \geq 2$;

(xv) $\mathcal{C}_r(a, r + 2; a + 1, a + r + 2)$ *for any* $a \geq 2$ *and* $r \geq 2$;

(xvi) $\mathcal{C}_r(a, a + 1; r + 2, a + r + 2)$ *for any* $a \geq 2$ *and* $r \geq 2$;

(xvii) $C_r(r-1, c+1; c, c+r-1)$ *for any* $c \geq r \geq 2$;

(xviii) $C_r(r-1, c+r-1; c, c+1)$, *for any* $c \geq r \geq 2$.   □

It is easy to verify that the χ-equivalence classes of the exceptional cases are as follows (see Omoomi and Peng (2003b)).

Theorem 5.3.2 *Each of the following is a χ-equivalence class:*

 (i) $C_r(1, b; c, d) \cup (C_{c-1}(1, b; r+1, d) \cup C_{d-1}(1, b; c, r+1)$, *where* $r \geq 1$ ($r \neq 2$) *and* $b, c, d \geq 2$;

 (ii) $C_0(2, b; b+1, b+2) \cup C_2(1, b+1; b, b+1) \cup C_{c-1}(1, b+1; 3, b+1) \cup C_c(1, b+1; b, 3)$;

 (iii) $C_1(2, 3; 3, 5) \cup C_3(2, 3; 2, 4)$;

 (iv) $C_1(3, 5; 5, 8) \cup C_5(2, 6; 4, 5)$;

 (v) $C_1(3, b; b+1, b+3) \cup C_3(2, b+1; b, b+2)$, *where* $b \geq 3$;

 (vi) $C_1(3, c+3; c, c+1) \cup C_3(2, c+2; c, c+1)$, *where* $c \geq 3$;

 (vii) $C_1(3, 3; c, c+2) \cup C_{c-1}(2, 4; 3, c+1)$, *where* $c \geq 3$;

 (viii) $C_1(3, b; 3, b+2) \cup C_{b-1}(2, b+1; 3, 4)$, *where* $b \geq 3$;

 (ix) $C_r(a, r+2; a+1, a+r+2) \cup C_{r+2}(r+1, a+1; a, a+r+1)$, *where* $a \geq r+2 \geq 2$;

 (x) $C_r(a, a+1; r+2, a+r+2) \cup C_{r+2}(a, a+1; r+1, a+r+1)$, *where* $a \geq r+2 \geq 2$.   □

Let G_1 and G_2 be two graphs and each contains a chain isomorphic to P_k. A graph obtained from G_1 and G_2 by indentifying the two chains is called a P_k-*gluing* of G_1 and G_2.

Problem 5.3.1 *Study the chromaticity of any P_k-gluing of $\theta(a_1, a_2, \cdots, a_k)$ and a cycle C_{k+d}, where $k \geq 4$, $d \geq 2$.*

The family $C_r(a, b; c, d)$ can also be viewed as the family of graphs obtained by joining two cycles C_{a+b} and C_{c+d} with two disjoint paths of lengths s and t such that $r = s + t$. Let us take two graphs G and H. Suppose there exist $u, v \in V(G)$ and $w, x \in V(H)$ such that $uv \notin E(G)$ and $wx \notin E(H)$. Now join u to w by a path P_s, and v to x by another path P_t, disjoint from P_s. Denote the resulting graph by $M_{wx}^{uv}(G, H, s, t)$ (see Figure 5.3).

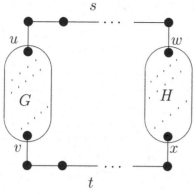

Figure 5.3

Problem 5.3.2 *For what g.p. trees G, H and $r \geq 1$, does*

$$\mathcal{M}_r = \{M_{wx}^{uv}(G, H, s, t) : s + t = r\}$$

form a χ-equivalence class?

Note that the answer to the question is in the affirmative when $G = K_4 - e$ and $H = C_q$, where $q \geq 4$ and $e \in E(G)$ (see Koh and K.L. Teo (1994)). The answer is also positive when $G = H = K_4 - e$, where $e \in E(G)$ (see Koh and C.P. Teo (1990a)). Let P_n^* denote the graph $K_1 + P_n$. For each $i, j \geq 4$, let u, v be the vertices of degree 2 in P_i^*, and w, x the vertices of degree 2 in P_j^*. Denote by $U_{i,j}(s, t)$ the graph $M_{wx}^{uv}(P_i^*, P_j^*, s, t)$ (see Figure 5.4).

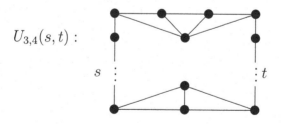

Figure 5.4

N.Z. Li and Whitehead (1993) conjectured that for $k \geq 1$, the family $\{U_{3,4}(s, t) : s + t = k\}$ is a χ-equivalence class. This was proved to be true by Chia, who further posed the following question in Chia (1996b).

Problem 5.3.3 *Let $k \geq 1$. For what values of $i, j \geq 4$, is the family $\{U_{i,j}(s, t) : s + t = k\}$ a χ-equivalence class?*

5.4 K_4-homeomorphs

Let G be 2-connected $(n, n+2)$-graph. Chao and Zhao (1983) showed that either G is a 4-bridge graph, a member of $\mathcal{C}_r(a, b; c, d)$ or a K_4-homeomorph. From the previous two sections we know that the chromaticity of each 2-connected $(n, n+2)$-graph that is not isomorphic to a K_4-homeomorph is completely settled. In this section focus on the chromaticity of K_4-homeomorphs.

A K_4-homeomorph is a subdivision of the complete graph K_4. Such a homeomorph will be denoted by $K_4(a_1, a_2, a_3, a_4, a_5, a_6)$ if the six edges of K_4 are replaced by six paths of length $a_1, a_2, \cdots, a_6$ respectively as shown in Figure 5.5.

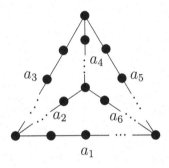

Figure 5.5

The six paths are the maximal chains of the homeomorph. The length of a chain P is denoted by $\ell(P)$.

The chromaticity of K_4-homeomorphs was first considered by S. Kahn (see Chao and Zhao (1993)). By computing the chromatic polynomials of all connected $(n, n+2)$-graphs, Chao and Zhao (1983) showed that any graph that is χ-equivalent to a K_4-homeomorph must itself be a K_4-homeomorph. In other words, we have:

Lemma 5.4.1   *The family of K_4-homeomorphs is χ-closed.*   □

The chromatic polynomial of a K_4-homeomorph can be found easily by using FRT, and is shown below (see Exercise 5.12 and Whitehead and Zhao (1984a)).

Lemma 5.4.2   *Let $G = K_4(a_1, a_2, \cdots, a_6)$. Then*

$$P(G, 1-x) = (-1)^m \frac{x}{(1-x)^2}(x^{m-1} + S(G, x) - (x+1)(x+2)), \quad (5.6)$$

where $x = 1 - \lambda$, $m = e(G)$ and

$$S(G, \lambda) = -\left(\sum_{i=1}^{5} x^{a_i + a_{i+1} + a_{i+2}} + \sum_{i=1}^{3} x^{a_i + a_{i+3}}\right) + (1 + x)\sum_{i=1}^{6} x^{a_i}, \quad (5.7)$$

with $a_7 = a_1$. □

It is obvious that if G and H are two K_4-homeomorphs, then $H \sim G$ if and only if $S(G, x) = S(H, x)$. We shall refer to $S(G, x)$ as the *essential polynomial* of G.

Not all K_4-homeomorphs are χ-unique. For example, $K_4(4, 1, 3, 1, 3, 1)$ and $K_4(2, 3, 1, 1, 5, 1)$ are χ-equivalent but non-isomorphic graphs. What is the proportion of χ-unique K_4-homeomorphs in the family of K_4-homeomorphs? This problem was considered by W.M. Li (1987). To give an answer to this question, he first established the following result.

Theorem 5.4.1 *If $\{a_1, a_2, \cdots, a_6\} = \{b_1, b_2, \cdots, b_6\}$ as multisets and*

$$K_4(a_1, a_2, \cdots, a_6) \sim K_4(b_1, b_2, \cdots, b_6),$$

then the two graphs are isomorphic. □

From this, W.M. Li derived the following result.

Corollary 5.4.1 *The graph $G = K_4(a_1, a_2, \cdots, a_6)$ is χ-unique if each of the following conditions is satisfied:*

(1) $\min\{a_1, a_2, \cdots, a_6\} > 1$;

(2) for any maximal chains P, Q, R in G such that P and Q are disjoint,

$$\ell(P) + \ell(Q) \neq \lceil \frac{1}{2}(\ell(P) + \ell(Q) + \ell(R)) \rceil;$$

(3) for any maximal chains P, Q, R, S in G such that P, Q and R have a vertex in common,

$$\ell(P) + \ell(Q) + \ell(R) \neq \lceil \frac{1}{2}(\ell(P) + \ell(Q) + \ell(R) + \ell(S)) \rceil.$$ □

Let $\mathcal{S}_0$ be the family of all K_4-homeomorphs of order n with the four vertices of degree 3 labeled. What is $|\mathcal{S}_0|$? It is known that the number of ways of distributing k identical objects into r distinct boxes is given by

$\binom{k+r-1}{r-1}$. Since each member of S_0 can be constructed by inserting $n - 4$ vertices into the six labeled edges of a given K_4,

$$|S_0| = \binom{n - 4 + (6 - 1)}{6 - 1} = \binom{n + 1}{5}.$$

Let S_j, where $j = 1, 2, 3$, be the family of those members of S_0 which violate condition (j) in the above corollary. Then it is not hard to see that $|S_1| \leq 4\binom{n}{4}$. Li further showed that each of $|S_2|$ and $|S_3|$ is bounded above by a polynomial in n of degree 4. Thus

$$\lim_{n \to \infty} \frac{|S_1| + |S_2| + |S_3|}{|S_0|} = 0.$$

This means that almost every K_4-homeomorph is χ-unique.

Theorem 5.4.2   *For any $n \geq 4$, let $f(n)$ (resp., $g(n)$) denote the number of K_4-homeomorphs (resp., χ-unique K_4-homeomorphs) of order n. Then*

$$\lim_{n \to \infty} g(n)/f(n) = 1. \qquad \qquad \square$$

So far, the method used to study the chromaticity of K_4-homeomorphs relies heavily on comparing coefficients of two essential polynomials (see Lemma 5.4.2). The following result of Whitehead Jr. and Zhao (1984a), is useful towards this end.

Theorem 5.4.3   *Suppose $K_4(a_1, a_2, \cdots, a_6) \sim K_4(b_1, b_2, \cdots, b_6)$. Then*

(i) $\min\{a_1, a_2, \cdots, a_6\} = \min\{b_1, b_2, \cdots, b_6\}$ (= ℓ, say);

(ii) $|\{i : a_i = \ell\}| = |\{i : b_i = \ell\}|$. $\square$

As a consequence of Theorem 5.3.1, W.M. Li (1987) proved the following useful result.

Lemma 5.4.3   *Let G be a K_4-homeomorph. If at most a pair of terms in the essential polynomial $S(G, x)$ can be cancelled, then G is χ-unique.* $\square$

In what follows, we shall let $G = K_4(a_1, a_2, \cdots, a_6)$, $A = \{a_1, a_2, \cdots, a_6\}$ (as a set), and for $s \in \mathbb{N}$, we let $I_s = \{i : a_i = s\}$. We summarize below some known results on the chromaticity of K_4-homeomorphs.

(1) (Whitehead Jr. and Zhao (1984a)) If $|A| \leq 2$, then G is χ-unique.

(2) (W.M. Li (1991)) If $\min A = s$ and $|I_s| \geq 4$, then G is χ-unique. (See also Ren and Zhang (2001).)

(3) (Zhang (2000)) If $\min A = s \geq 2$ and $|I_s| \geq 3$, then G is χ-unique. (See also Yin (2001).)

(4) (Ren (2002b))

    (i) If $\max A = s \geq 2$ and $|I_s| = 3$, then G is χ-unique, except $G = K_4(s, s, s-2, 1, s, 2)$, where $s \geq 3$. Moreover, for $s \geq 3$,

$$\{K_4(s, s, s-2, 1, s, 2), \ K_4(1, s, s+1, s-2, s-1, 2)\}$$

    is a χ-equivalence class.

    (ii) If $|I_s| = 3$, where $s \geq 2$, and two numbers in $\{a_1, a_2, \cdots, a_6\}$ (as a multiset) are less than s and one greater than s, then G is χ-unique, except $G = K_4(s, s, s, 2s-2, s-2, 1)$, where $s \geq 3$, or $G = K_4(s, s, 1, 2s, s, s-1)$, where $s \geq 2$.
Moreover,

$$\{K_4(s, s, 1, 2s, s, s-1), \ K_4(1, s, s-1, s+1, 2s-1, s) : \ s \geq 2\},$$

    and

$$\{K_4(s, s, s, 2s-2, s-2, 1), \ K_4(1, s, s-1, s-2, 2s-1, s) : \ s \geq 3\}$$

    are respectively χ-equivalence classes.
    (See also Ren and Zhang (2001).)

(5) (Ren and Liu (2001)) If $|I_s| = 3$, where $s \geq 2$, and there are two numbers in $\{a_1, a_2, \cdots, a_6\}$ (as a multiset) greater than s and one less than s, then G is χ-unique, except $G \in \{K_4(s, s, 1, 2s, s, s+2), \ K_4(s, s, s, 2s+1, s+1, 1)\}$. Moreover, these two exceptional graphs form a χ-equivalence class.

(6) (Guo and Whitehead Jr. (1997)) If $|I_1| = 3$, then G is χ-unique, except

$$G \in \{K_4(1, 1, 1, s+1, s, 3), K_4(1, 1, s, 2, 1, s+2)\},$$

$s \geq 2$. Moreover, the two exceptional graphs form a χ-equivalence class.
(See also Ren and Liu (2001), Peng (1994), Hu (1998), Zhang (2000), Chao and Zhao (1983), W.M. Li (1987), Chen and Ouyang (1987a).)

(7) If $|I_1| = 2$, then there are two situations.

  (i) (Guo and Whitehead Jr. (1997), S.J. Xu (1993))
      The graph $G = K_4(1, a_2, a_3, 1, a_5, a_6)$, where $a_2, a_3, a_5, a_6 \geq 2$,
      is χ-unique, except that G is included in

$$\{K_4(1, a+2, a, 1, 2, 2), \; K_4(1, b+1, b+3, 1, b, 2),$$
$$K_4(1, a+2, b, 1, a, 2)\},$$

where $a, b \geq 2$. Moreover, the following are respectively χ-equivalence classes:

$$\{K_4(1, a+2, a, 1, 2, 2), \; K_4(3, a+1, 1, 2, 1, a)\};$$
$$\{K_4(1, a+1, a+3, 1, a, 2), \; K_4(b+1, b+2, 1, a, 1, 3)\};$$
$$\{K_4(1, a+2, b, 1, a, 2), \; K_4(a+1, a, 1, b, 1, 3)\}.$$

  (ii) (Peng and Liu (2002))
       $G = K_4(a_1, a_2, 1, a_4, 1, a_6)$, where $a_1, a_2, a_4, a_6 \geq 2$, is χ-unique,
       except

$$G \in \{K_4(3, a+1, 1, 2, 1, a), \; K_4(a+1, a+2, 1, a, 1, 3),$$
$$K_4(a+1, a, 1, b, 1, 3), \; K_4(a, b+1, 1, a+b+1, 1, b),$$
$$K_4(a, a+b, 1, b, 1, b+2), \; K_4(a+1, a, 1, a+3, 1, 2),$$
$$K_4(a+2, a+2, 1, a, 1, 2)\}.$$

The χ-equivalence classes of the first three exceptional graphs are, respectively shown in (i). Each of the following is a χ-equivalence class:

$$\{K_4(a, b+1, 1, a+b+1, 1, b), \; K_4(a, a+b, 1, b, 1, b+2)\};$$
$$\{K_4(a+1, a, 1, a+3, 1, 2), \; K_4(a+2, a+2, 1, a, 1, 2)\}.$$

(See also Chen and Ouyang (1997a).)

(8) (i) (W.M. Li (1991)) For $a \geq 2$, $a > h$, $K_4(a, a+h, a+2h, a+5h, a+4h, a+3h)$ is χ-unique.
     (ii) (Zhang (2000)) For any $a, q \in \mathbb{N}$, $K_4(a, aq, aq^2, aq^5, aq^4, aq^3)$ is χ-unique.
     (iii) (W.M. Li (1991)) If $1 + \max A < 2 \min A$, then G is χ-unique.

From the above results, we see that the chromaticity of K_4-homeomorphs is complete for $|I_1| \geq 2$, and for $|I_s| \geq 3$, when $s \geq 2$. By extensive computer search, Whitehead Jr. (1986) has found many χ-equivalence pairs of K_4-homeomorphs with $|I_1| = 1$. Here are two tasks to be tackled.

Problem 5.4.1   *Study the chromaticity of K_4-homeomorphs for $|I_s| = 2$, $s \geq 2$, and for $|I_1| = 1$.*

Chen and Ouyang (1997a) determined the χ-equivalence classes of all K_4-homeomorphs with girth 5. In particular, the chromaticity of K_4-homeomorphs with girth 5 and $|I_1| = 1$ is therefore settled. We provide below their particular result.

Theorem 5.4.4   *The graph $K_4(x, y, z, 1, 2, 2)$, where $x, y, z \geq 2$, is χ-unique if and only if (x, y, z) is not a member of the following set:*

$$\bigcup_{i=1}^{\infty} \{(i+2, i, i+2), (i, i+1, i+3)\}.$$

Moreover, for each $i \geq 2$, the following family is a χ-equivalence class:

$$\{K_4(i, i+1, i+3, 1, 2, 2), K_4(i+2, i, i+2, 1, 2, 2)\}. \qquad \square$$

Very recently, Y.L. Peng (2003) determined the χ-equivalence classes of all K_4-homeomorphs with girth 6. Thus the chromaticity of all K_4-homeomorphs with girth 6 and $|I_1| = 1$ is settled. We provide this particular result of hers as follows.

Theorem 5.4.5   *The graph $K_4(a_1, a_2, a_3, 1, 2, 3)$, where $a_i \geq 2$, for $i = 1, 2, 3$, is χ-unique if and only if (a_1, a_2, a_3) is not one of the following:*

$$(i, i+1, i+4), (i+2, i+3, i), (i, i+4, i+1),$$
$$(i+3, i+2, i), (i, i+3, i+1), (i+2, i, i+2),$$

where $i \geq 2$. Moreover, for each $i \geq 2$, each of the following sets is a χ-equivalence class:

$$\{K_4(i, i+1, i+4, 1, 2, 3), K_4(i+2, i+3, i, 1, 2, 3)\},$$
$$\{K_4(i, i+4, i+1, 1, 2, 3), K_4(i+3, i+2, i, 1, 2, 3)\},$$
$$\{K_4(i, i+3, i+1, 1, 2, 3), K_4(i+2, i, i+2, 1, 2, 3)\}. \qquad \square$$

5.5 Chromaticity of uniform subdivisions of graphs

Theorem 5.2.8 implies that every uniform subdivision of $K(2, q)$ is χ-unique, where $q \geq 2$. For which graph, are its uniform subdivisions χ-unique? K.L.

Teo and Koh (1991) found a sufficient condition that gives an affirmative answer (see Theorem 5.5.3). Consequently, they found two families of graphs whose uniform subdivisions are χ-unique. These are extremal graphs that satisfy condition $(*)$ in the following lemma due to C.P. Teo and Koh (1992).

Lemma 5.5.1   *Let G be a 2-connected (n, m)-graph. Then*

$$n_G(C_g) \leq \begin{cases} \frac{m}{g}(m - n + 1), & \text{when } g \text{ is even} \\ \frac{n}{g}(m - n + 1), & \text{when } g \text{ is odd.} \end{cases}$$

Moreover, for even g, the equality holds if and only if G satisfies the condition

$(*)$ *every two edges of G are contained in a common shortest cycle.*
And for odd g, the equality holds if and only if G satisfies the condition

$(\#)$ *every vertex and every edge of G are contained in a common shortest cycle.*   □

We shall need a structural theorem on these extremal graphs. First of all, we state a theorem of Homobono and Peyrat (1989).

Theorem 5.5.1   *Every graph with even girth that satisfies condition $(*)$ is bipartite.*   □

Let G be a graph, which is not a cycle with $g(G) = 2h$ and satisfing condition $(*)$. Then there exists $ab \in E(G)$ with $d(a) \geq 3$. Let

$$s = min\{d(a, v) : d(v) \geq 3\}.$$

Note that *diam* $G = h$, and so $s \leq h$. K.L. Teo and Koh (1991) showed that $h = sk$, for some $k \in \mathbb{N}$. We refer (h, s, k) as the *structural triple* of G. Moreover, they provided some essential features of the structure of these extremal graphs.

Lemma 5.5.2   *Let G be a graph satisfying condition $(*)$ in Lemma 5.5.1 with $g(G) = 2h \geq 4$ and the structure triple (h, s, k).*

(i) If $k = 1$, then $G \cong \theta_q(h)$ for some $q \geq 2$.

(ii) If $k = 2$, then $G \cong H(s)$, where $H \cong K(p, q)$ for some p, q.

(iii) If $k \geq 2$, then $G \cong H(s)$ for a unique graph H.

(iv) If $k \geq 2$, then G contains some C_{g+2s} but no C_t with $g < t < g + 2s$.

$\square$

Two 2-connected graphs G and H are said to be n_g-*equivalent* if $v(G) = v(H)$, $e(G) = e(H)$, $g(G) = g(H)$ and $n_G(C_g) = n_H(C_g)$. A graph G is n_g-*unique* if $H \cong G$ whenever H is n_g-equivalent to G. It follows from Theorem 3.2.1 that every n_g-unique graph must be χ-unique. The converse is not true. The following two n_g-equivalent graphs are both χ-unique.

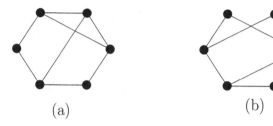

(a)   (b)

Figure 5.6

The following theorem of C.P. Teo and Koh (1992) provides an alternative proof of the χ-uniqueness of complete bipartite graphs.

Theorem 5.5.2   *For any $q \geq p \geq 2$, the complete bipartite graph $K(p,q)$ is n_g-unique.*

Proof.   Let G be n_g-equivalent to $K(p,q)$. Then

$$v(G) = p + q, \ e(G) = pq, \ g(G) = 4 \text{ and } n_G(C_4) = \binom{p}{2}\binom{q}{2}.$$

Let $n = v(G)$, $m = e(G)$. Then

$$\frac{m}{g}(m - n + 1) = \frac{1}{4}pq(pq - p - q + 1) = \binom{p}{2}\binom{q}{2} = n_G(C_4).$$

Thus, by Lemma 5.5.1, G satisfies condition $(*)$. By Theorem 5.5.1, G is bipartite. Condition $(*)$ now implies that G must be a complete bipartite graph. Let X, Y be the two partite sets of G with $|X| = p+r$ and $|Y| = q-r$ for some $r \geq 0$. Then

$$(p + r)(q - r) = |X||Y| = pq;$$

that is,

$$r(q - p - r) = 0.$$

It follows that $G \cong K(p,q)$.   □

Let $M(r,g)$ denote an r-regular Moore graph with girth g. Teo and Koh (1991) showed that $M(r,g)$ with even girth satisfies condition $(*)$. Moreover, they showed that $M(r,g)$ is n_g-unique if $r = p^c + 1$ for some prime p and $c \in \mathbb{N}$. Their next result implies that any uniform subdivision of the above Moore graphs and complete bipartite graphs is χ-unique.

Theorem 5.5.3   *If G is an n_g-unique graph with even girth satisfying condition $(*)$, then for each integer $f \geq 2$, the f-uniform subdivision $G(f)$ of G is χ-unique.*

Proof.   The result is true if $G \cong \theta_k(t)$ for some $k, t \geq 2$. Hence we may and shall assume that $G \not\cong \theta_k(t)$ for all $k, t \geq 2$. We may and will assume that $G \not\cong H(t)$ any graph H, since if $H(t)$ is n_g-unique and satisfies condition $(*)$, then H is also n_g-unique and satisfies condition $(*)$.

Evidently, $G(f)$ has even girth and satisfies condition $(*)$. Let $g' = g(G(f))$, and (h', s', k') be the structural triple of $G(f)$ as defined earlier. Note that $h' = s'k'$ and $s' = fs = f$, by Lemma 5.5.2 (iii).

By our assumption, $k' \geq 2$. By Lemma 5.5.2 (iv), $G(f)$ contains some $C_{g'+2f}$. Note that if $k' \geq 3$, then

$$g' = 2h' = 2fk' \geq 12,$$

and so

$$g' + 2f \leq g' + \frac{g'}{3} = \frac{3}{2}g' - \frac{g'}{6} \leq \frac{3g'}{2} - 2.$$

Thus we conclude that

$(**)$ if $k' \geq 3$, then $G(f)$ contains some $C_{g'+2f}$, where

$$g' < g' + 2f \leq \frac{3}{2}g' - 2.$$

We are now ready to show that $G(f)$ is χ-unique. Thus let $H^* \sim G(f)$. Then $g(H^*) = g'$ and H^* satisfies condition $(*)$. Let (h^*, s^*, k^*) be the structural triple of H^*. Then $g' = 2h^* = 2k^*s^*$. By the choice of our G, $k^* \neq 1$. We consider two cases.

Case (i): $k^* = 2$.

In this case, $g' = 4s^*$. By Lemma 5.5.2 (iv), H^* contains a cycle of length $g' + 2s^* = \frac{3}{2}g'$. Thus H^* does not contain any C_t with $g' < t < \frac{3}{2}g' - 2$. By Theorem 3.2.1 and statement $(**)$ above, we see that $k' = 2$.

Thus $s^* = s' = f$. By Lemma 5.5.2 (iii), $H^* \cong H(s^*) = H(f)$ for some graph H. In this case, it is clear that both G and H are complete bipartite graphs of same order and size. Thus $H \cong G$, and hence $H^* \cong G(f)$.

Case (ii): $k^* \geq 3$.

If $s^* \geq 2$, then again, H^* contains some $C_{g'+2s^*}$ with $g' + 2s^* \leq \frac{3}{2}g' - 2$. Since $G(f)$ (resp., H^*) does not contain any C_t with $g' < t < g'+2f$ (resp., $g' < t < g'+2s^*$), by Theorem 2.3.4, $g'+2f = g'+2s^*$, and so $s^* = f$, which implies that $k' = k^*$. Let H be the graph such that $H^* \cong H(s^*)$. Then both H and G are 2-connected and are n_g-equivalent. By the n_g-uniqueness of G, $H \cong G$, and thus $H^* \cong H(s^*) \cong G(f)$.

Assume now $s^* = 1$. If $k^* = 3$, then $g' = 6$. Since $g' = 2k'f$ and $f \geq 2$, we have $k' = 1$, which contradicts our choice of G.

Thus $k^* \geq 4$ and $g' = 2k^*s^* \geq 8$, which imply that

$$\frac{3}{2}g' - 2 \geq g' + 2s^*.$$

By an argument similar to that for $s^* \geq 2$, we have $s^* = f$ and $k^* = k'$, and hence arrive at the conclusion that $H^* \cong G(f)$. $\square$

It is easy to see that any uniform subdivision of K_4 is χ-unique (see Whitehead and Zhao (1984a)). We make the following conjecture.

Conjecture 5.5.1 *For all $n, t \in \mathbb{N}$ with $n \geq 4$, any t-uniform subdivision of K_n is χ-unique.*

The chromaticity of W_5-homeomorphs was studied by Giudici and Margaglio (1988).

Problem 5.5.1 *Study the chromaticity of uniform subdivisions of wheels.*

Brown and Hickman (2002b) derived an expression for a uniform subdivision of a graph G of recursive nature that is based on the underlying graph G.

5.6 Further results

Recall that the family of K_4-homeomorphs is χ-closed. However, this is not the case for general K_n-homeomorphs, where $n \geq 5$.

For example, the following two graphs (see Example 4.2.3(3) and Chee and Royle (1990)) are χ-equivalent. They both have chromatic polynomial

$$(\lambda - 1)(\lambda - 2)(\lambda - 3)\Big((\lambda - 1)^2(\lambda - 3) + 3\Big).$$

The first graph is a K_5-homeomorph while the second is not.

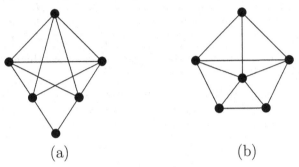

(a)   (b)

Figure 5.7

Let $K_n^{(s)}$ be the graph obtained from K_n by replacing an edge with a path of length s. Thus $K_n^{(1)} = K_n$. The first graph in Figure 5.7 is $K_5^{(2)}$. Read (1975) found that $K_6(2)$ is χ-equivalent to a chordal graph with chromatic polynomial

$$\lambda(\lambda - 1)(\lambda - 2)(\lambda - 3)^3(\lambda - 4),$$

and any such chordal graph is not a K_6-homeomorph (see Example 4.2.3).

The chromatic polynomial of $K_5^{(s)}$ has been computed by Chee and Royle (1990). In general, we have:

Lemma 5.6.1 *For any $n, s \in \mathbb{N}$ with $n \geq 5$,*

$$P(K_n^{(s)}, \lambda) = \frac{(\lambda)_{n-1}}{\lambda}\left((\lambda - 1)^s(\lambda - n + 2) + (-1)^s(n - 2)\right). \qquad (5.8)$$

Proof. By Theorem 1.3.3,

$$
\begin{aligned}
P(K_n^{(s)}, \lambda) &= \frac{1}{\lambda(\lambda - 1)}P(C_{s+1}, \lambda)P(K_n - e, \lambda) + (-1)^s P(K_{n-1}) \\
&= \frac{1}{\lambda(\lambda - 1)}\Big((\lambda - 1)^{s+1} + (-1)^{s+1}(\lambda - 1)\Big)\Big((\lambda)_n + (\lambda)_{n-1}\Big) \\
&\quad + (-1)^s(\lambda)_{n-1} \\
&= \frac{1}{\lambda}\Big((\lambda - 1)^s + (-1)^{s+1}\Big)\Big((\lambda)_n + (\lambda)_{n-1}\Big) + (-1)^s(\lambda)_{n-1} \\
&= \frac{(\lambda)_{n-1}}{\lambda}\Big((\lambda - 1)^s(\lambda - n + 2) + (-1)^s(n - 2)\Big). \qquad \square
\end{aligned}
$$

Chee and Royle (1990) showed that all $K_5^{(s)}$ are χ-unique except $K_5^{(2)}$.

Theorem 5.6.1   *For all $s \in \mathbb{N}$, $s \neq 2$, the graph $K_5^{(s)}$ is χ-unique. Moreover, the two graphs in Figure 5.7 form a χ-equivalence class (when $s = 2$).*

Chee and Royle (1990) made the following conjecture.

Conjecture 5.6.1   *All graphs $K_n^{(s)}$, $n \geq 6$, are χ-unique with finitely many exceptions.*

The χ-equivalence of subdivisions of the Petersen graph was considered by Shahmohamad and Whitehead Jr. (2002). Many χ-equivalent pairs of graphs have been found, but no χ-equivalence class has been determined.

Exercise 5

5.1 Let H be a connected (n, m)-graph. Show that

$$n_H(C_g) \le \binom{m - n + 2}{2},$$

where $g = g(H)$, and the equality holds if and only if $m = n$, or $g(H)$ is even and the subgraph H' generated by the vertex set V' is isomorphic to $\theta_{m-n+2}(g/2)$, where

$$V' = \{x \in V(H) : x \text{ is a vertex of } H\}.$$

Let $H \sim \theta_k(f)$. Show that $H' \cong \theta_k(f)$, and hence $H \cong \theta_k(f)$.

(See Dong (1993).)

5.2 Let $G = \theta(a_1, a_2, \cdots, a_k)$, where $k \ge 2$. Show that for any real number x,

$$P(G, 1 - x) = \frac{(-1)^{e(G)+1}}{(1 - x)^{e(G)-v(G)+1}} \, Q(x),$$

where

$$Q(x) = x \prod_{i=1}^{k}(x^{a_i} - 1) - \prod_{i=1}^{k}(x^{a_i} - x).$$

5.3 Show that a 2-connected graph is a g.p. tree if and only if it does not contain a subgraph isomorphic to a subdivision of K_4. Hence the family of g.p. trees is χ-closed.

(See Dirac (1952), Duffin (1965), Xu (1994).)

5.4 Let $r \in \mathbb{N}$. Show that the family of all g.p. trees, with r communication pairs, is χ-closed.

(See Xu (1994).)

5.5 Show that for any integers $b \ge a \ge 3$, $\theta(2, 2, 2, a, b)$ is χ-unique if and only if $(a, b) \ne (3, 4)$. Moreover, the following is a χ-equivalence class:

$$\{\theta(2, 2, 2, 3, 4)\} \cup \mathcal{G}_e(\theta(2, 2, 3), C_4, C_4).$$

Investigate the chromaticity of $\theta(a, a, a, b, c)$.

(See Li and Wei (2001).)

5.6   Show that if $\min\{a_1, a_2, \cdots, a_k\} \geq k - 1$, then $\theta(a_1, a_2, \cdots, a_k)$ is χ-unique for each $k \geq 4$.

(See Xu, Liu and Peng (1994).)

5.7   Let $k, f \geq 2$ and $p \geq 3$. Show that $\mathcal{G}_e(\theta_k(f), C_p)$ is a χ-equivalence class.

(See Xu, Liu and Peng (1994).)

5.8   Show that for $k \geq 2$, $p \geq 3$, the family $\mathcal{G}_e(\theta(a_1, a_2, \cdots, a_k), C_p)$ is a χ-equivalence class, if $a_1 \geq a_2 \geq \cdots \geq a_k > k$.

(See Xu, Liu and Peng (1994).)

5.9   Let G be a member of $\mathcal{C}_r(a, b; c, d)$ of order n, as defined in Section 5.3. Show that

$$P(G, 1 - x) = \frac{(-1)^n x}{(1 - x)^2} \, R(G, x),$$

for some polynomial $R(G, x)$.

(See Peng, Little, Teo and Wang (1997).)

5.10   Show that $\mathcal{C}_r(1, b; r + 1, r + 1)$ is a χ-equivalence class for all $r \in \mathbb{N}$, $b \geq 2$.

(See Xu, Liu and Peng (1994).)

5.11   For $s, t \in \mathbb{N}_0$, let $\mathcal{U}(s, t)$ be the graph shown below:

$U_{3,4}(s, t) :$

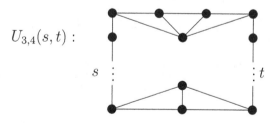

Show that for any $k \in \mathbb{N}$, $\{\mathcal{U}(s, t) : s + t = k\}$ is a χ-equivalence class.
(See Chia (1996b).)

5.12   Derive the formula for the chromatic polynomial of $K_4(a_1, a_2, \cdots, a_6)$, as in Lemma 5.4.2.

(See Whitehead and Zhao (1984a).)

5.13   Let S be the family of K_4-homeomorphs of order n that violate condition (2) or (3) in Corollary 5.4.1. Show that $|S|$ is bounded above by a polynomial in n of degree 4.

(See W.M. Li (1987).)

5.14   Suppose $K_4(a_1, a_2, \cdots, a_6) \sim K_4(b_1, b_2, \cdots, b_6)$. Show that

(1) $\min\{a_1, a_2, \cdots, a_6\} = \min\{b_1, b_2, \cdots, b_6\}$.

(2) Let $\ell = \min\{a_1, a_2, \cdots, a_6\}$. Show that

$$|\{i : a_i = \ell\}| = |\{i : b_i = \ell\}|.$$

(See Whitehead and Zhao (1984a).)

5.15   Suppose $G = K_4(a_1, a_2, \cdots, a_6)$ and $G' = K_4(b_1, b_2, \cdots, b_6)$ are χ-equivalent and $\{a_1, a_2, \cdots, a_6\} = \{b_1, b_2, \cdots, b_6\}$ as multisets. Show that $G \cong G'$.

(See W.M. Li (1987).)

5.16   Determine the χ-equivalence classes of all K_4-homeomorphs with girth 5.

(See Chen and Ouyang (1997a).)

5.17   Let G be a 2-connected (n, m)-graph and $g = g(G)$. Prove that

$$n_G(C_g) \leq \begin{cases} \frac{m}{g}(m - n + 1), & \text{when } g \text{ is even} \\ \frac{n}{g}(m - n + 1), & \text{when } g \text{ is odd} \end{cases}$$

(See C.P. Teo and Koh (1992).)

5.18   Show that if $K_5^{(s)}$ is the graph obtained from K_5 by replacing and edge with a path of length s, then $K_5^{(s)}$ is χ-unique for all $s \neq 2$.

(See Chee and Royle (1990).)

5.19   Show that for $k \geq 4$, $a \geq 2$, $\theta(k - 2, a, a + 1, \cdots, a + k - 2) \sim H$ for every $H \in \mathcal{G}_2(\theta(k - 1, a, a + 1, \cdots, a + k - 3), C_{a+k-2})$.

(See Dong, Teo, Little, Hendy and Koh (2004).)

5.20   Show that for $k \geq 5$,

$$\theta(2, 3, \cdots, k - 1, k, k - 3) \sim H,$$

for every $H \in \mathcal{G}_2(\theta(2, 3, \cdots, k - 1), C_{k-1}, C_k)$.

(See Dong, Teo, Little, Hendy and Koh (2004).)

Chapter 6

Graphs in Which any Two Colour Classes Induce a Tree (I)

6.1 Introduction

Dmitriev (1980), Chao, Li and Xu (1986), Vaderlind (1988) and Han (1988) showed, independently, that the family of q-trees of the same order forms a χ-equivalence class. Xu and Li (1984) proved that the wheel W_n is χ-unique for odd $n \geq 7$. This was then generalized by Dong (1990) who showed that the join $C_n + K_m$ is χ-unique for all $m \geq 1$ and even $n \geq 4$. While Chia (1990,1996a) (see also Dong, Liu and Koh (1997)) showed that the broken wheel $W(n, n-2)$ is χ-unique for $n = 6$ or $n \geq 8$, Dong and Liu (1998) proved that the broken wheel $W(n, n-3)$ is also χ-unique for $n \geq 5$. Chia (1995a) further proved that the join $W(n, n-2) + K_m$ is χ-unique for all $m \geq 1$ and even $n \geq 6$. Wanner (1989) showed that the family of graphs obtained from q-trees of the same order by deleting one edge that is contained in exactly $q-1$ triangles forms a χ-equivalence class. Borodin and Dmitriev (1991) proved a deep and more general result that for $2 \leq p \leq q$, the family of graphs of the same order obtained from a q-tree by adding a simplicial vertex of degree $q - p$ forms also a χ-equivalence class (see also Dong and Koh (1999b)).

Basically, in establishing the above results, the authors first considered the colour classes of the respective graphs and observed that the subgraph induced by any pair of colour classes possesses certain property such as forming a tree, a forest or a graph containing a unique cycle. They then proceeded to investigate the extremal value of some χ-invariants, usually the triangle number $t(G)$, to help complete their tasks.

Motivated by these, in the next four chapters, we shall unify and generalize these approaches by systematically setting some sets of conditions and introducing various families of graphs whose vertex sets could be partitioned into independent sets such that the subgraph induced by any pair of independent sets satisfies the required conditions of the respective sets. We shall then derive the expression for the triangle numbers of the graphs in each family, which, in turn, are used to help produce not only the χ-equivalence classes or χ-unique graphs that motivated our study, but also some unexpected new ones.

Consider the three graphs in Figure 6.1. Observe that their vertex sets are partitioned into independent sets in such a way that every two independent sets induce a tree. These are examples of graphs that we shall study in this and next chapter.

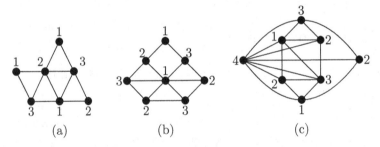

Figure 6.1

In general, given a graph G, an r-partition $\{A_1, \cdots, A_r\}$ of $V(G)$ is called an *r-independent partition* (or simply *independent partition*) of G if each A_i is a non-empty independent set of G for $i = 1, \cdots, r$. Obviously, a vertex colouring of G corresponds to an independent partition of G. A graph G is said to satisfy condition **(T)** if G possesses an r-independent partition $\{A_1, \cdots, A_r\}$, where $r \geq 3$, such that the subgraph $[A_i \cup A_j]$ is a tree for every pair i, j with $1 \leq i < j \leq r$.

Let $\mathcal{T}$ be the family of graphs satisfying **(T)**, and for $r \geq 3$, let $\mathcal{T}_r$ be the family of graphs G such that G satisfies **(T)** with an r-independent partition.

In Section 6.2 below, we shall derive an expression for the triangle number $t(G)$ of the graphs G in $\mathcal{T}_r$, and deduce from it that

$$t(G) \leq \frac{1}{3}(3v(G) - 2r)\binom{r-1}{2}.$$

The main objective in this chapter then is to characterize the family of graphs G in $\mathcal{T}_r$ such that the above equality for $t(G)$ holds. It turns out that this is indeed the χ-equivalence class of q-trees of the same order. However, before we get to this, we shall derive in Section 6.3 some useful information concerning the graphs in $\mathcal{T}_r$ and present in Section 6.4 some relevant results on chordal graphs. The results in Sections 6.2-6.4 will be found useful also in our next chapter.

6.2 The sizes and triangle numbers of graphs in $\mathcal{T}_r$

In this section, we shall aim at deriving an expression for $t(G)$, where $G \in \mathcal{T}_r$. A graph G is said to be *acyclic* r-*colourable* if G has an independent partition $\{A_1, \cdots, A_r\}$ such that $[A_i \cup A_j]$ is acyclic for all i and j with $1 \leq i < j \leq r$. The *acyclic chromatic number* $\chi^a(G)$ is the smallest positive integer r such that G is acyclic r-colourable.

We begin by providing a sharp upper bound for the size of an acyclic r-colourable graph.

Lemma 6.2.1 *Let G be an acyclic r-colourable graph. Then*

$$e(G) \leq (r-1)v(G) - \binom{r}{2}, \tag{6.1}$$

where equality holds if and only if G has an independent partition $\{A_1, \cdots, A_r\}$ such that $[A_i \cup A_j]$ is a tree for all i and j with $1 \leq i < j \leq r$.

Proof. By definition, G contains an independent partition $\{A_1, \cdots, A_r\}$ such that $[A_i \cup A_j]$ is acyclic for all i and j with $1 \leq i < j \leq r$. Thus

$$\begin{aligned}
e(G) &= \sum_{1 \leq i < j \leq r} e([A_i \cup A_j]) \\
&\leq \sum_{1 \leq i < j \leq r} (|A_i| + |A_j| - 1) \\
&= (r-1)\sum_{1 \leq i \leq r} |A_i| - \binom{r}{2}
\end{aligned}$$

$$= (r-1)v(G) - \binom{r}{2}.$$

Observe that $e(G) = (r-1)v(G) - \binom{r}{2}$ if and only if

$$e([A_i \cup A_j]) = |A_i| + |A_j| - 1,$$

i.e., $[A_i \cup A_j]$ is a tree for all i and j with $1 \le i < j \le r$. □

Lemma 6.2.2   *Let G be a graph and $r \ge 3$. Then the following statements are equivalent:*

(i) $G \in \mathcal{T}_r$;

(ii) $v(G) \ge r$ and G has an independent partition $\{A_1, \cdots, A_r\}$ such that $[A_i \cup A_j]$ is a tree for all i, j with $1 \le i < j \le r$;

(iii) $\chi^a(G) = r$ and $e(G) = (r-1)v(G) - \binom{r}{2}$.

Proof.   (i)$\Rightarrow$(ii) Assume that G satisfies condition **(T)** with an independent partition $\{A_1, \cdots, A_r\}$. As $A_i \ne \emptyset$ for all i, we have $v(G) \ge r$. Thus (ii) holds.

(ii)$\Rightarrow$(iii) By Lemma 6.2.1, we have $e(G) = (r-1)v(G) - \binom{r}{2}$. It is obvious that $\chi^a(G) \le r$. If $\chi^a(G) \le r-1$, then by Lemma 6.2.1,

$$\begin{aligned}
e(G) &\le (r-2)v(G) - \binom{r-1}{2} \\
&= (r-1)v(G) - \binom{r}{2} - (v(G) - r + 1) \\
&< (r-1)v(G) - \binom{r}{2},
\end{aligned}$$

a contradiction. Thus $\chi^a(G) = r$ and (iii) follows.

(iii)$\Rightarrow$(i) By Lemma 6.2.1, G has an independent partition $\{A_1, \cdots, A_r\}$ such that $[A_i \cup A_j]$ is a tree for all i and j with $1 \le i < j \le r$. As $\chi^a(G) = r$, no A_i is empty. Thus $G \in \mathcal{T}_r$. □

Corollary 6.2.1   *Let $G \in \mathcal{T}_r$. Then*
(i) $v(G) \ge r$,
(ii) $\chi^a(G) = r$ and
(iii) $e(G) = (r-1)v(G) - \binom{r}{2}$.   □

It follows from Corollary 6.2.1 that $\mathcal{T}_{r_1} \cap \mathcal{T}_{r_2} = \emptyset$ for any pair r_1 and r_2 with $r_1 \neq r_2$. Thus $\mathcal{T}$ is partitioned into $\mathcal{T}_2, \mathcal{T}_3, \cdots$.

Given $G \in \mathcal{T}$, let $\mathcal{I}_t(G)$ be the set of independent partitions of G with which G satisfies **(T)**. In this section, we always assume that $G \in \mathcal{T}_r$ and $\{A_1, \cdots, A_r\} \in \mathcal{I}_t(G)$. For simplicity, we denote by $G_{i,j}$ the induced subgraph $[A_i \cup A_j]$ of G, where $1 \leq i < j \leq r$. Let i, j, k be pairwise distinct integers in $\{1, 2, \cdots, r\}$. For $x \in A_k$, let $d(x, A_i, A_j)$ denote the degree of x in $[A_i \cup A_j]$ and let

$$\rho(x, A_i, A_j) = c([N(x) \cap (A_i \cup A_j)]) - 1.$$

It is clear that $\rho(x, A_i, A_j) \geq 0$, and $\rho(x, A_i, A_j) = 0$ if and only if $[N(x) \cap (A_i \cup A_j)]$ is a tree.

Lemma 6.2.3 *Let $G \in \mathcal{T}_r$ and $\{A_1, \cdots, A_r\} \in \mathcal{I}_t(G)$. For any i, j, k with $1 \leq i < j < k \leq r$, we have*

$$\sum_{x \in A_k} \rho(x, A_i, A_j) = \sum_{x \in A_i} \rho(x, A_j, A_k) = \sum_{x \in A_j} \rho(x, A_i, A_k)$$
$$= |A_i| + |A_j| + |A_k| - 2 - t([A_i \cup A_j \cup A_k]). \tag{6.2}$$

Proof. Any triangle in $[A_i \cup A_j \cup A_k]$ contains exactly one vertex in A_i, one vertex in A_j and one vertex in A_k. For any vertex $x \in A_k$, the number of triangles containing x in $[A_i \cup A_j \cup A_k]$ equals the number of edges in the subgraph $[N(x) \cap (A_i \cup A_j)]$. Since $[N(x) \cap (A_i \cup A_j)]$ is a forest with $1 + \rho(x, A_i, A_j)$ components, $e([N(x) \cap (A_i \cup A_j)]) = d(x, A_i, A_j) - 1 - \rho(x, A_i, A_j)$. Thus

$$
\begin{aligned}
t([A_i \cup A_j \cup A_k]) &= \sum_{x \in A_k} t([\{x\} \cup A_i \cup A_j]) \\
&= \sum_{x \in A_k} (d(x, A_i, A_j) - 1 - \rho(x, A_i, A_j)) \\
&= \sum_{x \in A_k} d(x, A_i, A_j) - |A_k| - \sum_{x \in A_k} \rho(x, A_i, A_j) \\
&= e(G_{i,k}) + e(G_{j,k}) - |A_k| - \sum_{x \in A_k} \rho(x, A_i, A_j) \\
&= |A_i| + |A_k| - 1 + |A_j| + |A_k| - 1 - |A_k| \\
&\quad - \sum_{x \in A_k} \rho(x, A_i, A_j) \\
&= |A_i| + |A_j| + |A_k| - 2 - \sum_{x \in A_k} \rho(x, A_i, A_j).
\end{aligned}
$$

Similarly,

$$
\begin{aligned}
t([A_i \cup A_j \cup A_k]) &= |A_i| + |A_j| + |A_k| - 2 - \sum_{x \in A_i} \rho(x, A_j, A_k) \\
&= |A_i| + |A_j| + |A_k| - 2 - \sum_{x \in A_j} \rho(x, A_i, A_k).
\end{aligned}
$$

The result thus follows.   □

We now arrive at an expression for the triangle number of a graph in $\mathcal{T}_r$.

Lemma 6.2.4   Let $G \in \mathcal{T}_r$ and $\{A_1, \cdots, A_r\} \in \mathcal{I}_t(G)$. Then

$$
t(G) = \frac{1}{3}(3v(G) - 2r)\binom{r-1}{2} - \sum_{1 \le i < j < k \le r} \sum_{x \in A_k} \rho(x, A_i, A_j). \qquad (6.3)
$$

Proof. By Lemma 6.2.3,

$$
\begin{aligned}
t(G) &= \sum_{1 \le i < j < k \le r} t([A_i \cup A_j \cup A_k]) \\
&= \sum_{1 \le i < j < k \le r} (|A_i| + |A_j| + |A_k| - 2) - \sum_{1 \le i < j < k \le r} \sum_{x \in A_k} \rho(x, A_i, A_j) \\
&= v(G)\binom{r-1}{2} - 2\binom{r}{3} - \sum_{1 \le i < j < k \le r} \sum_{x \in A_k} \rho(x, A_i, A_j) \\
&= \frac{1}{3}(3v(G) - 2r)\binom{r-1}{2} - \sum_{1 \le i < j < k \le r} \sum_{x \in A_k} \rho(x, A_i, A_j).
\end{aligned}
$$
                                                                                             □

Corollary 6.2.2   Let $G \in \mathcal{T}_r$. Then $t(G) \le \frac{1}{3}(3v(G) - 2r)\binom{r-1}{2}$.   □

Let $h(G)$ denote the difference between $t(G)$ and the upper bound; i.e.,

$$
h(G) = \frac{1}{3}(3v(G) - 2r)\binom{r-1}{2} - t(G). \qquad (6.4)
$$

Then $\mathcal{T}_r$ is partitioned into $\mathcal{T}_{r,0}, \mathcal{T}_{r,1}, \cdots$, where

$$
\mathcal{T}_{r,h} = \{G : G \in \mathcal{T}_r, h(G) = h\}, \quad h = 0, 1, \cdots. \qquad (6.5)
$$

By the above two lemmas, we have:

Lemma 6.2.5 *Let $G \in \mathcal{T}_r$ and $\{A_1, \cdots, A_r\} \in \mathcal{I}_t(G)$. Then*

$$
\begin{aligned}
h(G) &= \sum_{1 \leq i < j < k \leq r} \sum_{x \in A_k} \rho(x, A_i, A_j) \\
&= \sum_{1 \leq i < j < k \leq r} \sum_{x \in A_j} \rho(x, A_i, A_k) \\
&= \sum_{1 \leq i < j < k \leq r} \sum_{x \in A_i} \rho(x, A_j, A_k).
\end{aligned}
\tag{6.6}
$$

$\square$

6.3 Graphs in $\mathcal{T}_r$

In this section, we first introduce some graphs in $\mathcal{T}_r$ which are uniquely r-colourable and then proceed to present some properties of the graphs in $\mathcal{T}_r$.

By definition, $\mathcal{T}_2$ is the family of trees. For $r \geq 3$, which graphs are members in $\mathcal{T}_r$? Obviously, K_r belongs to $\mathcal{T}_r$. It can be checked that any $(r-1)$-tree of order at least r is contained in $\mathcal{T}_r$. A subfamily of $\mathcal{T}_r$ is given below.

Theorem 6.3.1 *Let G be a uniquely r-colourable graph with $e(G) = (r-1)v(G) - \binom{r}{2}$. Then $G \in \mathcal{T}_r$.*

Proof. Let $(A_1, \cdots, A_r)$ be an independent partition of G. Since G is uniquely r-colourable, we have $A_i \neq \emptyset$ for all i with $1 \leq i \leq r$ and $[A_i \cup A_j]$ is connected for all i, j with $1 \leq i < j \leq r$. Thus

$$
\begin{aligned}
e(G) &= \sum_{1 \leq i < j \leq r} e([A_i \cup A_j]) \\
&\geq \sum_{1 \leq i < j \leq r} (|A_i| + |A_j| - 1) \\
&= (r-1)v(G) - \binom{r}{2}.
\end{aligned}
$$

Since $e(G) = (r-1)v(G) - \binom{r}{2}$, we have $e([A_i \cup A_j]) = |A_i| + |A_j| - 1$, and so $[A_i \cup A_j]$ is a tree for all i, j with $1 \leq i < j \leq r$. Thus $G \in \mathcal{T}_r$. $\square$

Note Not every graph in $\mathcal{T}_r$ is uniquely r-colourable. For example, let G_0 be the graph in Figure 6.2 and let $A_1 = \{a, b'\}$, $A_2 = \{b, c'\}$ and $A_3 = \{c, a'\}$. Then all $[A_i \cup A_j]$'s are trees, and so $G_0 \in \mathcal{T}_3$. But G_0 is not uniquely 3-colourable.

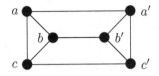

Figure 6.2

The graphs in $\mathcal{T}_r$ are obviously connected. We shall prove that they are, in fact, $(r-1)$-connected.

Lemma 6.3.1 *If a graph G has an independent partition $(B_1, \cdots, B_r)$ such that $[B_i \cup B_j]$ is connected for all i, j with $1 \leq i < j \leq r$, then G is $(r-1)$-connected.*

Proof. Suppose that S is a cut of G with $|S| \leq r - 2$ and that $G_1, G_2, \cdots, G_q$, where $q \geq 2$, are the components of $G - S$. (See Figure 6.3.)

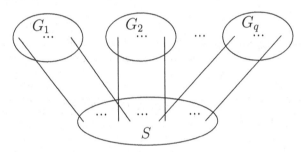

Figure 6.3

Since $|S| \leq r - 2$, there exist two of B_i's, say B_1 and B_2, such that $B_1 \cap S = B_2 \cap S = \emptyset$. As $[B_1 \cup B_2]$ is connected, $B_1 \cup B_2 \subseteq V(G_i)$ for some i, say $i = 1$. Thus $V(G_2) \subseteq B_3 \cup \cdots \cup B_r$. Assume $V(G_2) \cap B_3 \neq \emptyset$. Then $[B_1 \cup B_3]$ is disconnected, a contradiction. Thus G is $(r-1)$-connected. $\square$

Corollary 6.3.1 *Let $G \in \mathcal{T}_r$. Then*
  (i) G is $(r-1)$-connected and
  (ii) $d(x) = r - 1$ for any simplicial vertex x of G. $\square$

To end this section, we derive the following properties of graphs in $\mathcal{T}_r$.

Lemma 6.3.2 *Let $r \geq 2$ be an integer and suppose that G is not a complete graph and x is a simplicial vertex of G with $d(x) = r - 1$. Then*
  (i) $G \in \mathcal{T}_r$ if and only if $G - x \in \mathcal{T}_r$, and
  (ii) $h(G) = h(G - x)$ when $G \in \mathcal{T}_r$.

Proof. (i) [Necessity] Suppose that $G \in \mathcal{T}_r$ and $\{A_1, \cdots, A_r\} \in \mathcal{I}_t(G)$. Assume that $x \in A_1$. Then $\{A_1 \backslash \{x\}, A_2, \cdots, A_r\}$ is an independent partition of $G - x$.

Suppose that $A_1 = \{x\}$. As $[A_1 \cup A_j]$ is a tree for all j with $2 \leq j \leq r$, x is adjacent to all other vertices of G. The fact that x is a simplicial vertex implies that G is complete, a contradiction. Thus $A_1 \backslash \{x\} \neq \emptyset$. Since x is a simplicial vertex of G, x is an end-vertex of the tree $[A_1 \cup A_j]$, and so $[(A_1 \backslash \{x\}) \cup A_j]$ is a tree for each j with $2 \leq j \leq r$. It follows that $G - \{x\} \in \mathcal{T}_r$.

[Sufficiency] Suppose that $G - x \in \mathcal{T}_r$ and $\{B_1, \cdots, B_r\} \in \mathcal{I}_t(G - x)$. Since x is a simplicial vertex of G and $d(x) = r - 1$, there is exactly one i in $\{1, 2, \cdots, r\}$, say $i = 1$, such that $N(x) \cap B_1 = \emptyset$. Thus $\{B_1 \cup \{x\}, B_2, \cdots, B_r\}$ is an independent partition of G. Since $[B_1 \cup B_j]$ is a tree and $|N(x) \cap B_j| = 1$, $[B_1 \cup \{x\} \cup B_j]$ is a tree for each j with $2 \leq j \leq r$. Thus $G \in \mathcal{T}_r$.

(ii) Since x is a simplicial vertex of degree $r - 1$,

$$t(G) = t(G - x) + \binom{r-1}{2}.$$

Thus

$$
\begin{aligned}
h(G - x) &= \frac{1}{3}(3(v(G) - 1) - 2r)\binom{r-1}{2} - t(G - x) \\
&= \frac{1}{3}(3v(G) - 2r)\binom{r-1}{2} - \binom{r-1}{2} - t(G - x) \\
&= \frac{1}{3}(3v(G) - 2r)\binom{r-1}{2} - t(G) \\
&= h(G).
\end{aligned}
$$

$\square$

6.4 Chordal graphs

In this section, we shall present some useful properties of chordal graphs; study a condition under which a vertex is contained in a pure cycle; and finally provide two sufficient conditions for a graph to be chordal. These results will be found useful later.

Theorem 6.4.1 (Dirac (1961)) *Suppose that G is a chordal graph. If G is not complete, then G contains two non-adjacent simplicial vertices.* $\square$

Corollary 6.4.1   *Let G_1 and G_2 be two graphs and let r be an integer with $1 \leq r \leq \min\{\omega(G_1), \omega(G_2)\}$. Let $G \in \mathcal{G}[G_1 \cup_r G_2]$. If G_1 is a chordal graph and $G_1 \not\cong K_r$, then there is an x in $V(G_1)\backslash V(G_2)$ such that x is a simplicial vertex of G.*   □

The proof of Corollary 6.4.1 is left to the reader (see Exercise 6.2).

Suppose that $v(G) = n$. An ordering: $x_1, x_2, \cdots, x_n$ of vertices in G is called a *perfect elimination ordering* of G if for each $i = 1, 2, \cdots, n - 1$, x_i is a simplicial vertex in the graph $[V_i]$, where $V_i = \{x_i, x_{i+1}, \cdots, x_n\}$. The following characterization of chordal graphs follows easily from Theorem 6.4.1.

Theorem 6.4.2 (Fulkerson and Gross (1965))   *A graph G is chordal if and only if G has a perfect elimination ordering.*   □

Given a vertex x in G, how can we know whether x is contained in a pure cycle of G? We shall answer this question in Lemma 6.4.2. First of all, we have the following observation.

Lemma 6.4.1   *Let x be a vertex in a graph G with $N(x) = \{u, v\}$. If $G - x$ is connected and $uv \notin E(G)$, then x, u and v are contained in a common pure cycle of G.*

Proof. Let $P(u, v)$ be a shortest u-v path in the connected graph $G - x$. Then $xP(u, v)x$ is a pure cycle of G.   □

For $S \subseteq V(G)$, let $N_G^-(S)$, or simply $N^-(S)$, be the set

$$\{x : x \in V(G)\backslash S, xy \in E(G) \text{ for some } y \in S\};$$

i.e., $N^-(S) = \left(\bigcup_{x \in S} N(x) \right) \backslash S$. Also we write $N[x]$ for $N(x) \cup \{x\}$.

Lemma 6.4.2   *Suppose that G is a connected graph of order n and that x is a vertex in G with $d(x) < n - 1$. Then x is contained in a pure cycle of G if and only if there exists a component H of the graph $G - (N[x])$ such that the set $N^-(V(H))$ is not a clique in G.*

Proof. [Necessity] Suppose that $xu_1u_2 \cdots u_kx$, where $k \geq 3$, is a pure cycle of G. Then $G - (N[x])$ contains a component H such that $u_1, u_k \in N^-(V(H))$ and $\{u_2, \cdots, u_{k-1}\} \subseteq V(H)$ (see Figure 6.4). Then $u_1u_k \notin E(G)$, implying that $N^-(V(H))$ is not a clique in G.

[Sufficiency] Suppose that $N^-(V(H))$ is not a clique of G for some component H of $G - (N[x])$. Then there are two vertices u and v in $N^-(V(H))$ such that $uv \notin E(G)$. Obviously, $\{u, v\} \subseteq N(x)$. Since $[\{u, v\} \cup V(H)]$ is connected, by Lemma 6.4.1, x, u and v are contained in a common pure cycle of G. $\square$

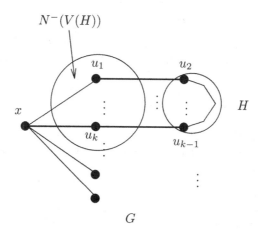

Figure 6.4

Recall that a cut S of G is called a *clique-cut* of G if S is a clique in G. By Lemma 6.4.2, we have the following result.

Corollary 6.4.2 *Let G be a connected graph and let $x \in V(G)$. If x is contained in no pure cycle of G, then one of the following statements holds:*
(i) $d(x) = v(G) - 1$;
(ii) G contains a clique-cut. $\square$

The proof is left to the reader (see Exercise 6.3).

In the remainder of this section, we give two sufficient conditions for a graph to be chordal. These results were established by Dong and Koh (1997b).

Theorem 6.4.3 *Let G be an r-colourable graph, where $r \geq 3$, and $A_1, \cdots, A_r$ be the colour classes of an r-colouring of G. Then G is a chordal graph if the following three conditions hold:*
(i) $G - A_r$ is a chordal graph,

    (ii) $[A_i \cup A_r]$ *is a forest for each* $i = 1, 2, \cdots, r - 1$, *and*

    (iii) for any $x \in A_r$ *and any pair* i, j *with* $1 \le i < j \le r - 1$, *either* x *is an isolated vertex in* $[A_i \cup A_j \cup A_r]$ *or* $[N(x) \cap (A_i \cup A_j)]$ *is a tree.* $\square$

Theorem 6.4.4   *Let* G *be an* r-*colourable graph, where* $r \ge 3$, *and* $A_1, \cdots, A_r$ *be the colour classes of an* r-*colouring of* G. *Then* G *is a chordal graph if the following conditions hold:*

    (i) $[A_i \cup A_j]$ *is a forest for every pair* i *and* j *with* $1 \le i < j \le r$ *and*

    (ii) for any i, j, k *with* $1 \le i < j < k \le r$ *and any* $x \in A_k$, *either* x *is an isolated vertex in* $[A_i \cup A_j \cup A_k]$ *or* $[N(x) \cap (A_i \cup A_j)]$ *is a tree.* $\square$

6.5 q-trees

In this section, we shall establish three characterizations of q-trees, which were obtained independently by Dmitriev (1982), Chao, Li and Xu (1986), Vaderlind (1986) and Han (1986).

    It is clear that every q-tree is a chordal graph. Note that K_q is the unique q-tree of order q and K_{q+1} is the unique q-tree of order $q + 1$. The following results follow directly from the definition of q-trees.

Lemma 6.5.1   *Let* G *be a* q-*tree of order* n, *where* $n \ge q + 1$. *Then*
(i) $\omega(G) = \chi(G) = q + 1$,
(ii) G *is uniquely* $(q + 1)$-*colourable*,
(iii) $e(G) = \binom{q}{2} + (n - q)q$ *and* $t(G) = \binom{q}{3} + (n - q)\binom{q}{2}$,
(iv) G *is* q-*connected*,
(v) every cut of cardinality q *induces a* K_q,
(vi) for every $1 \le k \le q$, *every* k-*clique is contained in a* $(q + 1)$-*clique, and*
(vii) $P(G, \lambda) = \lambda(\lambda - 1) \cdots (\lambda - q + 1)(\lambda - q)^{n-q}$. $\square$

    While every q-tree is chordal, not every chordal graph is a q-tree. The following result characterizes q-trees within the family of chordal graphs.

Theorem 6.5.1   *Let* G *be a chordal graph of order* n, *where* $n \ge q$ *for some* $q \in \mathbb{N}$. *Then* G *is a* q-*tree if and only if* G *is* q-*connected with* $\chi(G) \le q + 1$.

Proof.   The necessity follows from Lemma 6.5.1. We now prove the sufficiency. If $n = q$, then as G is q-connected, $G \cong K_q$. Suppose that $n > q$. By Theorem 6.4.1, G contains a simplicial vertex x. Since G is q-connected and $\chi(G) \leq q+1$, it is clear that $d(x) = q$. Observe that $G - x$ is a chordal graph such that $\chi(G-x) \leq q+1$ and $G-x$ is q-connected. By the induction hypothesis, $G - x$ is a q-tree. Thus G is a q-tree.   □

Corollary 6.5.1 *If G is a 2-connected chordal graph with $\chi(G) \leq 3$, then G is a 2-tree.*   □

Lemma 6.5.2   *Let $G \in \mathcal{T}_r$. Then $G \in \mathcal{T}_{r,0}$ if and only if G is a chordal graph.*

Proof.   Suppose that $\{A_1, \cdots, A_r\} \in \mathcal{I}_t(G)$.

[Necessity] If $G \in \mathcal{T}_{r,0}$, by Lemma 6.2.5, then $\rho(x, A_i, A_j) = 0$ for all distinct i, j, k from $\{1, 2, \cdots, r\}$ and any $x \in A_k$. By Theorem 6.4.4, G is a chordal graph.

[Sufficiency] Suppose that $G \notin \mathcal{T}_{r,0}$; i.e., $h(G) > 0$. Then, by Lemma 6.2.5, there exist $x \in V(G)$ and integers i, j with $i \neq j$ and $x \notin A_i \cup A_j$ such that

$$\rho(x, A_i, A_j) > 0.$$

Thus $[N(x) \cap (A_i \cup A_j)]$ is disconnected. But then there exists a pure cycle in $[\{x\} \cup A_i \cup A_j]$, and so G is not a chordal graph.   □

We are now ready to establish our main result.

Theorem 6.5.2   *Let G be a graph of order n and r be an integer with $n \geq r \geq 3$. Then the following statements are equivalent:*

(i) $G \in \mathcal{T}_{r,0}$;

(ii) G is an $(r-1)$-tree;

(iii) $P(G, \lambda) = \lambda(\lambda-1) \cdots (\lambda-r+2)(\lambda-r+1)^{n-(r-1)}$;

(iv) G is a uniquely r-colourable graph with $e(G) = \binom{r}{2} + (n-r)(r-1)$ and $t(G) = \binom{r}{3} + (n-r)\binom{r-1}{2}$.

Proof.   (i)⇒(ii) Let G be a graph in $\mathcal{T}_{r,0}$. If $n = r$, then $G \cong K_r$, which is an $(r-1)$-tree.

Assume that $n \geq r+1$. By Lemma 6.5.2, G is a chordal graph. Let x be a simplicial vertex of G. By Corollary 6.3.1, $d(x) = r - 1$. Since $G \in \mathcal{T}_{r,0}$, by Lemma 6.3.2, $G - x \in \mathcal{T}_{r,0}$. By the induction hypothesis, $G - x$ is an $(r-1)$-tree. Thus G is an $(r-1)$-tree by definition.

(ii)⇒(iii) This follows from Lemma 6.5.1.

(iii)$\Rightarrow$(iv) As $n \geq r$, $P(G, \lambda) > 0$ if and only if $\lambda \geq r$. Thus $\chi(G) = r$. Since $P(G, r) = r!$, G is uniquely r-colourable. Let H be an $(r - 1)$-tree of order n. Then $H \sim G$. By Theorem 3.2.1, we have $e(G) = e(H)$ and $t(G) = t(H)$. By Lemma 6.5.1,

$$e(G) \;=\; e(H) = \binom{r}{2} + (r - 1)(n - r - 1)$$

$$t(G) \;=\; t(H) = \binom{r}{3} + (n - r - 1)\binom{r - 1}{2}.$$

(iv)$\Rightarrow$(i) By Theorem 6.3.1, $G \in \mathcal{T}_r$. Since $t(G) = \binom{r}{3} + (n - r)\binom{r-1}{2}$, $G \in \mathcal{T}_{r,0}$ by definition. $\qquad\square$

Corollary 6.5.2 *The family of q-trees of order n, where $n \geq q \geq 2$, forms a χ-equivalence class.* $\qquad\square$

Exercise 6

6.1 Let $G \in \mathcal{T}_r$. Prove that

(i) G is $(r-1)$-connected and

(ii) $d(x) = r - 1$ for any simplicial vertex x of G.

6.2 Let G_1 and G_2 be two graphs and let r be an integer with $1 \le r \le \min\{\omega(G_1), \omega(G_2)\}$. Let $G \in \mathcal{G}[G_1 \cup_r G_2]$ and G_1 be a chordal graph with $G_1 \not\cong K_r$. Prove that there is an x in $V(G_1) \backslash V(G_2)$ such that x is a simplicial vertex of G.

6.3 Let G be a connected graph and $x \in V(G)$. Prove that if x is contained in no pure cycle of G, then one of the following statements holds:

(i) $d(x) = v(G) - 1$;

(ii) G contains a clique-cut.

6.4 Prove that if G is a 2-connected chordal graph with $\chi(G) \le 3$, then G is a 2-tree.

6.5 Let G be a q-tree of order n, where $n \ge q + 1$. Prove that

(i) $\omega(G) = \chi(G) = q + 1$,

(ii) G is uniquely $(q+1)$-colourable,

(iii) $e(G) = \binom{q}{2} + (n-q)q$ and $t(G) = \binom{q}{3} + (n-q)\binom{q}{2}$,

(iv) G is q-connected,

(v) every cut of cardinality q induces a K_q,

(vi) for every $1 \le k \le q$, every k-clique is contained in a $(q+1)$-clique, and

(vii) $P(G, \lambda) = \lambda(\lambda - 1) \cdots (\lambda - q + 1)(\lambda - q)^{n-q}$.

6.6 Let G be a 2-tree. Assume that G has only two simplicial vertices u and v. Prove that

(i) if $d(u, v) = 2$, then G is a fan, as shown in Figure 6.5(a); and

(ii) if $d(u, v) = 3$, then G is a graph in Figure 6.5(b) or (c).

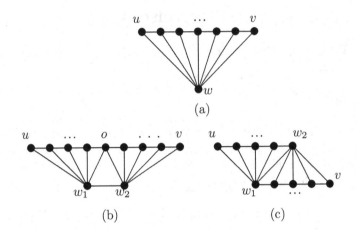

Figure 6.5

Chapter 7

Graphs in Which any Two Colour Classes Induce a Tree (II)

7.1 Introduction

In the preceding chapter, we characterize the family $\mathcal{T}_{r,0}$ as the χ-equivalence class of $(r-1)$-trees. Based on the basic results developed therein, we shall proceed in this chapter to study the structure and hence the chromaticity of the graphs in $\mathcal{T}_{r,1}$. Our main objective is to show that a graph G in $\mathcal{T}_{r,1}$ is χ-unique if and only if $G + K_m$ is χ-unique for each $m \geq 1$. Some new families of χ-unique graphs in $\mathcal{T}_{3,1}$ will be presented at the end of this chapter.

We begin in what follows with some observations on the properties of the graphs in $\mathcal{T}_{r,1}$, which will be useful in characterizing the structure of graphs in $\mathcal{T}_{r,1}$.

For convenience, we always assume throughout this chapter that G is a graph in $\mathcal{T}_{r,1}$ and $\{A_1, \cdots, A_r\} \in \mathcal{I}_t(G)$, unless otherwise stated. By Lemma 6.2.5,

$$\sum_{1 \leq i < j < k \leq r} \sum_{x \in A_k} \rho(x, A_i, A_j) = 1.$$

Thus there exist i_0, j_0, k_0 with $1 \leq i_0 < j_0 < k_0 \leq r$ such that for i, j, k

with $1 \leq i < j < k \leq r$,

$$\sum_{x \in A_k} \rho(x, A_i, A_j) = \begin{cases} 1, & \text{if } (i, j, k) = (i_0, j_0, k_0), \\ 0, & \text{otherwise.} \end{cases} \quad (7.1)$$

By Lemma 6.2.3, for i, j, k with $1 \leq i < j < k \leq r$,

$$\sum_{x \in A_i} \rho(x, A_j, A_k) = \begin{cases} 1, & \text{if } (i, j, k) = (i_0, j_0, k_0), \\ 0, & \text{otherwise;} \end{cases} \quad (7.2)$$

$$\text{and} \quad \sum_{x \in A_j} \rho(x, A_i, A_k) = \begin{cases} 1, & \text{if } (i, j, k) = (i_0, j_0, k_0), \\ 0, & \text{otherwise.} \end{cases} \quad (7.3)$$

We may assume that $i_0 = 1$, $j_0 = 2$ and $k_0 = 3$.

Lemma 7.1.1   *There exist three vertices $x_1 \in A_1$, $x_2 \in A_2$ and $x_3 \in A_3$ such that for any pairwise distinct integers i, j, k with $1 \leq i, j, k \leq r$ and any $x \in A_k$,*

$$\rho(x, A_i, A_j) = \begin{cases} 1, & \text{if } \{i, j, k\} = \{1, 2, 3\} \text{ and } x = x_k, \\ 0, & \text{otherwise.} \end{cases}$$

Proof.   By (7.1), (7.2) and (7.3), there exist $x_1 \in A_1$, $x_2 \in A_2$ and $x_3 \in A_3$ such that for i, j, k with $\{i, j, k\} = \{1, 2, 3\}$,

$$\rho(x, A_i, A_j) = \begin{cases} 1, & \text{if } x = x_k, \\ 0, & \text{if } x \in A_k \backslash \{x_k\}. \end{cases}$$

For all distinct i, j, k with $\{i, j, k\} \neq \{1, 2, 3\}$, we have $\sum_{x \in A_k} \rho(x, A_i, A_j) = 0$, and thus $\rho(A_i, A_j, x) = 0$ for any $x \in A_k$. $\square$

In the remainder of this section, let x_1, x_2 and x_3 denote the three special vertices in G having the properties mentioned above. By Lemma 7.1.1, we have the following result.

Corollary 7.1.1   *For all distinct i, j, k with $1 \leq i, j, k \leq r$ and any $x \in A_k$, if $\{i, j, k\} = \{1, 2, 3\}$ and $x = x_k$, then $[N(x) \cap (A_i \cup A_j)]$ is a forest with two components, and a tree otherwise.* $\square$

Lemma 7.1.2   *$G - x_i$ is a chordal graph for each $i = 1, 2, 3$.*

Proof. Consider the independent partition $\{A_1 \backslash \{x_1\}, A_2, \cdots, A_r\}$ of $G - x_1$.

If we treat $A_1 \backslash \{x_1\}$ as B_r and A_i as B_{i-1} for $i = 2, \cdots, r$, then $B_1, B_2, \cdots, B_r$ are the colour classes of an r-colouring of $G - x_1$. By Theorem 6.4.4, $G - x_1$ is a chordal graph. Likewise, $G - x_2$ and $G - x_3$ are chordal graphs, too. □

Corollary 7.1.2 *Any pure cycle of G contains x_1, x_2 and x_3. In particular, $x_i x_j \notin E(G)$ for some i, j with $1 \leq i < j \leq 3$.* □

Lemma 7.1.3 *G contains a simplicial vertex if and only if G contains a clique-cut.*

Proof. Obviously, G is not a complete graph. If x is a simplicial vertex of G, then $N(x)$ is a clique-cut of G.

Suppose now that G contains a clique-cut S. Then there are two proper subgraphs G_1 and G_2 of G such that $G \in \mathcal{G}[G_1 \cup_m G_2]$ for some $m \geq 1$ and $V(G_1) \cap V(G_2) = S$. Since S is a clique in G, any pure cycle of G is either in G_1 or in G_2. Suppose that G_1 contains pure cycles. Then $\{x_1, x_2, x_3\} \subseteq V(G_1)$ by Corollary 7.1.2. Since $[\{x_1, x_2, x_3\}] \not\cong K_3$, $\{x_1, x_2, x_3\} \not\subseteq S$ and thus $\{x_1, x_2, x_3\} \not\subseteq V(G_2)$. By Corollary 7.1.2 again, G_2 contains no pure cycle, and thus it is a chordal graph. By Corollary 6.4.1, G contains a simplicial vertex in G_2. □

7.2 The number $s_3(H)$

For a graph H and $r \in \mathbb{N}$, define (see Exercise 1.14)

$$s_r(H) = \frac{P(H,r)}{r!}. \tag{7.4}$$

When $r = \chi(H)$, $s_r(H)$ counts the number of different families $\{B_1, \cdots, B_r\}$ where $\{B_1, \cdots, B_r\} \in \mathcal{I}_t(H)$. If $s_r(H) = 1$, where $r = \chi(H)$, then H is uniquely r-colourable.

In this section, we shall present a result on $s_3(H)$ for a 3-colourable graph H, which will be used to study the structure of graphs in $\mathcal{T}_{3,1}$ in next section.

Lemma 7.2.1   *Let H be a 3-colourable graph in Figure 7.1(a), where $\{x, y\}$ is a cut of H. If $xy \in E(H)$, then*

$$s_3(H) = s_3(H_1 + xy)s_3(H_2 + xy);$$

and if $xy \notin E(H)$, then

$$s_3(H) = s_3(H_1 + xy)s_3(H_2 + xy) + 2s_3(H_1 \cdot xy)s_3(H_2 \cdot xy),$$

where H_1 and H_2 are as shown in Figure 7.1(a).

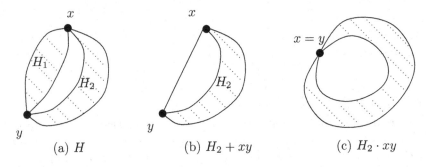

(a) H (b) $H_2 + xy$ (c) $H_2 \cdot xy$

Figure 7.1

Proof.   (i) Assume that $xy \in E(H)$. Then $\{x, y\}$ is a clique-cut of H. Thus

$$P(H, \lambda) = \frac{P(H_1 + xy, \lambda)P(H_2 + xy, \lambda)}{\lambda(\lambda - 1)},$$

and so

$$
\begin{aligned}
s_3(H) &= \frac{P(H_1 + xy, 3)P(H_2 + xy, 3)}{3 \times 2 \times 3!} \\
&= s_3(H_1 + xy)s_3(H_2 + xy).
\end{aligned}
$$

(ii) Assume that $xy \notin E(H)$. We have

$$
\begin{aligned}
P(H, \lambda) &= P(H + xy, \lambda) + P(H \cdot xy, \lambda) \\
&= \frac{P(H_1 + xy, \lambda)P(H_2 + xy, \lambda)}{\lambda(\lambda - 1)} + \frac{P(H_1 \cdot xy, \lambda)P(H_2 \cdot xy, \lambda)}{\lambda},
\end{aligned}
$$

and so

$$s_3(H) = s_3(H_1 + xy)s_3(H_2 + xy) + 2s_3(H_1 \cdot xy)s_3(H_2 \cdot xy). \qquad \square$$

Corollary 7.2.1   *Let H be a 3-colourable graph in Figure 7.1 (a), where $\{x, y\}$ is a cut of H and H_1 is uniquely 3-colourable.*

(i) *If x, y are in different colour classes of a 3-colouring of H_1, then $s_3(H) = s_3(H_2 + xy)$.*

(ii) *If x, y are in the same colour class of a 3-colouring of H_1, then $s_3(H) = 2s_3(H_2 \cdot xy)$.*   □

7.3   The family $T_{3,1}$

As we shall see, the structure of the graphs in $T_{r,1}$ is completely determined by the structure of the graphs in $T_{3,1}$. We thus focus our attention first in this section on the case when $r = 3$.

Given a graph H and $x \in V(H)$, let $H \ominus x$ denote the graph obtained from H by deleting x, adding c new vertices $w_1, \cdots, w_c$, where $c = c(H[N(x)])$, and joining w_i to all vertices in H_i for $i = 1, 2, \cdots, c$, where $H_1, H_2, \cdots, H_c$ are the components of $H[N(x)]$. If $H[N(x)]$ is connected, then $H \cong H \ominus x$. An example is shown in Figure 7.2.

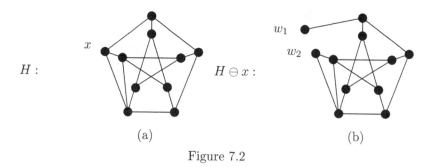

Figure 7.2

Remember that in this section, G (unless otherwise stated) is a graph in $T_{3,1}$ and x_1, x_2 and x_3 are vertices having the properties stated in Lemma 7.1.1.

Lemma 7.3.1   *The graph G has the structure of Figure 7.3(a), where*

(i) *G_i is a 2-tree for $i = 1, 2, 3$,*

(ii) *$V(G) = V(G_1) \cup V(G_2) \cup V(G_3)$, $E(G) = E(G_1) \cup E(G_2) \cup E(G_3)$,*

(iii) *$V(G_i) \cap V(G_j) = \{x_k\}$ for all i, j, k with $\{i, j, k\} = \{1, 2, 3\}$, and*

(iv) with respect to any 3-colouring of G_i, x_j and x_k are in different colour classes for all i, j and k such that $\{i, j, k\} = \{1, 2, 3\}$.

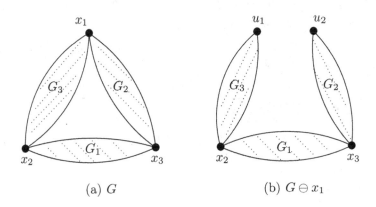

(a) G (b) $G \ominus x_1$

Figure 7.3

Proof. Let $H = G \ominus x_1$. Observe that $B_2 = A_2, B_3 = A_3$ and $B_1 = V(H)\backslash(A_2 \cup A_3)$ are the colour classes of a 3-colouring of H. Since any two colour classes of this colouring induce a forest and $H[N(x)]$ is a tree for any $x \in B_1$, by Theorem 6.4.4, H is a chordal graph. Observe that H is connected and 3-colourable. By Corollary 6.5.1, each block is a 2-tree. Let $H_1, H_2, \cdots, H_b$ be the blocks of H. Then

$$
\begin{aligned}
e(H) &= \sum_{1 \leq i \leq b(H)} e(H_i) = \sum_{1 \leq i \leq b(H)} (2v(H_i) - 3) \\
&= 2v(H) + 2(b(H) - 1) - 3b(H) = 2v(H) - b(H) - 2.
\end{aligned}
$$

Since $e(H) = e(G) = 2v(G) - 3 = 2v(H) - 5$, we have $b(H) = 3$. By Corollary 7.1.1, only x_2 and x_3 are possibly the cut-vertices of H. Thus H is the graph in Figure 7.3(b). Since $G = H \cdot u_1 u_2$, G is a graph in Figure 7.3(a), and clearly, (i), (ii) and (iii) hold.

Finally, we prove (iv). Let $B'_i = V(G_3) \cap A_i$ for $i = 1, 2, 3$. Then B'_1, B'_2, B'_3 are the colour classes of a 3-colouring of G_3. Since G_3 is a 2-tree, either $G_3 \cong K_2$ or G_3 is uniquely 3-colourable. As $x_1 \in B'_1$ and $x_2 \in B'_2$, x_1 and x_2 must be in different colour classes of any 3-colouring of G_3. Similarly, G_1 and G_2 have the same property. This proves (iv) and the lemma holds. $\square$

Our main result in this section, which characterizes the structure of the graphs in $\mathcal{T}_{3,1}$, is now given below.

Theorem 7.3.1   *Let G be a graph. The following three statements are equivalent:*

   (i) $G \in \mathcal{T}_{3,1}$;

   (ii) G is a uniquely 3-colourable graph with $e(G) = 2v(G) - 3$ and $t(G) = v(G) - 3$;

   (iii) G is a graph in Figure 7.3 (a) satisfying

     (1) $[\{x_1, x_2, x_3\}] \not\cong K_3$,

     (2) G_i is a 2-tree for each $i = 1, 2, 3$, and

     (3) with respect to any 3-colouring of G_i, x_j and x_k are in different colour classes for all i, j and k such that $\{i, j, k\} = \{1, 2, 3\}$.

Proof.   We prove the theorem in the following way: (ii)$\Rightarrow$(i)$\Rightarrow$(iii)$\Rightarrow$(ii).

(ii)$\Rightarrow$(i). As G is a uniquely 3-colourable graph with $e(G) = 2v(G) - 3$, by Theorem 6.3.1, $G \in \mathcal{T}_3$. Since $t(G) = v(G) - 3$, $G \in \mathcal{T}_{3,1}$ by definition.

(i)$\Rightarrow$(iii). This follows directly from Lemma 7.3.1.

(iii)$\Rightarrow$(ii). By Corollary 7.2.1(i), $s_3(G) = s_3(G') = 1$, where G' is the graph in Figure 7.4, and thus G is uniquely 3-colourable. Since G_1, G_2 and G_3 are 2-trees, we have

$$e(G) = \sum_{1 \leq i \leq 3} e(G_i) = \sum_{1 \leq i \leq 3} (2v(G_i) - 3) = 2v(G) - 3$$

and

$$t(G) = \sum_{1 \leq i \leq 3} t(G_i) = \sum_{1 \leq i \leq 3} (v(G_i) - 2) = v(G) - 3. \qquad \square$$

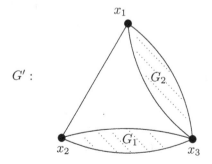

Figure 7.4

7.4 The structure of graphs in $\mathcal{T}_{r,1}$ $(r \geq 4)$

We now proceed to consider the case when $r \geq 4$. We first prove the following useful result due to Dong (1997). An $x - y$ path P in a graph G is said to be *pure* if $[V(P)] \cong P$ or $[V(P)] \cong P + xy$.

Theorem 7.4.1   Let $G \in \mathcal{T}_r$, $\{A_1, \cdots, A_r\} \in \mathcal{I}_t(G)$ and $k \in \mathbb{N}$ with $k \leq r$. Then $V(C) \cap A_k = \emptyset$ for all pure cycles C in G if and only if

$$\sum_{\substack{1 \leq i < j \leq r \\ i \neq k, j \neq k}} \sum_{x \in A_k} \rho(x, A_i, A_j) = 0. \qquad (7.5)$$

Proof.   By definition, $\rho(x, A_i, A_j) \geq 0$. Since $\{A_1, \cdots, A_r\} \in \mathcal{I}_t(G)$, $[A_i \cup A_j]$ is a tree for all i, j with $1 \leq i < j \leq r$. If $\rho(x, A_i, A_j) > 0$, then there is a pure cycle in $[A_i \cup A_j \cup \{x\}]$. Hence the necessity holds.

By Theorem 6.4.4, the sufficiency holds for $r = 3$. Assume that the sufficiency holds for $r \leq m - 1$ where $m \geq 4$. We now consider the case that $r = m \geq 4$. We may further assume that $k = r$. Then the given condition is that

$$\rho(x, A_i, A_j) = 0$$

for all i, j with $1 \leq i < j < r$ and all $x \in V_r$. Our aim is to prove that $V(C) \cap A_r = \emptyset$ for any pure cycle C in G.

Let $\Phi = \{C : C$ is a pure cycle of G and $V(C) \cap A_r \neq \emptyset\}$, and suppose that $\Phi \neq \emptyset$. It suffices to prove that this supposition leads to a contradiction.

By the induction hypothesis that the theorem holds for $r \leq m - 1$, we have $V(C) \cap A_i \neq \emptyset$ for all $C \in \Phi$ and all i with $1 \leq i \leq r - 1$. Let

$$p_1 = \min \{|V(C) \cap A_1| : C \in \Phi\},$$
$$p_2 = \min \{v(C) : C \in \Phi, |V(C) \cap A_1| = p_1\}$$
and $$\Phi_0 = \{C : C \in \Phi, |V(C) \cap A_1| = p_1, v(C) = p_2\}.$$

Then $\Phi_0 \neq \emptyset$. Let $p_3 = \max \{|V(C) \cap A_r| : C \in \Phi_0\}$ and let C be any member in Φ_0 such that $|V(C) \cap A_r| = p_3$. Let $w \in V(C) \cap A_1$ and $x, y \in V(C) \cap N(w)$. Denote the path $C - w$ by $x u_1 u_2 \cdots u_s y$, where $s = p_2 - 3 \geq 1$.

Claim 1:   There exists a pure $x - y$ path $x v_1 v_2 \cdots v_t y$ in $[N(w)]$ such that either $\{v_1, v_2, \cdots, v_t\} \cap A_r \neq \emptyset$ or $x, y \in A_r$ and $t = 1$.

By Theorem 6.4.4, $[A_i \cup A_j \cup A_r]$ is chordal for all i, j with $1 \leq i < j < r$. Thus, for each $x \in A_i$ (or resp., $x \in A_j$), $[N(x) \cap (A_j \cup A_r)]$ (or resp., $[N(x) \cap (A_i \cup A_r)])$ is a tree.

To prove Claim 1, we just need to consider these cases on x, y: (1) $x, y \in A_r$; (2) $x \in A_i$ and $y \in A_i \cup A_r$, where $2 \leq i < r$ and (3) $x \in A_i$ and $y \in A_j$ for some i, j with $1 \leq i < j < r$. Other symmetric cases can be settled similarly.

If $x, y \in A_r$, then $[N(w) \cap (A_2 \cup A_r)]$ is a tree and so there exists an $x - y$ path (only one) $x v_1 v_2 \cdots v_t y$ in this tree. As $x, y \in A_r$, t must be odd and so $\{v_1, v_2, \cdots, v_t\} \cap A_r \neq \emptyset$ if $t \geq 2$. If $x \in A_i$ and $y \in A_i \cup A_r$ for some i with $i \neq r$, then $i \neq 1$ and $[N(w) \cap (A_i \cup A_r)]$ is a tree and so there exists an $x - y$ path (only one) $x v_1 v_2 \cdots v_t y$ in this tree. It is clear $v_1 \in A_r$. The last case is that $x \in A_i$ and $y \in A_j$ for some i, j with $2 \leq i < j < r$. Since $[A_1 \cup A_r]$ is a tree, there exists $z \in A_r \cap N(w)$. So both $[N(w) \cap (A_i \cup A_r)]$ and $[N(w) \cap (A_j \cup A_r)]$ are trees. There exist an $x - z$ path $x v_1 v_2 \cdots v_{t_1} z$ in $[N(w) \cap (A_i \cup A_r)]$ and a $z - y$ path $z w_1 w_2 \cdots w_{t_2} y$ in $[N(w) \cap (A_j \cup A_r)]$. Both paths are pure. If the path $x v_1 v_2 \cdots v_{t_1} z w_1 w_2 \cdots w_{t_2} y$ is pure, then the claim holds; if it is not pure, then the following is a pure $x - y$ path

$$x v_1 v_2 \cdots v_p w_q \cdots w_{t_2} y,$$

where p is the minimum index such that $N(v_p) \cap \{w_1, w_2, \cdots, w_{t_2}\} \neq \emptyset$ (where $v_0 = x$) and with p fixed, q is the maximum index such that $v_p w_q \in E(G)$ (where $w_{t_2+1} = y$). Since $xy \notin E(G)$, it is not possible that $p = 0$ and $q = t_2 + 1$, i.e., either $p \geq 1$ or $q \leq t_2$. Observe that $v_1, w_{t_2} \in A_1$. Thus $\{v_1, v_2, \cdots, v_p, w_q, \cdots, w_{t_2}\} \cap A_r \neq \emptyset$. Hence Claim 1 holds.

Claim 2: There is no pure $x - y$ path $x v_1 v_2 \cdots v_t y$ in $[N(w)]$ such that $\{v_1, v_2, \cdots, v_t\} \cap A_r \neq \emptyset$.

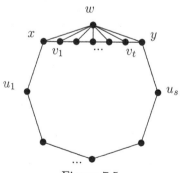

Figure 7.5

Assume that there is a pure $x - y$ path $xv_1v_2 \cdots v_t y$ in $[N(w)]$ such that $\{v_1, v_2, \cdots, v_t\} \cap A_r \neq \emptyset$. Since $wu_i \notin E(G)$ for $j = 1, 2, \cdots, s$, we have $\{v_1, v_2, \cdots, v_t\} \cap \{u_1, u_2, \cdots, u_s\} = \emptyset$, as shown in Figure 7.5.

If $t = 1$, then $v_1 \in A_r$. Since $V(C) \cap A_r \neq \emptyset$, we have $u_p \in A_r$ for some p with $1 \leq p \leq s$, and so $v_1 u_p \notin E(G)$. Thus there is a pure cycle $C' = v_1 u_f u_{f+1} \cdots u_h v_1$, where $0 \leq f \leq p \leq h \leq s + 1$ (where $u_0 = x$ and $u_{s+1} = y$). So $C' \in \Phi$. But $|V(C') \cap A_1| < p_1$, a contradiction.

Hence we have $t \geq 2$. Assume that $v_l \in A_r$. If $v_l u_i \notin E(G)$ for $i = 1, 2, \cdots, s$, then by Lemma 6.4.1, v_l is contained in a pure cycle C'' in the subgraph induced by $\{v_1, v_2, \cdots, v_t\} \cup \{u_1, u_2, \cdots, u_s\}$. But $C'' \in \Phi$ and $|V(C'') \cap A_1| < p_1$, a contradiction.

Thus $v_l u_i \in E(G)$ for some i with $1 \leq i \leq s$. Since $t \geq 2$, without loss of generality, we assume that $l \geq 2$ (if $l = 1$, we can switch x with y). Let p be the minimum index such that $v_l u_p \in E(G)$, where $1 \leq p \leq s$. Then $C''' = wxu_1 \cdots u_p v_l w$ is a pure cycle. Observe that $|V(C''') \cap A_1| \leq p_1$, $v(C''') \leq p_2$. Thus $v(C''') = p_2$, implying that $p = s$. If $y \in A_r$, then $C_0 : wv_l u_s yw$ is a pure cycle. But, as $r \geq 4$, we have $V(C_0) \cap A_j = \emptyset$ for some j with $2 \leq j \leq r - 1$, a contradiction. Hence $y \notin A_r$, implying that $|V(C''') \cap A_r| = p_3 + 1$, a contradiction. Thus Claim 2 holds.

By Claims 1 and 2, $x, y \in A_r$. This also means that for every vertex of $V(C) \cap A_1$, its two neighbours on C belong to A_r. But, as C is not a cycle in $[A_1 \cup A_r]$, the vertex $w \in V(C) \cap A_1$ can be selected such that $u_1 \notin A_1$ or $u_s \notin A_1$, say $u_1 \notin A_1$. Assume $u_1 \in A_2$ without loss of generality. As $\rho(w, A_2, A_r) = 0$, the subgraph $[N(w) \cap (A_2 \cup A_r)]$ is a tree and so there is an $x - y$ path $xv_1v_2 \cdots v_t y$ in this tree. It is a pure path. By Claim 2, $t = 1$. So $v_1 \in A_2$ and $v_1 u_1 \notin E(G)$. Let p be the minimum index such that $v_1 u_p \in E(G)$ (where $u_{s+1} = y$). It is clear $p \geq 2$. Then $C^{(4)} = yu_1 \cdots u_p v_1 y$ is a pure cycle. As $y \in A_1$, we have $C^{(4)} \in \Phi$. But $|V(C^{(4)}) \cap A_1| < p_1$, a contradiction.

Therefore we conclude that $\Phi = \emptyset$. This completes the proof. $\qquad \square$

Let $\mathcal{T}_{r,1}^*$ denote the family of the graphs in $\mathcal{T}_{r,1}$ that contain no simplicial vertex. The following result follows directly from Theorem 7.4.1 and Lemma 7.1.1.

Lemma 7.4.1　*Given any G in $\mathcal{T}_{r,1}$ and any pure cycle C in G, $V(C) \subseteq A_1 \cup A_2 \cup A_3$.* $\qquad \square$

Lemma 7.4.2   *For any G in $T_{r,1}$, either $G \cong H + K_{r-3}$ for some $H \in T_{3,1}^*$ or there exists a sequence of vertices $y_1, y_2, \cdots, y_s$ in G, where $s \geq 1$, such that*

(1) $G - \{y_1, \cdots, y_s\} \cong H + K_{r-3}$ for some $H \in T_{3,1}^*$ and

(2) y_i is a simplicial vertex of degree $r-1$ in the graph $G - \{y_1, \cdots, y_{i-1}\}$ for $i = s, s-1, \cdots, 1$.

Proof.   **Case 1:** G contains no simplicial vertex.

By Lemma 7.1.3, G does not contain a clique-cut. For $i = 4, 5, \cdots, r$, let x_i be a vertex in A_i. By Lemma 7.4.1, x_i is not contained in any pure cycle of G for $i = 4, 5, \cdots, r$. Then, by Corollary 6.4.2, $d_G(x_i) = v(G) - 1$ for $i = 4, 5, \cdots, r$. Thus $A_i = \{x_i\}$ for $i = 4, 5, \cdots, r$. Let $H = [A_1 \cup A_2 \cup A_3]$. Then $G \cong H + K_{r-3}$. It is obvious that $H \in T_{3,1}$. As G contains no simplicial vertex, H contains no simplicial vertex and so $H \in T_{3,1}^*$.

Case 2: G contains a simplicial vertex.

If x is a simplicial vertex of G, then $d(x) = r - 1$, and by Lemma 6.3.2, $G - x \in T_{r,1}$. Hence, there is a sequence of vertices $y_1, y_2, \cdots, y_s$ in G such that y_i is a simplicial vertex of degree $r - 1$ in the graph $G - \{y_1, \cdots, y_{i-1}\}$ for $i = 1, 2, \cdots, s$, and $G - \{y_1, \cdots, y_s\} \in T_{r,1}^*$. By Case 1, $G - \{y_1, \cdots, y_s\} \cong H + K_{r-3}$ for some $H \in T_{3,1}^*$. $\square$

We are now in a position to characterize the graphs in $T_{r,1}$.

Theorem 7.4.2   *Let G be a graph and $r \in \mathbb{N}$ with $r \geq 4$. The following statements are equivalent:*

(i) $G \in T_{r,1}$;

(ii) G is a uniquely r-colourable graph with $e(G) = (r-1)v(G) - \binom{r}{2}$ and $t(G) = \frac{1}{3}(3v(G) - 2r)\binom{r-1}{2} - 1$;

(iii) either $G \cong H + K_{r-3}$ for some $H \in T_{3,1}^*$ or there exists a sequence of vertices $y_1, y_2, \cdots, y_s$ in G, $s \geq 1$, such that

     (1) $G - \{y_1, \cdots, y_s\} \cong H + K_{r-3}$ for some $H \in T_{3,1}^*$ and

     (2) for $i = s, s-1, \cdots, 1$, y_i is a simplicial vertex of degree $r-1$ in the graph $G - \{y_1, \cdots, y_{i-1}\}$.

Proof.   We prove the theorem in the following way: (ii)$\Rightarrow$(i)$\Rightarrow$(iii)$\Rightarrow$(ii).

(ii)$\Rightarrow$(i) By Theorem 6.3.1, $G \in T_r$. By definition, $G \in T_{r,1}$.

(i)$\Rightarrow$(iii) This follows directly from Lemma 7.4.2.

(iii)⇒(ii) Suppose that G satisfies (iii). We first show that G is uniquely r-colourable. If G contains no simplicial vertex, then $G \cong H + K_{r-3}$, where $H \in T^*_{3,1}$. Since H is uniquely 3-colourable, G is uniquely r-colourable. If x is a simplicial vertex of G, then $d(x) = r - 1$. By Lemma 6.3.2, $G - x \in T_{r,1}$. By the induction hypothesis, $G - x$ is uniquely r-colourable. Since $P(G, \lambda) = (\lambda - r + 1)P(G - x, \lambda)$, $s_r(G) = s_r(G - x)$. Thus G is uniquely r-colourable.

If $G \cong H + K_{r-3}$ for some $H \in T_{3,1}$, then

$$
\begin{aligned}
e(G) &= e(H) + \binom{r-3}{2} + (r-3)v(H) \\
&= 2v(H) - 3 + \binom{r-3}{2} + (r-3)v(H) \\
&= (r-1)v(H) - 3 + \binom{r-3}{2} \\
&= (r-1)(v(H) + r - 3) - \binom{r}{2} \\
&= (r-1)v(G) - \binom{r}{2}
\end{aligned}
$$

and

$$
\begin{aligned}
t(G) &= t(H) + \binom{r-3}{3} + (r-3)e(H) + v(H)\binom{r-3}{2} \\
&= v(H) - 3 + \binom{r-3}{3} + (r-3)(2v(H) - 3) + v(H)\binom{r-3}{2} \\
&= (v(H) + r - 3)\binom{r-1}{2} - \frac{2r}{3}\binom{r-1}{2} - 1 \\
&= v(G)\binom{r-1}{2} - \frac{2r}{3}\binom{r-1}{2} - 1.
\end{aligned}
$$

Thus (ii) holds if $G \cong H + K_{r-3}$ for some $H \in T_{3,1}$.

If x is a simplicial vertex of G of degree $r-1$, then $e(G) = r-1+e(G-x)$ and $t(G) = \binom{r-1}{2} + t(G - x)$. By the induction hypothesis, we have

$$
e(G - x) = (r - 1)(v(G) - 1) - \binom{r}{2}
$$

and

$$
t(G - x) = (v(G) - 1)\binom{r-1}{2} - \frac{2r}{3}\binom{r-1}{2} - 1.
$$

Hence (ii) holds again.   □

7.5 Chromatically unique graphs in $\mathcal{T}_{r,1}$

In this section, we shall apply the characterization theorems established in the above sections to investigate the chromaticity of the graphs in $\mathcal{T}_{r,1}$. We shall prove that for any $G \in \mathcal{T}_{r,1}$, if G is χ-unique, then $G \cong H + K_{r-3}$ for some $H \in \mathcal{T}_{3,1}^*$; and given $G \in \mathcal{T}_{3,1}$ and $m \geq 1$, $G + K_m$ is χ-unique if and only if G is χ-unique. To get to this, we first prove the following lemma.

Lemma 7.5.1 *For any $r \geq 3$, the family $\mathcal{T}_{r,1}$ is χ-closed.*

Proof. Let $G \in \mathcal{T}_{r,1}$ and H be a graph such that $H \sim G$. By Theorems 7.3.1 and 7.4.2, G is uniquely r-colourable, and so H is also uniquely r-colourable.

Since $v(H) = v(G)$, $e(H) = e(G)$ and $t(H) = t(G)$, by Theorems 7.3.1 and 7.4.2 again, $H \in \mathcal{T}_{r,1}$. $\qquad\square$

Lemma 7.5.2 *Given any graph $G \in \mathcal{T}_{r,1}$, where $r \geq 4$, there exists a graph $H \in \mathcal{T}_{3,1}$ such that $G \sim H + K_{r-3}$, and $G \in \mathcal{T}_{r,1}^*$ if and only if $H \in \mathcal{T}_{3,1}^*$.*

Proof. **Case 1**: $G \in \mathcal{T}_{r,1}^*$. By Theorem 7.4.2, there exists $H \in \mathcal{T}_{3,1}^*$ such that $G = H + K_{r-3}$.

Case 2: $G \in \mathcal{T}_{r,1} \backslash \mathcal{T}_{r,1}^*$.

By Theorem 7.4.2, there is a sequence of vertices $y_1, \cdots, y_s$ in G, where $s \geq 1$, such that y_i is a simplicial vertex of degree $r - 1$ in the graph $G - \{y_1, \cdots, y_{i-1}\}$, for $i = 1, \cdots, s$, and $G - \{y_1, \cdots, y_s\} \cong H_0 + K_{r-3}$, where $H_0 \in \mathcal{T}_{3,1}^*$. Now we construct a graph H from H_0 by adding s new vertices and joining each of them to u and v, where uv is a fixed edge in $E(H_0)$. Observe that $H \in \mathcal{T}_{3,1} \backslash \mathcal{T}_{3,1}^*$ and

$$
\begin{aligned}
& P(H + K_{r-3}, \lambda) \\
= {} & \lambda(\lambda - 1) \cdots (\lambda - (r-3) + 1) P(H, \lambda - (r-3)) \\
= {} & \lambda(\lambda - 1) \cdots (\lambda - (r-3) + 1)(\lambda - (r-3) - 2)^s P(H_0, \lambda - (r-3)) \\
= {} & (\lambda - r + 1)^s \lambda(\lambda - 1) \cdots (\lambda - (r-3) + 1) P(H_0, \lambda - (r-3)).
\end{aligned}
$$

On the other hand,

$$
\begin{aligned}
P(G, \lambda) &= (\lambda - r + 1)^s P(G - \{y_1, \cdots, y_s\}, \lambda) \\
&= (\lambda - r + 1)^s \lambda(\lambda - 1) \cdots (\lambda - (r-3) + 1) P(H_0, \lambda - (r-3)).
\end{aligned}
$$

Thus $G \sim H + K_{r-3}$, where $H \in \mathcal{T}_{3,1} \backslash \mathcal{T}_{3,1}^*$. $\qquad\square$

Lemma 7.5.3　　*If G is a χ-unique graph in $T_{r,1}$, then*

(i) when $r = 3$, $G \in T_{3,1}^*$;

(ii) when $r \geq 4$, $G = H + K_{r-3}$, where H is a χ-unique graph in $T_{3,1}^*$.

Proof.　　(i) Assume that $r = 3$.

Claim 1: There do not exist two vertices y_1 and y_2 in G such that y_1 is a simplicial vertex of G and y_2 is a simplicial vertex of $G - y_1$.

Suppose otherwise; and let $y_1, y_2, \cdots, y_s$, where $s \geq 2$, be a sequence of vertices in G such that y_i is a simplicial vertex of $G - \{y_1, \cdots, y_{i-1}\}$ for $i = 1, 2, \cdots, s$, and $G - \{y_1, \cdots, y_s\}$ contains no simplicial vertices. Then

$$P(G, \lambda) = (\lambda - 2)^s P(G - \{y_1, \cdots, y_s\}, \lambda).$$

Let $G_1 = G - \{y_1, \cdots, y_s\}$ and G_2 be a 2-tree of order $s + 2$ having exactly two simplicial vertices. Then for any $H \in \mathcal{G}[G_1 \cup_2 G_2]$, we have $P(H, \lambda) = P(G, \lambda)$. Since G is χ-unique, $G \cong H$ for any $H \in \mathcal{G}[G_1 \cup_2 G_2]$. But this is certainly absurd as some graphs in the family $\mathcal{G}[G_1 \cup_2 G_2]$ have exactly one simplicial vertex while others have two.

Claim 2: G contains no simplicial vertex.

Suppose on the contrary that G contains a simplicial vertex x. By Claim 1, $G - x$ contains no simplicial vertex. Observe that $d(x) = 2$ and $G - x \in T_{3,1}$ by Lemma 6.3.2. Thus $G - x$ has the structure as stated in Theorem 7.3.1(iii).

Let e be any edge of $G - x$, and let G_e be the graph obtained from $G - x$ by adding a new vertex and joining it to the two vertices incident with e. Then $G_e \sim G$. Since G is χ-unique, $G \cong G_e$ for every edge $e \in G - x$. But this, again, is absurd as $G_e \not\cong G_f$ for some edges e and f in $G - x$. We thus conclude that $G \in T_{3,1}^*$.

(ii) Assume that $r \geq 4$. By Lemma 7.5.2, $G \sim H + K_{r-3}$ for some $H \in T_{3,1}$ such that $G \in T_{r,1}^*$ if and only if $H \in T_{3,1}^*$. As G is χ-unique, $G \cong H + K_{r-3}$ and H is also χ-unique. By (i), we have $H \in T_{3,1}^*$.　　□

We are now ready to establish our main result in this chapter.

Theorem 7.5.1　　*Let G be a graph in $T_{r,1}$, where $r \geq 3$, and let $m \in \mathbb{N}$. Then G is χ-unique if and only if $G + K_m$ is χ-unique.*

Proof.　　The sufficiency is obvious (see Theorem 1.5.1). We shall prove the necessity. By Lemma 7.5.3, it suffices to consider the case when $r = 3$.

Assume that G is χ-unique. By Lemma 7.5.3, $G \in T_{3,1}^*$. Evidently, $G + K_m \in T_{m+3,1}$. Let H be a graph such that $H \sim G + K_m$. We shall show that $H \cong G + K_m$. By Lemma 7.5.1, $H \in \mathcal{T}_{m+3,1}$. By Lemma 7.5.2, there is a graph $H_0 \in \mathcal{T}_{3,1}$ such that $H_0 + K_m \sim H$, and $H_0 \in \mathcal{T}_{3,1}^*$ if and only if $H \in \mathcal{T}_{m+3,1}^*$. As $H_0 + K_m \sim H \sim G + K_m$, we have $H_0 \sim G$. The fact that G is χ-unique implies that $H_0 \cong G$. Thus $H_0 \in \mathcal{T}_{3,1}^*$, and so $H \in \mathcal{T}_{m+3,1}^*$. By Theorem 7.4.2, $H \cong H' + K_m$ for some $H' \in \mathcal{T}_{3,1}^*$, which implies that $G + K_m \sim H \cong H' + K_m$, and so $G \sim H'$. The fact that G is χ-unique again implies that $H' \cong G$. Hence $H \cong H' + K_m \cong G + K_m$, as was to be shown. $\qquad \square$

The procedure of arguments in the proof are summarized in the following flow chart.

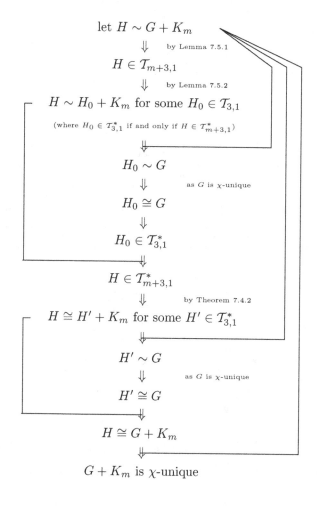

7.6 Further results

To search for χ-unique graphs in $\mathcal{T}_{r,1}$, by Lemma 7.5.3, we may confine ourselves to the family $\mathcal{T}_{3,1}^*$. Which graphs in $\mathcal{T}_{3,1}^*$ are χ-unique? This is, by no means, a simple problem.

For a non-chordal graph G, let $g_p(G)$ denote the length of a shortest pure cycle in G. The family $\mathcal{T}_{3,1}^*$ is thus divided into $\mathcal{T}_{3,1,4}^*$, $\mathcal{T}_{3,1,5}^*$, $\cdots$, where

$$\mathcal{T}_{3,1,q}^* = \{G \in \mathcal{T}_{3,1}^* : g_p(G) = q\}.$$

Let $W(n, n-2)$ denote the broken wheel of Figure 7.6. It can be checked that for even $n \geq 6$, $G = W(n, n-2)$ is a graph in $\mathcal{T}_{3,1}^*$ while $g_p(G) = 4$.

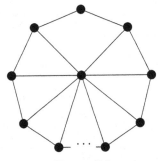

Figure 7.6

Let $W_{n,2k}^{(1)}$, $W_{n,2k}^{(2)}$ and $W_{n,2k}^{(3)}$ denote, respectively, the graphs in Figures 7.7, 7.8 and 7.9, where n and k satisfy the respective conditions stated therein. It can be checked that these graphs G belong to $\mathcal{T}_{3,1}^*$ with $g_p(G) = 5$.

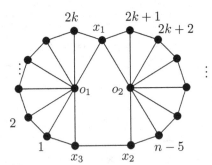

Figure 7.7. $W_{n,2k}^{(1)}$ of order n, where $k \geq 1$ and $n \geq 2k+7$ is odd

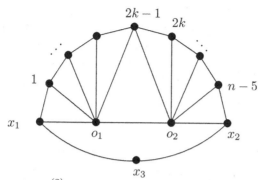

Figure 7.8. $W_{n,2k}^{(2)}$ of order n, where $k \geq 1$ and $n \geq 2k + 4$

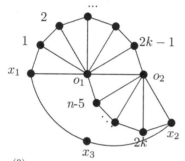

Figure 7.9. $W_{n,2k}^{(3)}$ of order n, where $k \geq 1$ and $n \geq 2k + 5$

Indeed, we have the following result due to Dong and Koh (1997b).

Lemma 7.6.1   (i) $\mathcal{T}_{3,1,4}^* = \{W(n, n - 2) : n \geq 6 \text{ and } n \text{ is even}\}$.
(ii)

$$
\begin{aligned}
\mathcal{T}_{3,1,5}^* = \quad & \{W_{n,2k}^{(1)} : k \geq 1 \text{ and } n \geq 2k + 7 \text{ is odd }\} \\
\cup \quad & \{W_{n,2k}^{(2)} : k \geq 1, n \geq 2k + 4, n \geq 7\} \\
\cup \quad & \{W_{n,2k}^{(3)} : k \geq 1, n \geq 2k + 7\}.
\end{aligned}
$$

$\square$

While we do not know whether every graph in $\mathcal{T}_{3,1}^*$ is χ-unique, we do have the following result due to Dong and Koh (1997b).

Theorem 7.6.1   *Every graph in $\mathcal{T}_{3,1,4}^* \cup \mathcal{T}_{3,1,5}^*$ is χ-unique.* $\square$

Chia (1990,1995a) proved that the graph $W(n, n-2) + K_m$ is χ-unique for any two $n, m \in \mathbb{N}$, where $n \geq 6$ is even and $m \geq 1$. By Theorems 7.5.1 and 7.6.1, we have the following more general result (see Exercise 7.5).

Corollary 7.6.1   *For any graph G in $\mathcal{T}^*_{3,1,4} \cup \mathcal{T}^*_{3,1,5}$ and any $m \in \mathbb{N}$, the graph $G + K_m$ is χ-unique.* $\square$

To end this chapter, we would like to propose two problems.

Problem 1   *Is every graph G in $\mathcal{T}^*_{3,1}$ always χ-unique?*

Problem 2   *If G is a graph in $\mathcal{T}^*_{3,1}$ such that $(\lambda - 2)^2$ is not a factor of $P(G, \lambda)$, is G χ-unique?*

Exercise 7

7.1 Prove that for $G \in \mathcal{T}_{r,1}$, if $(\lambda - r + 1)^2 \nmid P(G, \lambda)$, then $G \in \mathcal{T}_{r,1}^*$.

7.2 Prove that

(i) $\mathcal{T}_{3,1,4}^* = \{W(n, n - 2) : n \geq 6 \text{ and n is even}\}$.

(ii)

$$
\begin{aligned}
\mathcal{T}_{3,1,5}^* \ = \ & \{W_{n,2k}^{(1)} : k \geq 1 \text{ and } n \geq 2k + 7 \text{ is odd } \} \\
& \cup \{W_{n,2k}^{(2)} : k \geq 1, n \geq 2k + 4, n \geq 7\} \\
& \cup \{W_{n,2k}^{(3)} : k \geq 1, n \geq 2k + 7\}.
\end{aligned}
$$

7.3 Prove that every graph in $\mathcal{T}_{3,1,4}^* \cup \mathcal{T}_{3,1,5}^*$ is χ-unique.

7.4 For $i = 1, 2, 3$, $W_{n,2k}^{(i)}$ are the graphs shown in Figures 7.7, 7.8 and 7.9. Show that

$$
\begin{aligned}
P(W_{n,2k}^{(1)}, \lambda) \ = \ & \frac{\lambda}{\lambda - 1} \left((\lambda - 2)^{n-2}(\lambda - 2)^2 + 2(\lambda - 2) + 2 \right) \\
& + (\lambda - 2)^{2k+2} + (\lambda - 2)^{n-2k-3} - (\lambda - 2)^3 \\
& - 2(\lambda - 2)^2 \Big) ; \\
P(W_{n,2k}^{(2)}, \lambda) \ = \ & \frac{\lambda}{\lambda - 1} \left((\lambda - 2)^{n-2}(\lambda - 2)^2 + 2(\lambda - 2) + 2 \right) \\
& + (-1)^{n-1}(\lambda - 2)^{2k} - (\lambda - 2)^{n-2k-2} \\
& + (-1)^n(\lambda - 2)((\lambda - 2)^2 + (\lambda - 2) - 1) \Big) ; \\
P(W_{n,2k}^{(3)}, \lambda) \ = \ & \frac{\lambda}{\lambda - 1} \left((\lambda - 2)^{n-2}(\lambda - 2)^2 + 2(\lambda - 2) + 2 \right) \\
& + (-1)^n(\lambda - 2)^{2k+1} - (\lambda - 2)^{n-2k-2} \\
& - (-1)^{n-1}(\lambda - 2) \Big) .
\end{aligned}
$$

7.5 Show that for any graph G in $\mathcal{T}_{3,1,4}^* \cup \mathcal{T}_{3,1,5}^*$ and any $m \in \mathbb{N}$, the graph $G + K_m$ is χ-unique.

(For the above problems, see Dong and Koh (1997b).)

Chapter 8

Graphs in Which All but One Pair of Colour Classes Induce Trees (I)

8.1 Introduction

In the preceding two chapters, we study the structure and chromaticity of graphs whose vertex sets can be partitioned into colour classes $\{A_1, \cdots, A_r\}$ satisfying condition (**T**). In this chapter, we shall modify a condition in (**T**) and study the structure and chromaticity of graphs whose vertex sets can be partitioned into colour classes $\{A_1, \cdots, A_r\}$ satisfying the modified conditions.

Consider the three graphs of Figure 8.1 (a)-(c) and their independent partitions $A_1, A_2, \cdots$ of the vertex sets as shown. We observe that, in each case, any two independent sets induce a tree except a pair, i.e., A_1 and A_2, which instead induce a forest with two components.

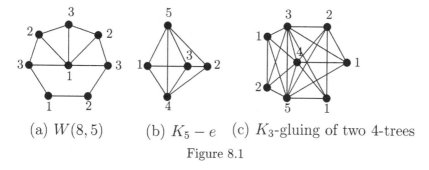

(a) $W(8,5)$ (b) $K_5 - e$ (c) K_3-gluing of two 4-trees

Figure 8.1

In general, a graph G is said to satisfy **(F)** if $V(G)$ possesses an independent partition $\{A_1, \cdots, A_r\}$, where $r \geq 3$, such that $[A_1 \cup A_2]$ is a forest with two components and $[A_i \cup A_j]$ is a tree for all i and j with $1 \leq i < j \leq r$ and $(i,j) \neq (1,2)$. Let $\mathcal{F}$ be the family of graphs satisfying **(F)** and for each $r \geq 3$, let $\mathcal{F}_r$ be the family of graphs satisfying **(F)** with an r-independent partition. Also, given $G \in \mathcal{F}_r$, let $\mathcal{I}_f(G)$ denote the family of r-independent partitions with which G satisfies **(F)**.

In Section 8.2, we shall show that for each $G \in \mathcal{F}_r$,

$$t(G) \leq \frac{1}{3}(3v(G) - 2r)\binom{r-1}{2} - (r-2).$$

We shall then proceed to Section 8.3 to study those graphs G in $\mathcal{F}_r$ which have their $t(G)$ attaining the upper bound. Wanner (1989) showed that the family of graphs of the same order which are subgraphs of an $(r-1)$-tree obtained by deleting an edge that is contained in exactly $r-2$ triangles forms a χ-equivalence class. Borodin and Dmitriev (1991) proved, independently, that this is precisely the χ-equivalence class of graphs which are K_{r-2}-gluings of two $(r-1)$-trees (see Figure 8.1(c)). Our main objective in this chapter is to show that the χ-equivalence class studied by Wanner, and Borodin and Dmitriev is indeed the family of graphs G in $\mathcal{F}_r$ with $t(G) = \frac{1}{3}(3v(G) - 2r)\binom{r-1}{2} - (r-2)$. A general result in this area will be mentioned at the end of this chapter.

8.2   The triangle number and an upper bound

In this section, we shall derive an expression for $t(G)$ and obtain the sharp upper bound for $t(G)$ as stated in Section 8.1 for G in $\mathcal{F}_r$.

We first compute the size of each graph in $\mathcal{F}_r$.

Lemma 8.2.1   *Let G be a graph in $\mathcal{F}_r$ of order n. Then*

$$e(G) = (r-1)n - \binom{r}{2} - 1. \tag{8.1}$$

Proof.   Let $\{A_1, \cdots, A_r\} \in \mathcal{I}_f(G)$. Then

$$
\begin{aligned}
e(G) &= \sum_{1 \leq i < j \leq r} e([A_i \cup A_j]) \\
&= |A_1| + |A_2| - 2 + \sum_{(i,j) \neq (1,2)} (|A_i| + |A_j| - 1)
\end{aligned}
$$

$$= -1 + (r-1) \sum_{k=1}^{r} |A_k| - \binom{r}{2}$$

$$= (r-1)n - \binom{r}{2} - 1.$$

$\square$

It follows from Lemma 8.2.1 that $\mathcal{F}_{r_1} \cap \mathcal{F}_{r_2} = \emptyset$ for all r_1 and r_2 with $r_1 > r_2 \geq 3$. Throughout this chapter, we shall always assume that $\{A_1, \cdots, A_r\} \in \mathcal{I}_f(G)$, where $G \in \mathcal{F}_r$ and $r \geq 3$.

Lemma 8.2.2   *For $1 \leq i < j \leq r$ and $x \in A_k$, where $k \neq i,j$,*

$$\sum_{x \in A_k} \rho(x, A_i, A_j)$$

$$= \begin{cases} |A_1| + |A_2| + |A_j| - 3 - t([A_1 \cup A_2 \cup A_j]), & \text{if } \{i,k\} = \{1,2\}; \\ |A_i| + |A_j| + |A_k| - 2 - t([A_i \cup A_j \cup A_k]), & \text{otherwise.} \end{cases}$$

Proof.   Indeed,

$$t([A_i \cup A_j \cup A_k]) = \sum_{x \in A_k} (d(x, A_i, A_j) - c([N_x \cap (A_i \cup A_j)]))$$

$$= \sum_{x \in A_k} (d(x, A_i, A_j) - 1 - \rho(x, A_i, A_j))$$

$$= e([A_i \cup A_k]) + e([A_j \cup A_k]) - |A_k|$$

$$\qquad - \sum_{x \in A_k} \rho(x, A_i, A_j),$$

and thus the result follows.

$\square$

Corollary 8.2.1   *For any graph G in $\mathcal{F}_r$,*

$$\sum_{1 \leq i < j < k \leq r} \sum_{x \in A_k} \rho(x, A_i, A_j) = r - 2 + \sum_{1 \leq i < j < k \leq r} \sum_{x \in A_j} \rho(x, A_i, A_k)$$

$$= r - 2 + \sum_{1 \leq i < j < k \leq r} \sum_{x \in A_i} \rho(x, A_j, A_k).$$

In particular,

$$\sum_{1 \leq i < j < k \leq r} \sum_{x \in A_k} \rho(x, A_i, A_j) \geq r - 2.$$

$\square$

The proof of Corollary 8.2.1 is left to the reader (see Exercise 8.2).

Lemma 8.2.3   *For any graph G of order n in $\mathcal{F}_r$,*

$$t(G) = \frac{1}{3}(3n - 2r)\binom{r-1}{2} - \sum_{1 \leq i < j < k \leq r} \sum_{x \in A_k} \rho(x, A_i, A_j). \qquad (8.2)$$

Proof. By Lemma 8.2.2,

$$
\begin{aligned}
t(G) &= \sum_{1 \leq i < j < k \leq r} t([A_i \cup A_j \cup A_k]) \\
&= \sum_{1 \leq i < j < k \leq r} \left(|A_i| + |A_j| + |A_k| - 2 - \sum_{x \in A_k} \rho(x, A_i, A_j) \right) \\
&= n\binom{r-1}{2} - 2\binom{r}{3} - \sum_{1 \leq i < j < k \leq r} \sum_{x \in A_k} \rho(x, A_i, A_j),
\end{aligned}
$$

and thus (8.2) follows.   □

Corollary 8.2.2   *For any graph G of order n in $\mathcal{F}_r$,*

$$t(G) \leq \tfrac{1}{3}(3n - 2r)\binom{r-1}{2} - (r - 2).$$ □

Given $G \in \mathcal{F}_r$, let

$$h(G) = \frac{1}{3}(3n - 2r)\binom{r-1}{2} - (r - 2) - t(G);$$

and for $h = 0, 1, 2, \cdots$, let

$$\mathcal{F}_{r,h} = \{G \in \mathcal{F}_r : h(G) = h\}.$$

Corollary 8.2.3   *For $G \in \mathcal{F}_{r,h}$,*

$$
\begin{cases}
\displaystyle \sum_{1 \leq i < j < k \leq r} \sum_{x \in A_i} \rho(x, A_j, A_k) = h \\
\displaystyle \sum_{1 \leq i < j < k \leq r} \sum_{x \in A_j} \rho(x, A_i, A_k) = h \\
\displaystyle \sum_{1 \leq i < j < k \leq r} \sum_{x \in A_k} \rho(x, A_i, A_j) = h + r - 2.
\end{cases} \qquad (8.3)
$$ □

8.3   Graphs in $\mathcal{F}_r$ having maximum triangle numbers

Given $r \in \mathbb{N}$ with $r \geq 3$, by definition, $\mathcal{F}_{r,0}$ consists of those graphs in $\mathcal{F}_r$ of order n whose triangle number attains the equality in Corollary 8.2.2. In this section, we shall establish our main result of this chapter by proving the equivalences of the following characterizations of the family $\mathcal{F}_{r,0}$. While the equivalence of (iii) and (iv) was obtained by Wanner [1989], that of (ii) and (iv) was due to Borodin and Dmitriev [1991].

Theorem 8.3.1 *Let G be a graph of order n, and r an integer with $n \geq r \geq 3$. The following statements are equivalent:*

(i) $G \in \mathcal{F}_{r,o}$;

(ii) *G is K_{r-2}-gluing of two $(r-1)$-trees (see Figure 8.2) and $(r-2)$ is the smallest integer with this property;*

(iii) *$G = H - e$, where H is an $(r-1)$-tree and e is an edge in H that is contained in exactly $(r-2)$ triangles of H;*

(iv)

$$P(G, \lambda) = \lambda(\lambda-1)\cdots(\lambda-(r-3))(\lambda-(r-2))^2(\lambda-(r-1))^{n-r}. \quad (8.4)$$

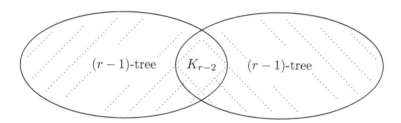

Figure 8.2

To ease our proof of Theorem 8.3.1, we first make some observations as given below.

Lemma 8.3.1 *Let $G \in \mathcal{F}_r$. Then $h(G) = 0$ if and only if G is a chordal graph.*

Proof. Assume $h(G) = 0$. By (8.3),

$$\sum_{1 \leq i < j < k \leq r} \sum_{x \in A_i} \rho(x, A_j, A_k) = 0.$$

Thus $\rho(x, A_j, A_k) = 0$, i.e., $[N(x) \cap (A_j \cup A_k)]$ is a tree for every $x \in A_i$ and any i, j, k with $1 \leq i < j < k \leq r$. By Theorem 6.4.4, G is a chordal graph.

Assume $h(G) \geq 1$. Then by (8.3) again,

$$\sum_{1 \leq i < j < k \leq r} \sum_{x \in A_i} \rho(x, A_j, A_k) \geq 1.$$

Thus $\rho(x, A_j, A_k) \geq 1$ for some i, j, k with $1 \leq i < j < k \leq r$ and some $x \in A_i$. Since $[A_j \cup A_k]$ is a tree, there exists a pure cycle in $[\{x\} \cup (A_j \cup A_k)]$, which implies that G is not chordal. $\qquad \square$

Corollary 8.3.1   *Let G be a graph of order $r \geq 3$. Then $G \in \mathcal{F}_{r,0}$ if and only if $G = K_r - e$ for some edge e in K_r.* $\qquad \square$

The proof of Corollary 8.3.1 is left to the reader (see Exercise 8.3).

Our next observation (see Exercise 8.4) follows directly from the definitions of $\mathcal{T}$ and $\mathcal{F}$.

Lemma 8.3.2   *Let G be a graph of order n in $\mathcal{F}_r$. If x is a simplicial vertex in G, then $r - 2 \leq d(x) \leq r - 1$, and*

$$\begin{cases} G - x \cong K_{n-1}, & \text{when } d(x) = r - 2, n = r, \\ G - x \in \mathcal{T}_r, & \text{when } d(x) = r - 2, n > r, \\ G - x \in \mathcal{F}_r, & \text{when } d(x) = r - 1, n > r. \end{cases} \qquad \square$$

We are now in a position to prove Theorem 8.3.1. We shall prove the implications: (i) $\Rightarrow$ (ii) $\Rightarrow$ (iv) $\Rightarrow$ (i) and (ii) $\Leftrightarrow$ (iii).

Proof of Theorem 8.3.1. (i) $\Rightarrow$ (ii).

Let G be a graph of order n in $\mathcal{F}_{r,0}$. We shall prove (ii) by induction on n. By Lemma 8.3.1, G is a chordal graph.

Assume that $n = r$. By Corollary 8.3.1, G is a K_{n-2}-gluing of two copies of K_{n-1}'s.

Assume that $n > r$. By Theorem 6.4.1, G contains two non-adjacent simplicial vertices u and v. By Lemma 8.3.2, $r - 2 \leq d(x) \leq r - 1$, where $x \in \{u, v\}$. If $d(u) = r - 2$, then by Lemma 8.3.2, $G - u \in \mathcal{T}_r$, and so $d(v) = r - 1$. Thus either $d(u) = r - 1$ or $d(v) = r - 1$, say the latter. By Lemma 8.3.2, $G - v \in \mathcal{F}_r$. Since $G - v$ is a chordal graph, by Lemma 8.3.1, $G - v \in \mathcal{F}_{r,0}$. By the induction hypothesis, $G - v$ is a K_{r-2}-gluing of two $(r - 1)$-trees, where $r - 2$ is the least integer having such a property. As v is a simplicial vertex of G, the same result holds for G.

(ii) $\Rightarrow$ (iv).

This follows from Theorem 1.3.2 and Theorem 6.5.2.

(ii) $\Rightarrow$ (iii).

Assume that G is a K_{r-2}-gluing of two $(r-1)$-trees G_1 and G_2, and let Q be the $(r-2)$-clique in common. By Lemma 6.5.1(vi), let $Q_1 = Q \cup \{u_1, u_2\}$ and $Q_2 = Q \cup \{v_1, v_2\}$ be, respectively, an r-clique in G_1 and G_2 containing Q. Let θ_i be an r-colouring of Q_i, $i = 1, 2$, such that both θ_1 and θ_2 agree on Q. Clearly, there is a pair $\{u_i, v_j\}$, say $\{u_1, v_1\}$ such that $\theta_1(u_1) \neq \theta_2(v_1)$. Let e be a new edge joining u_1 and v_1, and let $H = G + e$. Observe that a vertex x and 'e' induce a triangle in H if and only if $x \in Q$. Thus 'e' is contained in exactly $r - 2$ triangles in H. It is also easy to see that H is chordal and $(r-1)$-connected. In addition, the combination of θ_1 and θ_2 yields an r-colouring of H. By Theorem 6.5.1, H is an $(r-1)$-tree and we have (iii).

(iii) $\Rightarrow$ (ii).

The proof is by induction on $n \geq r$.

When $n = r$, it is clear that $H = K_r$, and so $G = K_r - e$ is a K_{r-2}-gluing of two copies of K_{r-1}'s, which are $(r-1)$-trees. Let H be an $(r-1)$-tree of order $n \geq r+1$ and e be an edge in H contained in exactly $r-2$ triangles of H.

By Theorem 6.4.1, H contains a simplicial vertex, say x, not incident to e. We assert that $e \notin E([N(x)])$. Indeed, as $H - x$ is also an $(r-1)$-tree, the $(r-1)$-clique $N(x)$ is contained in an r-clique in $H - x$ by Lemma 6.5.1, and so each edge in $[N(x)]$ is contained in at least $r-1$ triangles in H. Thus $e \notin E([N(x)])$ by assumption, as asserted. Now $H - x$ is an $(r-1)$-tree and e is an edge in $H - x$ which is contained in exactly $r-2$ triangles of $H - x$. By the induction hypothesis, $(H - x) - e$ is a K_{r-2}-gluing of two $(r-1)$-trees, where $r-2$ is the least integer having such a property. As x is a simplicial vertex of H, the same result holds for $H - e$.

(iv) $\Rightarrow$ (i).

Let G be a graph having its $P(G, \lambda)$ given in (8.4). We shall show that $G \in \mathcal{F}_{r,0}$.

If $n = r$, then

$$P(G, \lambda) = \lambda(\lambda - 1) \cdots (\lambda - (n - 3))(\lambda - (n - 2))^2,$$

and by Theorem 6.5.2, G is an $(r-2)$-tree of order r, i.e., $G = K_r - e$. Thus $G \in \mathcal{F}_{r,0}$.

Assume now that $n \geq r + 1$. As $P(G, \lambda)$ is given by (8.4), by Corollary 2.3.1, we have $\chi(G) = r$, $P(G, r)/r! = 2$ and

$$e(G) = h_1 = \sum_{i=1}^{r-1} i + (r-2) + (n-r-1)(r-1) = (r-1)n - \binom{r}{2} - 1.$$

Let $B_1, B_2, \cdots, B_r$ be the r colour classes of G. As $P(G, r)/r! = 2$, there exists at most one pair of colour classes B_i and B_j such that $[B_i \cup B_j]$ consists of 2 components. If $[B_i \cup B_j]$ is connected for each pair $\{i, j\}$ with $1 \leq i < j \leq r$, then

$$e(G) = \sum_{1 \leq i < j \leq r} e([B_i \cup B_j]) \geq \sum_{1 \leq i < j \leq r} (|B_i| + |B_j| - 1) = (r-1)n - \binom{r}{2},$$

a contradiction. Thus all the $[B_i \cup B_j]$'s are trees except one which is a forest with two components. By definition, $G \in \mathcal{F}_r$. Finally, we show that $h(G) = 0$. To see this, construct a new graph G_0 of order n as follows: starting with a K_{r-2}, add two new vertices x_1 and x_2 such that x_i is adjacent to each vertex in K_{r-2} for $i = 1, 2$, and then add another $n - r$ new vertices $y_1, y_2, \cdots, y_{n-r}$ such that y_i is adjacent to x_2 and each vertex in K_{r-2} (see Figure 8.3).

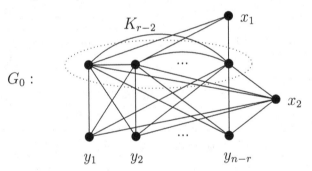

Figure 8.3

It is easy to see that $P(G_0, \lambda)$ is also given by (8.4); i.e., $G_0 \sim G$. Thus, by Theorem 3.2.1,

$$
\begin{aligned}
t(G) &= t(G_0) = \binom{r-2}{3} + 2\binom{r-2}{2} + (n-r)\binom{r-1}{2} \\
&= \frac{1}{3}(3n - 2r)\binom{r-1}{2} - (r-2).
\end{aligned}
$$

By Corollary 8.2.2, $h(G) = 0$ and so $G \in \mathcal{F}_{r,0}$. $\qquad\square$

8.4 A more general result

Let G be a graph of order n. In Theorem 6.5.2, a number of characterizations of G with

$$P(G, \lambda) = \lambda(\lambda - 1) \cdots (\lambda - q + 1)(\lambda - q)^{n-q}$$

are given. In Theorem 8.3.1, some characterizations of G with

$$P(G, \lambda) = \lambda(\lambda - 1) \cdots (\lambda - q + 1)^2(\lambda - q)^{n-q-1}$$

are presented.

A natural question one would like to ask is: What can be said about the structure of G with

$$P(G, \lambda) = \lambda(\lambda - 1) \cdots (\lambda - i)^2 \cdots (\lambda - q + 1)(\lambda - q)^{n-q-1}, \qquad (8.5)$$

where $1 \le i \le q - 2$?

This problem had indeed been solved by Han (1988) and Borodin and Dmitriev (1991).

Given $r, m \in \mathbb{N}$ with $r \ge 3$ and $m \ge 3$, let $\mathcal{G}^r(m)$ denote the family of connected graphs G with $\chi^a(G) \le r$ and $e(G) = m$, where $\chi^a(G)$ is the acyclic chromatic number of G (see Section 6.2); and let

$$t(r, m) = \max \{ t(G) : G \in \mathcal{G}^r(m) \}.$$

Dong and Koh (1997a) determined the quantity $t(r, m)$ and characterized also those graphs H in $\mathcal{G}^r(m)$ with $t(H) = t(r, m)$. As an application of their results, they solved also the above problem.

Theorem 8.4.1 *A graph G of order n has its $P(G, \lambda)$ given by (8.5) if and only if G is obtained from a q-tree of order $n-1$ by adding a simplicial vertex v with $d(v) = i$, where $1 \le i \le q - 2$.* $\qquad \square$

It was shown in Example 1.2.3 that if G is a chordal graph of order n, then

$$P(G, \lambda) = \lambda^{r_0}(\lambda - 1)^{r_1}(\lambda - 2)^{r_2} \cdots (\lambda - k)^{r_k} \qquad (8.6)$$

for some $r_i \in \mathbb{N}$, $0 \le i \le k$, such that $\sum_{i=0}^{k} r_i = n$. On the other hand, it follows from Theorems 8.3.1 and 8.4.1 that, for $i = 1, 2, \cdots, q-1$, any graph G having (8.5) as its $P(G, \lambda)$ is chordal. Another question arises: If G is a graph of order n having its $P(G, \lambda)$ given by (8.6), must G be chordal? This problem will be discussed in more details in Chapter 13.

Exercise 8

8.1 Let G be a connected graph. Show that

$$P(G, \lambda) = \lambda(\lambda - 1)^r(\lambda - 2)^s$$

for some $r \in \mathbb{N}$ and $s \in \mathbb{N}_0$ if and only if G is chordal and K_4-free.
(See Vaderlind (1988).)

8.2 Let G be any graph in $\mathcal{F}_r$. Prove that

$$
\begin{aligned}
\sum_{1 \leq i < j < k \leq r} \sum_{x \in A_k} \rho(x, A_i, A_j) &= r - 2 + \sum_{1 \leq i < j < k \leq r} \sum_{x \in A_j} \rho(x, A_i, A_k) \\
&= r - 2 + \sum_{1 \leq i < j < k \leq r} \sum_{x \in A_i} \rho(x, A_j, A_k),
\end{aligned}
$$

where $\rho(x, A_i, A_j) = c([N(x) \cap (A_i \cup A_j)]) - 1$.

8.3 Let G be a graph of order $r \geq 3$. Prove that $G \in \mathcal{F}_{r,0}$ if and only if $G = K_r - e$ for some edge e in K_r.

8.4 Let G be a graph of order n in $\mathcal{F}_r$ and $x \in V(G)$. Prove that if x is a simplicial vertex in G, then $r - 2 \leq d(x) \leq r - 1$, and

$$
\begin{cases}
G - x \cong K_{n-1}, & \text{when } d(x) = r - 2, n = r, \\
G - x \in \mathcal{T}_r, & \text{when } d(x) = r - 2, n > r, \\
G - x \in \mathcal{F}_r, & \text{when } d(x) = r - 1, n > r.
\end{cases}
$$

Chapter 9

Graphs in Which All but One Pair of Colour Classes Induce Trees (II)

9.1 Introduction

Xu and Li (1984) showed that for every odd $n \geq 5$, the wheel W_n (= $C_{n-1} + K_1$) is χ-unique. Dong (1990) generalized this result and showed that for every $k \geq 1$ and even $n \geq 4$, the graph $C_n + K_k$ is always χ-unique. It is observed that for $k \geq 1$ and even $n \geq 4$, the graph $C_n + K_k$ is $(k+2)$-colourable, and any pair of colour classes induces a tree except one, which induces a connected graph having a unique cycle (i.e., *unicyclic*).

Motivated by this observation, we introduce a family of graphs which satisfy the condition (**CT**) as stated below. A graph G is said to satisfy condition (**CT**) if G possesses an independent partition $\{A_1, \cdots, A_r\}$ such that

(i) $[A_1 \cup A_2]$ is a connected unicyclic graph and

(ii) $[A_i \cup A_j]$ is a tree for all i, j with $1 \leq i < j \leq r$ and $(i, j) \neq (1, 2)$.

Let $\mathcal{CT}$ be the family of graphs having an r-independent partition that satisfies condition (**CT**), and for $r \geq 3$, let $\mathcal{CT}_r$ be the family of graphs G that satisfy condition (**CT**) with an r-independent partition.

We shall first prove that for any $G \in \mathcal{CT}_r$,

$$e(G) = (r-1)v(G) - \binom{r}{2} + 1.$$

Accordingly, $\mathcal{CT}_{r_1} \cap \mathcal{CT}_{r_2} = \emptyset$ for all $r_1 \geq 3$ and $r_2 \geq 3$ with $r_1 \neq r_2$. Thus $\mathcal{CT}$ is partitioned into $\mathcal{CT}_3$, $\mathcal{CT}_4$, $\cdots$. We shall then proceed to show that for $G \in \mathcal{CT}_r$, where $r \geq 3$,

$$t(G) \leq r - 2 + \frac{1}{3}(3v(G) - 2r)\binom{r-1}{2}.$$

Thus for any $r \geq 3$, the family $\mathcal{CT}_r$ is further partitioned into $\mathcal{CT}_{r,0}$, $\mathcal{CT}_{r,1}$, $\cdots$, where for $h = 0, 1, 2, \cdots$,

$$\mathcal{CT}_{r,h} = \left\{ G \in \mathcal{CT}_r : t(G) = r - 2 + \frac{1}{3}(3v(G) - 2r)\binom{r-1}{2} - h \right\}.$$

In this chapter, our main objective is to characterize the graphs in $\mathcal{CT}_{r,0}$ in terms of their chromatic polynomials and deduce from it that the graph $C_n + K_k$ is χ-unique for all $k \geq 1$ and even $n \geq 4$. Some further results on the chromaticity of graphs in $\mathcal{CT}_{3,1}$ and the join of graphs will be mentioned in the final section.

9.2 Classification of graphs satisfying (CT)

In this section, we shall derive an expression for $e(G)$ and $t(G)$, respectively, for a graph G satisfying (**CT**), and then use them to classify such graphs.

Given $G \in \mathcal{CT}$, let $\mathcal{I}_c(G)$ be the family of independent partitions with which G satisfies condition (**CT**). We shall assume throughout this chapter that $\{A_1, \cdots, A_r\} \in \mathcal{I}_c(G)$ for $G \in \mathcal{CT}_r$.

Lemma 9.2.1   *Let $G \in \mathcal{CT}_r$. Then*
  (i) $v(G) \geq r + 2$,
  (ii) $e(G) = (r-1)v(G) - \binom{r}{2} + 1$ and
  (iii) $e(H) \leq (r-1)v(H) - \binom{r}{2} + 1$ for any subgraph H of G.

Proof.   Since $[A_1 \cup A_2]$ is a connected unicyclic graph, we have $|A_1| \geq 2$ and $|A_2| \geq 2$. For $i \geq 3$, we have $|A_i| \geq 1$. Thus (i) follows.

By definition,

$$\begin{aligned}
e(G) &= \sum_{1 \leq i < j \leq r} e([A_i \cup A_j]) = 1 + \sum_{1 \leq i < j \leq r} (|A_i| + |A_j| - 1) \\
&= 1 + (r-1)\sum_{1 \leq i \leq r} |A_i| - \binom{r}{2} = (r-1)v(G) - \binom{r}{2} + 1,
\end{aligned}$$

which is (ii). The inequality in (iii) can be proved in a similar way. $\square$

Corollary 9.2.1   If $r_1 \neq r_2$, then $\mathcal{CT}_{r_1} \cap \mathcal{CT}_{r_2} = \emptyset$.   □

Thus $\mathcal{CT}$ is partitioned into $\mathcal{CT}_2, \mathcal{CT}_3, \cdots$. In what follows, let C^* be the cycle in $[A_1 \cup A_2]$. We then have (see Exercise 9.2):

Lemma 9.2.2   Let $G \in \mathcal{CT}_r$. For any integer k with $3 \leq k \leq r$, there exists at most one vertex $x_k^* \in A_k$ such that $V(C^*) \subseteq N(x_k^*)$.   □

Let i and j be integers with $1 \leq i < j \leq r$ and $x \in V(G) \backslash (A_i \cup A_j)$. If $\{i, j\} = \{1, 2\}$ and $V(C^*) \subseteq N(x)$, then there are exactly

$$d(x, A_i, A_j) - \rho(x, A_i, A_j)$$

triangles in $[A_i \cup A_j \cup \{x\}]$; otherwise, there are exactly

$$d(x, A_i, A_j) - 1 - \rho(x, A_i, A_j)$$

triangles in $[A_i \cup A_j \cup \{x\}]$.

Lemma 9.2.3   Let $G \in \mathcal{CT}_r$. For any i, j, k with $1 \leq i < j < k \leq r$,

$$\sum_{x \in A_i} \rho(x, A_j, A_k) = \sum_{x \in A_j} \rho(x, A_i, A_k)$$

$$= \begin{cases} |A_i| + |A_j| + |A_k| - 1 - t([A_i \cup A_j \cup A_k]), & \text{if } (i, j) = (1, 2); \\ |A_i| + |A_j| + |A_k| - 2 - t([A_i \cup A_j \cup A_k]), & \text{if } (i, j) \neq (1, 2); \end{cases}$$

and if $(i, j) = (1, 2)$ and there is a vertex $x \in A_k$ such that $V(C^*) \subseteq N(x)$, then

$$\sum_{x \in A_k} \rho(x, A_i, A_j) = |A_i| + |A_j| + |A_k| - 1 - t([A_i \cup A_j \cup A_k]);$$

otherwise,

$$\sum_{x \in A_k} \rho(x, A_i, A_j) = |A_i| + |A_j| + |A_k| - 2 - t([A_i \cup A_j \cup A_k]).$$

Proof.   Observe that

$$t([A_i \cup A_j \cup A_k])$$

$$= \sum_{x \in A_i} t([\{x\} \cup A_j \cup A_k])$$

$$= \sum_{x \in A_i} (d(x, A_j, A_k) - 1 - \rho(x, A_j, A_k))$$

$$= e([A_i \cup A_j]) + e([A_i \cup A_k]) - |A_i| - \sum_{x \in A_i} \rho(x, A_j, A_k).$$

Likewise, we have

$$t([A_i \cup A_j \cup A_k]) = e([A_i \cup A_j]) + e([A_j \cup A_k]) - |A_j| - \sum_{x \in A_j} \rho(x, A_i, A_k).$$

Note that $[A_i \cup A_k]$ and $[A_j \cup A_k]$ are trees. When $(i,j) = (1,2)$, $e([A_i \cup A_j]) = |A_i| + |A_j|$; when $(i,j) \neq (1,2)$, $e([A_i \cup A_j]) = |A_i| + |A_j| - 1$. Thus the equalities in the first part follow.

Similarly, it can be shown that
(i) if there is a vertex $x \in A_k$ such that $V(C^*) \subseteq N(x)$, then

$$t([A_1 \cup A_2 \cup A_k]) = |A_1| + |A_2| + |A_k| - 1 - \sum_{x \in A_k} \rho(x, A_1, A_2);$$

(ii) if $V(C^*) \not\subseteq N(x)$ for all $x \in A_k$, then

$$t([A_1 \cup A_2 \cup A_k]) = |A_1| + |A_2| + |A_k| - 2 - \sum_{x \in A_k} \rho(x, A_1, A_2); \text{ and}$$

(iii) if $(i,j) \neq (1,2)$, then

$$t([A_i \cup A_j \cup A_k]) = |A_i| + |A_j| + |A_k| - 2 - \sum_{x \in A_k} \rho(x, A_i, A_j).$$

Thus the equalities in the second part hold. $\square$

Lemma 9.2.4   *Let $G \in \mathcal{CT}_r$. Then*

$$t(G) = r - 2 + \frac{1}{3}(3v(G) - 2r)\binom{r-1}{2} - \sum_{1 \leq i < j < k \leq r} \sum_{x \in A_i} \rho(x, A_j, A_k);$$

and

$$t(G) \leq r - 2 + \frac{1}{3}(3v(G) - 2r)\binom{r-1}{2} - \sum_{1 \leq i < j < k \leq r} \sum_{x \in A_k} \rho(x, A_i, A_j),$$

where equality holds if and only if for any $k \geq 3$, there is a vertex x_k^* in A_k such that $V(C^*) \subseteq N(x_k^*)$.

Proof. As the proofs are similar, we shall only prove the first result. By Lemma 9.2.3,

$$t(G) = \sum_{1 \leq i < j < k \leq r} t([A_i \cup A_j \cup A_k])$$

$$
\begin{aligned}
&= \sum_{3\le k\le r} t([A_1 \cup A_2 \cup A_k]) + \sum_{\substack{(i,j)\ne(1,2)\\1\le i<j<k\le r}} t([A_i \cup A_j \cup A_k])\\
&= \sum_{3\le k\le r}\left(|A_1| + |A_2| + |A_k| - 1 - \sum_{x\in A_1}\rho(x, A_2, A_k)\right)\\
&\quad + \sum_{\substack{(i,j)\ne(1,2)\\1\le i<j<k\le r}}\left(|A_i| + |A_j| + |A_k| - 2 - \sum_{x\in A_i}\rho(x, A_j, A_k)\right)\\
&= r - 2 + \sum_{1\le i<j<k\le r}(|A_i| + |A_j| + |A_k| - 2)\\
&\quad - \sum_{1\le i<j<k\le r}\sum_{x\in A_i}\rho(x, A_j, A_k)\\
&= r - 2 + \frac{1}{3}(3v(G) - 2r)\binom{r-1}{2} + \sum_{1\le i<j<k\le r}\sum_{x\in A_i}\rho(x, A_j, A_k),
\end{aligned}
$$

as required. $\square$

Corollary 9.2.2 *Let $G \in \mathcal{CT}_r$. Then*

$$
t(G) \le r - 2 + \tfrac{1}{3}(3v(G) - 2r)\binom{r-1}{2}.
$$
$\square$

The above corollary gives an upper bound for $t(G)$, where $G \in \mathcal{CT}_r$, which suggests that $\mathcal{CT}_r$ can be partitioned into $\mathcal{CT}_{r,0}$, $\mathcal{CT}_{r,1}$, $\mathcal{CT}_{r,2}, \cdots$, where for $h = 0, 1, 2, \cdots$,

$$
\mathcal{CT}_{r,h} = \left\{G \in \mathcal{CT}_r : t(G) = r - 2 + \frac{1}{3}(3v(G) - 2r)\binom{r-1}{2} - h\right\}.
$$

The following result now follows directly from Lemmas 9.2.3 and 9.2.4.

Lemma 9.2.5 *Let $G \in \mathcal{CT}_{r,h}$, where $h \ge 0$. Then*

$$
\sum_{1\le i<j<k\le r}\sum_{x\in A_i}\rho(x, A_j, A_k) = \sum_{1\le i<j<k\le r}\sum_{x\in A_j}\rho(x, A_i, A_k) = h
$$

and

$$
\sum_{1\le i<j<k\le r}\sum_{x\in A_k}\rho(x, A_i, A_j) \le h.
$$
$\square$

9.3   Graphs in $\mathcal{CT}$

It can be checked that $C_n + K_{r-2}$ is a graph in $\mathcal{CT}_r$ for all $r \geq 3$ and even $n \geq 4$. More generally, we have:

Theorem 9.3.1   *Let G be a uniquely r-colourable graph with $e(G) = (r-1)v(G) - \binom{r}{2} + 1$, where $r \geq 3$. Then $G \in \mathcal{CT}_r$.*

Proof. Let $\{A_1, \cdots, A_r\}$ be an independent partition of G. Since G is uniquely r-colourable, $A_i \neq \emptyset$ for all i and $[A_i \cup A_j]$ is connected for all i, j with $1 \leq i < j \leq r$. Thus

$$
\begin{aligned}
e(G) &= \sum_{1 \leq i < j \leq r} e([A_i \cup A_j]) \\
&\geq \sum_{1 \leq i < j \leq r} (|A_i| + |A_j| - 1) \\
&= (r-1)v(G) - \binom{r}{2}.
\end{aligned}
$$

Since $e(G) = 1 + (r-1)v(G) - \binom{r}{2}$, there is a pair (i_0, j_0) such that $e([A_{i_0} \cup A_{j_0}]) = |A_{i_0}| + |A_{j_0}|$ and $e([A_i \cup A_j]) = |A_i| + |A_j| - 1$ for all i and j with $(i, j) \neq (i_0, j_0)$. Thus $G \in \mathcal{CT}_r$ by definition.   □

Two simple properties of the graphs in $\mathcal{CT}_r$ are given below. By Lemma 6.3.1, we have:

Lemma 9.3.1   *For $G \in \mathcal{CT}_r$, G is $(r-1)$-connected.*   □

Lemma 9.3.2   *Let $G \in \mathcal{CT}_{r,h}$, where $h \geq 0$. If x is a simplicial vertex of G, then $d(x) = r - 1$ and $G - x \in \mathcal{CT}_{r,h}$.*

Proof. Assume that $x \in A_k$ for some k. By Lemma 9.3.1, $d(x) \geq r - 1$. Since x is a simplicial vertex, $|N(x) \cap A_i| \leq 1$ for all $i \neq k$. Thus $d(x) \leq r-1$, and so $d(x) = r - 1$.

If $A_k = \{x\}$, then $k \geq 3$ and $[A_i \cup A_k]$ is a star for all i with $i \neq k$. As x is a simplicial vertex, $|A_i| = |N(x) \cap A_i| = 1$ for all $i \neq k$, contradicting the fact that $|A_1| \geq 2$. Thus $|A_k| \geq 2$. Let $H = G - x$, and let $B_k = A_k \backslash \{x\}$ and $B_j = A_j$ when $j \neq k$. Then $(B_1, \cdots, B_r)$ is an independent partition of H. Since x is an end-vertex of the graph $[A_i \cup A_k]$ for all i with $i \neq k$, $H[B_1 \cup B_2]$ is a connected unicyclic graph and $H[B_i \cup B_j]$ is a tree when $\{i, j\} \neq \{1, 2\}$. Thus $H \in \mathcal{CT}_r$. As $t(H) = t(G) - \binom{r-1}{2}$ and $v(H) = v(G) - 1$, we have $H \in \mathcal{CT}_{r,h}$ by definition.   □

9.4   Graphs containing exactly one pure cycle

In order to study the structure of graphs in $\mathcal{CT}_{r,0}$, we need to obtain some useful information about the graphs which contain exactly one pure cycle. In this section, we shall first reveal the structure of such graphs, and then furnish a sufficient condition for an r-colourable graph to have a unique pure cycle.

Obviously, both the graphs C_n and $C_n + K_k$ contain exactly one pure cycle, where $n \geq 4$ and $k \geq 1$. We shall show that if a graph G contains exactly one pure cycle and no simplicial vertices, then either $G \cong C_n$ or $G \cong C_n + K_k$ for some $n \geq 4$ and $k \geq 1$.

Lemma 9.4.1   *Suppose that G contains exactly one pure cycle. Then G contains a simplicial vertex if and only if G contains a clique-cut.*

Proof.   If x is a simplicial vertex of G, then $N(x)$ is a clique-cut of G.

Assume that G contains a clique-cut. Then there are two proper subgraphs G_1 and G_2 of G such that $G \in \mathcal{G}[G_1 \cup_k G_2]$ for some integer k. Since G contains exactly one pure cycle, the cycle must be in G_1 or G_2. Thus at least one of G_1 and G_2 is a chordal graph. By Corollary 6.4.1, G contains a simplicial vertex.   □

The next result reveals the structures of connected graphs that have exactly one pure cycle.

Theorem 9.4.1   *Suppose that G is a connected graph containing exactly one pure cycle. Then one of the following holds:*

(i) $G \cong C_n$ for some $n \geq 4$,

(ii) $G \cong C_n + K_k$ for some $n \geq 4$ and $k \geq 1$, and

(iii) G contains a sequence of vertices $x_1, x_2, \cdots, x_s$, where $s \geq 1$, such that x_i is a simplicial vertex of the graph $G - \{x_1, \cdots, x_{i-1}\}$ for $i = s, s-1, \cdots, 1$, and either $G - \{x_1, \cdots, x_s\} \cong C_n$ for some $n \geq 4$ or $G - \{x_1, \cdots, x_s\} \cong C_n + K_k$ for some $n \geq 4$ and $k \geq 1$.

Proof.   Assume that C_n is the pure cycle in G. Obviously, $v(G) \geq n \geq 4$. When $v(G) = n$, (i) holds. Now assume that $v(G) \geq n + 1$.

If $d(x) = v(G) - 1$ for every $x \in V(G) \backslash V(C_n)$, then $G \cong C_n + K_k$, where $k = v(G) - n$, and (ii) holds.

If there is a vertex $z \in V(G) \backslash V(C_n)$ such that $d(z) < v(G) - 1$, then by Corollary 6.4.2, G contains a clique-cut, and thus G contains a simplicial vertex by Lemma 9.4.1. If G contains a simplicial vertex x, then $G - x$ contains exactly one pure cycle, and (iii) holds by induction. □

We shall now present a sufficient condition for an r-colourable graph to have exactly one pure cycle.

Theorem 9.4.2 *Let G be an r-colourable graph and $A_1, A_2, \cdots, A_r$ be the colour classes of an r-colouring of G. Suppose that the following conditions hold:*

(i) *$[A_1 \cup A_2]$ contains exactly one cycle, denoted by C^*, and $[A_i \cup A_j]$ is a forest for any pair i, j with $1 \leq i < j \leq r$ and $(i, j) \neq (1, 2)$,*

(ii) *for each $i = 3, 4, \cdots, r$, there exists a vertex $x_i \in A_i$ such that $V(C^*) \subseteq N(x_i)$, and*

(iii) *for all i, j, k with $1 \leq i < j < k \leq r$ and any vertex $x \in A_k$, either x is an isolated vertex in $[A_i \cup A_j \cup A_k]$ or $[N(x) \cap (A_i \cup A_j)]$ is connected.*

Then G contains exactly one pure cycle, namely C^*, and contains $C^* + K_{r-2}$ as a subgraph.

Proof. Clearly, C^* is a pure cycle of G. Suppose on the contrary that G contains more than one pure cycle. Let Φ be the set of pure cycles of G excluding C^*. Then $\Phi \neq \emptyset$.

By Theorem 6.4.4, both $G - A_1$ and $G - A_2$ are chordal graphs. So $V(C) \cap A_i \neq \emptyset$ for any $i = 1, 2$ and any $C \in \Phi$. If $V(C) \cap A_i = \emptyset$ for some $C \in \Phi$ and some $i \geq 3$, then the graph $G - A_i$ still satisfies the three conditions stated in the theorem. We may thus assume that $V(C) \cap A_i \neq \emptyset$ for any $C \in \Phi$ and any $i \geq 3$. So $V(C) \cap A_i \neq \emptyset$ for any $C \in \Phi$ and any i with $1 \leq i \leq r$. Let

$$p = \min \{|V(C) \cap A_r| : C \in \Phi\}.$$

Choose a member, say C, in Φ such that $|V(C) \cap A_r| = p$. Let $w \in V(C) \cap A_r$. Then $C - w$ is a path of order at least three, say $C - w = x u_1 \cdots u_s y$, where $s \geq 1$. We may assume that $x, y \in A_i \cup A_j$ for some i and j with $1 \leq i < j \leq r - 1$. Since w is not an isolated vertex in $[A_i \cup A_j \cup A_r]$, by (iii), $[N(w) \cap (A_i \cup A_j)]$ is connected. Let $x v_1 \cdots v_t y$ be one of the shortest x-y paths, where $t \geq 1$, in $[A_i \cup A_j]$ so that $w v_i \in E(G)$ for each $i = 1, \cdots, t$, as shown in Figure 9.1(a). Then $\{u_1, u_2, \cdots, u_s\} \cap \{v_1, v_2, \cdots, v_t\} = \emptyset$.

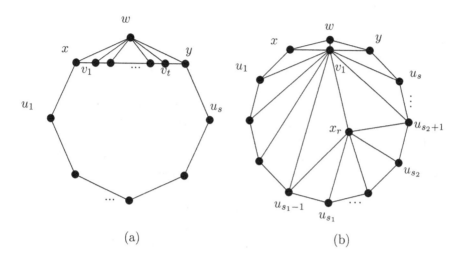

Figure 9.1

Let $P_1(x, y)$ and $P_2(x, y)$ denote the paths $x u_1 \cdots u_s y$ and $x v_1 \cdots v_t y$, respectively, and let $S = \{x, y, u_1, \cdots, u_s, v_1, \cdots, v_t\}$. Since $|S \cap A_r| = |V(P_1(x, y)) \cap A_r| = p - 1$, $[S]$ contains no pure cycles in Φ. So $[S]$ is either a chordal graph or a graph having exactly one pure cycle, namely C^*.

The value of t depends on the cycle C and the vertex w, where $C \in \Phi$ with $|V(C) \cap A_r| = p$ and $w \in A_r \cap V(C)$. Assume that C is a pure cycle in Φ with $|V(C) \cap A_r| = p$ and w is a vertex in $A_r \cap V(C)$ such that the corresponding number t reaches the minimum.

Claim 1: $t \geq 2$.

Suppose that the claim is not true. Then $t = 1$. If v_1 is adjacent to all vertices on $P_1(x, y)$, then $V(C) \cap A_i = \emptyset$ or $V(C) \cap A_j = \emptyset$, a contradiction. So v_1 is not adjacent to some vertex on $P_1(x, y)$, and thus $[S]$ is not a chordal graph. By the above argument, $[S]$ contains exactly one pure cycle, namely C^*. Then $P_1(x, y) - N_G(v_1)$ is a path $u_{s_1} \cdots u_{s_2}$ and $v_1 u_{s_1-1} \cdots u_{s_2+1} v_1$ is the cycle C^*. By (ii), x_r is adjacent to all vertices on this cycle, as shown in Figure 9.1(b). Consider the subgraph $[S']$, where $S' = \{w, x, y, x_r, u_1, \cdots, u_{s_1-1}, u_{s_2+1}, \cdots, u_s\}$. Since $w, x_r \in A_r$, $w x_r \notin E(G)$. As the path $u_{s_1-1}, \cdots, u_1, x, w, y, u_s, \cdots, u_{s_2+1}$ is a pure path, $[S']$ contains a pure cycle C' such that w and x_r are on C'. It is obvious that $C' \in \Phi$. Since $S' \subseteq N(v_1)$, either $V(C') \cap A_i = \emptyset$ or $V(C') \cap A_j = \emptyset$, a contradiction. Claim 1 thus follows.

Claim 2: $v_l u_m \notin E(G)$ for all $l = 1, 2, \cdots, t$ and $m = 1, 2, \cdots, s$.

Suppose that $v_l u_m \in E(G)$ for some l and m. If $l < t$, let m be the maximum index such that $v_l u_m \in E(G)$. Observe that C'' : $wv_l u_m \cdots u_s yw$ is a pure cycle. Since $w \in A_r$ and $w \in V(C'')$, we have $C'' \in \Phi$. However, $|V(C'') \cap A_r| \le p$ and the number 't' for the cycle C'' at the vertex w becomes smaller, a contradiction. If $l = t$, let m be the minimum index such that $v_t u_m \in E(G)$. Observe that C''' : $wv_t u_m u_{m-1} \cdots u_1 xw$ is a pure cycle in Φ. Similarly, $|V(C''') \cap A_r| \le p$ and the number 't' for this cycle at the vertex w becomes smaller, a contradiction. Hence Claim 2 holds.

By Claim 2, $[S]$ is a cycle, namely C^*. By (ii), x_r is adjacent to all vertices on C^*. Since $xv_1 v_2 \cdots v_t y$ is a path in $[A_i \cup A_j]$, either $\{x, v_2\} \subseteq A_i$ or $\{x, v_2\} \subseteq A_j$. Thus $wx x_r v_2 w$ is a cycle in either $[A_i \cup A_r]$ or $[A_j \cup A_r]$, a contradiction.

We thus conclude that $\Phi = \emptyset$ and that G contains exactly one pure cycle, namely C^*. Let $A = V(C^*) \cup \{x_3, \cdots, x_r\}$. If $x_i x_j \notin E(G)$ for some i, j with $3 \le i < j \le r$, then by (ii), $x_i u x_j v x_i$ is a pure cycle of G for some vertices u, v on C^*, a contradiction. Thus $x_i x_j \in E(G)$ for all i, j with $3 \le i < j \le r$. By (ii) again, $[A] \cong C^* + K_{r-2}$.   □

9.5   The main results

By definition, $\mathcal{CT}_{r,0}$ is the family of graphs G in $\mathcal{CT}_r$ with

$$t(G) = r - 2 + \left(v(G) - \frac{2r}{3}\right)\binom{r-1}{2}.$$

In this section, we shall characterize the graphs in $\mathcal{CT}_{r,0}$ via their chromatic polynomials in general, and prove, in particular, that the join $C_n + K_k$ is χ-unique for any $k \ge 1$ and even $n \ge 4$. As a reminder, we have $r \ge 3$ and $\{A_1, \cdots, A_r\} \in \mathcal{I}_c(G)$ for $G \in \mathcal{CT}_r$.

Theorem 9.5.1   *Let $G \in \mathcal{CT}_{r,0}$. Then either*

(i) $G \cong C_p + K_{r-2}$ *for some even $p \ge 4$, or*

(ii) G *contains a sequence of vertices $x_1, x_2, \cdots, x_s$, where $s \ge 1$, such that x_i is a simplicial vertex of the graph $G - \{x_1, \cdots, x_{i-1}\}$ for $i = s, s-1, \cdots, 1$, and either $G - \{x_1, \cdots, x_s\} \cong C_n$ for some $n \ge 4$ or $G - \{x_1, \cdots, x_s\} \cong C_n + K_k$ for some $n \ge 4$ and $k \ge 1$.*

Proof.   Let C^* be the cycle in $[A_1 \cup A_2]$. By Lemma 9.2.5, $\rho(x, A_i, A_j) = 0$ for all i and j with $1 \le i < j \le r$ and all $x \in V(G) \backslash (A_i \cup A_j)$. By

Lemma 9.2.4, there is a vertex x_k^* in A_k such that $V(C^*) \subseteq N(x_k^*)$ for all $k \geq 3$. Thus, by Theorem 9.4.2, G contains exactly one pure cycle, namely C^*. The result now follows by Theorem 9.4.1. $\qquad\square$

Theorem 9.5.2 *Let G be a graph and r be an integer with $r \geq 3$. Then the following statements are equivalent:*

(i) $G \in \mathcal{CT}_{r,0}$;

(ii) $P(G, \lambda) = (\lambda - r + 1)^{v(G) - p - r + 3} \lambda (\lambda - 1) \cdots (\lambda - r + 3)((\lambda - r + 1)^{p-1} + 1)$, *for some even $p \geq 4$;*

(iii) G *is a uniquely r-colourable graph with $e(G) = (r-1)v(G) - \binom{r}{2} + 1$ and $t(G) = r - 2 + \left(v(G) - \frac{2r}{3}\right)\binom{r-1}{2}$.*

Proof. (i)$\Rightarrow$(ii) If G contains no simplicial vertex, then by Theorem 9.5.1, $G \cong C_p + K_{r-2}$ for some even $p \geq 4$, and we have (ii). If x is a simplicial vertex of G, then $d(x) = r - 1$ and $G - x \in \mathcal{CT}_{r,0}$. By the induction hypothesis,

$$P(G - x, \lambda) = (\lambda - r + 1)^{v(G) - p - r + 2} \lambda(\lambda - 1) \cdots (\lambda - r + 3)((\lambda - r + 1)^{p-1} + 1),$$

for some even $p \geq 4$, and thus (ii) holds again.

(ii)$\Rightarrow$(iii) Observe that $P(G, r) = r!$. So G is uniquely r-colourable. By Corollary 2.3.1, we have the expressions of $e(G)$ and $t(G)$ as stated in (iii).

(iii)$\Rightarrow$(i) By Theorem 9.3.1 and definition, $G \in \mathcal{CT}_{r,0}$. $\qquad\square$

Theorem 9.5.3 *For any $k \geq 1$ and even $p \geq 4$, the graph $C_p + K_k$ is χ-unique.*

Proof. Let $G \sim C_p + K_k$. Then $v(G) = p + k$ and

$$P(G, \lambda) = \lambda(\lambda - 1) \cdots (\lambda - k + 1)((\lambda - k - 1)^p + (\lambda - k - 1)).$$

By Theorem 9.5.2, $G \in \mathcal{CT}_{k+2,0}$.

If G contains a simplicial vertex x, then by Lemma 9.3.2, $d(x) = k + 1$ and $G - x \in \mathcal{CT}_{k+2,0}$, and so $(\lambda - k - 1)^2$ is a factor of $P(G, \lambda)$, a contradiction. By Theorem 9.5.1 again, $G \cong C_{p'} + K_k$ for some even p'. Clearly $p' = p$ and the result follows. $\qquad\square$

Remark Chao and Whitehead Jr. (1979b) observed that $W_6(= C_5 + K_1)$ is not χ-unique, and there is only one graph, denoted by W' (see

Example 3.2.3(2)(b)), such that $W' \sim W_6$ and $W' \not\cong W_6$. Thus $C_5 + K_k$ is not χ-unique for any $k \geq 2$, since $C_5 + K_k \sim W' + K_{k-1}$. Petriso Gută (1997) proved that for any $k \geq 2$, $W' + K_{k-1}$ is the only graph which is χ-equivalent to $C_5 + K_k$, but not isomorphic to it.

9.6　Further results

(I) Chromaticity of graphs in $\mathcal{CT}_{3,1}$.

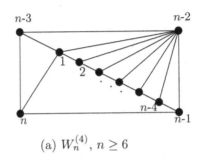

(a) $W_n^{(4)}$, $n \geq 6$

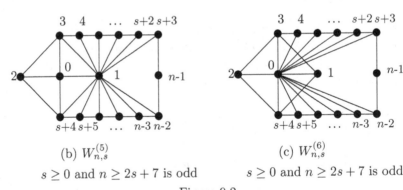

(b) $W_{n,s}^{(5)}$

$s \geq 0$ and $n \geq 2s + 7$ is odd

(c) $W_{n,s}^{(6)}$

$s \geq 0$ and $n \geq 2s + 7$ is odd

Figure 9.2

By definition, $\mathcal{CT}_{3,1}$ is the family of graphs in $\mathcal{CT}_3$ having $t(G) = v(G) - 2$. We shall further introduce a special subfamily of $\mathcal{CT}_{3,1}$. Define $\mathcal{CT}_{3,1}^0$ to be the family of graphs G in $\mathcal{CT}_{3,1}$ such that

　(i) G contains at least two pure cycles of order 4 and

　(ii) $(\lambda - 2)^2$ is not a factor of $P(G, \lambda)$.

Some examples of graphs in $\mathcal{CT}_{3,1}^0$ are given below.

Let $W_n^{(4)}$ be the graph in Figure 9.2 (a), where $n \geq 6$; let $W_{n,s}^{(5)}$ and $W_{n,s}^{(6)}$ be the graphs in Figure 9.2 (b) and (c) respectively, where $s \geq 0$,

$n \geq 2s + 7$ and n is odd. It can be checked that $W_n^{(4)} \in \mathcal{CT}_{3,1}$ for all $n \geq 6$ and $W_{n,s}^{(5)}, W_{n,s}^{(6)} \in \mathcal{CT}_{3,1}$ for all $s \geq 0$ and odd $n \geq 2s + 7$.

We now compute the chromatic polynomials of the graphs $W_n^{(4)}$, $W_{n,s}^{(5)}$ and $W_{n,s}^{(6)}$. Let x and y be respectively the vertices '$n - 3$' and '$n - 2$' in $W_n^{(4)}$, and let u and v be respectively the vertices '2' and '$s + 4$' in $W_{n,s}^{(5)}$. By FRT, we have

$$
\begin{aligned}
P(W_n^{(4)}, \lambda) &= P(W_n^{(4)} - xy, \lambda) - P(W_n^{(4)} \cdot xy, \lambda) \\
&= (\lambda - 2)P(W(n - 1, n - 3), \lambda) - P(W_{n-1}, \lambda) \\
&= \lambda(\lambda - 2)\left((\lambda - 2)^{n-2} + (\lambda - 2)^{n-4} + (-1)^n(\lambda - 3)\right) \\
&\quad \text{where } n \geq 6, \\
P(W_{n,s}^{(5)}, \lambda) &= P(W_{n,s}^{(5)} - uv, \lambda) - P(W_{n,s}^{(5)} \cdot uv, \lambda) \\
&= (\lambda - 2)P(W(n - 1, n - 3), \lambda) \\
&\quad -(\lambda - 3)P(W(n - 2, n - 4), \lambda) \\
&= \lambda(\lambda - 2)\left((\lambda - 2)^{n-2} + (\lambda - 2)^{n-4} + (\lambda - 2)^{n-6} - 2\lambda + 5\right) \\
&\quad \text{where } s \geq 0, n \geq 2s + 7 \text{ and } n \text{ is odd,} \\
P(W_{n,s}^{(6)}, \lambda) &= P(W_{n,s}^{(5)}, \lambda).
\end{aligned}
$$

$$(9.1)$$
$$(9.2)$$

Thus these graphs are members in $\mathcal{CT}_{3,1}^0$. Indeed, Dong (1997) proved the following result.

(1)

$$
\begin{aligned}
\mathcal{CT}_{3,1}^0 &= \left\{W_n^{(4)} : n \geq 6\right\} \cup \left(\bigcup_{\substack{n \geq 8 \text{ is even}}} \mathcal{G}[W(n - 2, n - 4) \cup_3 W_5]\right) \\
&\quad \cup \left\{W_{n,s}^{(5)}, W_{n,s}^{(6)} : s \geq 0, n \geq 2s + 7 \text{ and } n \text{ is odd}\right\}.
\end{aligned}
$$

He also observed that the families $\mathcal{CT}_{3,1}$ and $\mathcal{CT}_{3,1}^0$ are, respectively, χ-closed, and used them to establish the following:

(2) (i) For any integer n with $n = 6$ or $n \geq 9$, the graph $W_n^{(4)}$ is χ-unique.

(ii) The family $\{W_7^{(4)}, W_{7,0}^{(5)}, W_{7,0}^{(6)}\}$ is a χ-equivalence class.

(iii) The family $\{W_8^{(4)}\} \cup \mathcal{G}[W(6, 4) \cup_3 W_5]$ is a χ-equivalence class.

(iv) For any odd $n \geq 9$, $\left\{W_{n,s}^{(5)}, W_{n,s}^{(6)} : 0 \leq s \leq \frac{n-7}{2}\right\}$ is a χ-equivalence class.

(v) For any even $n \geq 10$, $\mathcal{G}[W(n-2, n-4) \cup_3 W_5]$ is a χ-equivalence class.

(II) Chromaticity of joins of graphs

We have seen that $P(G+H, \lambda)$ can be expressed in terms of $P(G, \lambda)$ and $P(H, \lambda)$, as shown in Theorem 1.5.1 for any two graphs G and H. It follows that if $G_1 \sim G_2$ and $H_1 \sim H_2$, then $G_1 + H_1 \sim G_2 + H_2$. Thus, if $G + H$ is χ-unique, then both G and H must be χ-unique. Conversely, is $G + H$ χ-unique if both G and H are χ-unique? Observe that if $G = K_m$ and $H = K_n$ (both are χ-unique), then $G + H = K_{m+n}$ is χ-unique; and if $G = O_m$ and $H = O_n$, where $m \geq n \geq 2$ (both are χ-unique), then $G + H = K(m, n)$ is also χ-unique. However, if $G = K_m$ and $H = O_n$, then $G + H$ is an m-tree of order $m + n$, which is clearly not χ-unique. Thus, in general, $G + H$ need not be χ-unique even if G and H are χ-unique.

For which χ-unique graphs G and H, their join $G+H$ is χ-unique? Until now, no work has been reported on this problem except the case when one of the graphs is complete. Results in this case are summarized below.

(1) (Dong (1990))

     (i) $C_n + K_m$ is χ-unique for even $n \geq 4$;

     (ii) $\theta(d, e, f) + K_m$ is χ-unique if d, e, f have the same parity.

(2) (Chia (1995a))

     (i) Let G be a uniquely r-colourable graph such that each colour class has exactly 2 vertices. Then $G + K_m$ is χ-unique if and only if G is so;

     (ii) $W(n, n-2) + K_m$ is χ-unique for even $n \geq 4$.

(3) (Dong and Koh (1997b))

     (i) For $G \in \mathcal{T}_{r,1}$, G is χ-unique if and only if $G + K_m$ is χ-unique (see Theorem 7.5.1);

     (ii) For any graph $G \in \mathcal{T}_{3,1,4}^* \cup \mathcal{T}_{3,1,5}^*$, $G + K_m$ is χ-unique (see Corollary 7.6.1).

Exercise 9

9.1 Prove that if $r_1 \neq r_2$, then $\mathcal{CT}_{r_1} \cap \mathcal{CT}_{r_2} = \emptyset$.

9.2 Let $G \in \mathcal{CT}_r$ and $\{A_1, \cdots, A_r\}$ be an independent partition of G with which G satisfies condition (CT). Let C^* be the only cycle in $[A_1 \cup A_2]$. Prove that for any integer k with $3 \leq k \leq r$, there exists at most one vertex $x_k^* \in A_k$ such that $V(C^*) \subseteq N(x_k^*)$.

9.3 Prove that for any integer n with $n = 6$ or $n \geq 9$, the graph $W_n^{(4)}$ is χ-unique.

9.4 Prove that $W_7^{(4)}, W_{7,0}^{(5)}$ and $W_{7,0}^{(6)}$ are χ-equivalent.

9.5 Prove that $W_{n,s}^{(5)}$ and $W_{n,s}^{(6)}$ are χ-equivalent for all odd n with $n \geq 9$ and $0 \leq s \leq (n-7)/2$.

9.6 Let $G \in \mathcal{CT}_r$. Prove that $e(H) \leq (r-1)v(H) - \binom{r}{2} + 1$ for any subgraph H of G.

9.7 Let $G \in \mathcal{CT}_r$. Prove that

$$t(G) \leq r - 2 + \frac{1}{3}(3v(G) - 2r)\binom{r-1}{2} - \sum_{1 \leq i < j < k \leq r} \sum_{x \in A_k} \rho(x, A_i, A_j),$$

where equality holds if and only if for any $k \geq 3$, there is a vertex x_k^* in A_k such that $V(C^*) \subseteq N(x_k^*)$.

Chapter 10

Chromaticity of Extremal 3-Colourable Graphs

10.1 Introduction

In Chapters 6 to 9, we study the chromaticity of graphs which are uniquely 3-colourable or nearly uniquely 3-colourable. In this chapter, we shall study the chromaticity of a family of 3-colourable graphs which are not uniquely 3-colourable.

Let G be a 3-colourable connected graph. By a proof similar to that of Theorem 6.3.1, it can be proved that if G is uniquely 3-colourable, then $e(G) \geq 2v(G) - 3$ (see Exercise 10.9). Recall that $s_r(G) = \frac{P(G,r)}{r!}$ for any $r \in \mathbb{N}$. In Section 10.2, we shall show that if $e(G) = 2v(G) - k$, where $k \geq 3$, then $s_3(G) \geq 2^{k-3}$. In particular, G is not uniquely 3-colourable if $e(G) \geq 2v(G) - 4$. We shall further prove that if $e(G) = 2v(G) - k$ and $s_3(G) < 2^{k-2}$, where $k \geq 3$, then $t(G) \leq v(G) - k + 1$, where $t(G)$ is the number of triangles in G, and equality holds if and only if G is chordal.

In Section 10.3, we shall characterize the structure of any 3-colourable connected graph G with $e(G) = 2v(G) - k$, $t(G) = v(G) - k$ and $s_3(G) < 2^{k-2}$, where $k \geq 3$, by showing that G is isomorphic to $H \cdot xy$ for some 3-colourable connected chordal graph H and two vertices x and y in H with $d_H(x, y) \geq 4$. It then follows that if G is 2-connected, then G is a graph obtained from a cycle by replacing each edge by a 2-tree.

Note that $W(n, s)$, where $1 \leq s \leq n - 2$, is the graph obtained from the wheel W_n by deleting all but s consecutive spokes, as shown in Figure 10.1.

Koh and Teo (1990) proposed the following problem on the chromaticity of $W(n, s)$:

Problem 10.1.1   *For any $n, s \in \mathbb{N}$ with $n \geq 5$ and $2 \leq s \leq n - 2$, is $W(n, s)$ χ-unique?*

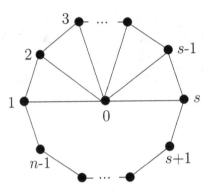

Figure 10.1

The graph $W(n, 1)$ is a vertex-gluing of C_{n-1} and K_2, and so it is χ-unique by Theorem 3.5.1. The graph $W(n, 2)$ is a special θ-graph, and so it is χ-unique. Chao and Whitehead Jr. (1979b) showed that the graphs $W(n, 3)$, where $n \geq 5$, and $W(n, 4)$, where $n \geq 6$, are also χ-unique; but pointed out that $W(7, 5)$ is not so (see Example 3.2.3(3)). Koh and Teo (1991a), on the other hand, proved that the graph $W(n, 5)$ is χ-unique if $n \geq 8$. Answering Problem 2 in Koh and Teo (1990), Li and Whitehead Jr. (1993) proceeded one step further to show that the graph $W(n, 6)$ is χ-unique for all $n \geq 8$. The graph $W(7, 5)$ is also a graph of the form $W(n, n-2)$. Chia (1990) proved that the graph $W(n, n-2)$ is χ-unique for all even $n \geq 6$. Later, Chia (1996) and Dong, Liu and Koh (1997) both showed that $W(n, n-2)$ is also χ-unique for odd $n \geq 9$ via different approaches. Dong and Liu (1998) further proved that the graph $W(n, n-3)$ is χ-unique for all $n \geq 6$. The above problem for the remaining case was eventually solved by Dong, Koh and Teo (2001). In Section 10.4, we shall introduce their proof, which makes use of the results developed in Section 10.3, in settling the problem.

10.2   3-colourable graphs

Let G be a graph of order n. By Theorem 1.4.1,

$$P(G, \lambda) = \sum_{k=1}^{n} \alpha(G, k)(\lambda)_k, \tag{10.1}$$

where $\alpha(G, k)$ is the number of ways of partitioning $V(G)$ into k independent sets. For any $r \in \mathbb{N}$, define

$$s_r(G) = P(G, r)/r!. \tag{10.2}$$

It is obvious that $s_r(G) = 0$ if $r < \chi(G)$. For $r \geq \chi(G)$, it can be verified by (10.1) (see Exercise 10.1) that

$$s_r(G) = \alpha(G, r) + \alpha(G, r-1) + \sum_{k=\chi(G)}^{r-2} \frac{\alpha(G, k)}{(r-k)!}. \tag{10.3}$$

The next result then follows immediately.

Lemma 10.2.1   *Let G be a graph and $r \in \mathbb{N}$ with $\chi(G) \leq r \leq \chi(G) + 1$. Then $s_r(G) \in \mathbb{N}$.*   □

By (10.2), if G is uniquely r-colourable, then $s_r(G) = 1$. The converse is not true; for instance, $s_r(K_{r-1}) = 1$ but K_{r-1} is not uniquely r-colourable. In general, we have (see Exercise 1.14):

Lemma 10.2.2   *Let G be a graph and $r \in \mathbb{N}$ with $r \leq v(G) + 1$. Then $s_r(G) = 1$ if and only if either $G \cong K_{r-1}$ or G is uniquely r-colourable.*   □

Our next result follows directly from FRT and the definition of $s_r(G)$.

Lemma 10.2.3   *Let G be a graph and $r \in \mathbb{N}$. Then for any two non-adjacent vertices u and v in G,*

$$s_r(G) = s_r(G + uv) + s_r(G \cdot uv).$$ □

We shall now confine our consideration to 3-colourable graphs. We first introduce the following result (see Exercise 10.2).

Lemma 10.2.4   *Let G be a 3-colourable connected graph with $e(G) \leq 2v(G) - 4$. Then there exist two non-adjacent vertices u and v in G such that both $G + uv$ and $G \cdot uv$ are 3-colourable and $N(u) \cap N(v) \neq \emptyset$.*   □

We next establish a lower bound of $s_3(G)$ for the family of 3-colourable connected graphs G.

Theorem 10.2.1   *Let G be a 3-colourable connected graph with $e(G) = 2v(G) - k$, where $k \geq 3$. Then $s_3(G) \geq 2^{k-3}$.*

Proof.   Since G is 3-colourable, we have $s_3(G) \geq 1$ by (10.3). Thus the theorem holds for $k = 3$.

Assume that $k \geq 4$. Since $e(G) = 2v(G) - k \leq 2v(G) - 4$, by Lemma 10.2.4, G contains two non-adjacent vertices u and v such that both $G + uv$ and $G \cdot uv$ are 3-colourable and $N(u) \cap N(v) \neq \emptyset$. Observe that

$$e(G + uv) = e(G) + 1 = 2v(G + uv) - (k - 1).$$

Since $G + uv$ is a 3-colourable connected graph, by the induction hypothesis, $s_3(G + uv) \geq 2^{k-4}$.

We now assert that $s_3(G \cdot uv) \geq 2^{k-4}$. Since $N(u) \cap N(v) \neq \emptyset$, we have

$$e(G \cdot uv) \leq e(G) - 1 = 2v(G) - k - 1 = 2v(G \cdot uv) - (k - 1).$$

If $e(G \cdot uv) = 2v(G \cdot uv) - (k - 1)$, then $s_3(G \cdot uv) \geq 2^{k-4}$ by the induction hypothesis. Assume that $e(G \cdot uv) = 2v(G \cdot uv) - k'$, where $k' > k - 1 \geq 3$. By applying Lemma 10.2.4 repeatedly, a 3-colourable connected graph G' can be obtained from $G \cdot uv$ by adding $k' - (k - 1)$ edges. Since

$$e(G') = e(G \cdot uv) + k' - (k - 1) = 2v(G') - (k - 1),$$

by the induction hypothesis, $s_3(G') \geq 2^{k-4}$. Thus $s_3(G \cdot uv) \geq s_3(G') \geq 2^{k-4}$ by Lemma 10.2.3.

Finally, by Lemma 10.2.3, $s_3(G) = s_3(G + uv) + s_3(G \cdot uv) \geq 2^{k-3}$. $\square$

We next state a property of 3-colourable connected chordal graphs (see Exercise 10.4).

Lemma 10.2.5   *Let G be a 3-colourable connected chordal graph. If $e(G) = 2v(G) - k$, where $k \geq 3$, then $t(G) = v(G) - k + 1$.* $\square$

Now we consider the case when G is non-chordal.

Theorem 10.2.2   *Let G be a 3-colourable connected non-chordal graph with $e(G) = 2v(G) - k$, where $k \geq 3$. If $s_3(G) < 2^{k-2}$, then $t(G) \leq v(G) - k$.*

Proof. Let $k = 3$. Then $s_3(G) < 2^{k-2} = 2$, and so $s_3(G) = 1$. Since G is non-chordal, G is uniquely 3-colourable by Lemma 10.2.2. Thus, by Corollary 6.2.2, $t(G) \le v(G) - 2$. Further, by Theorem 6.5.2, if $t(G) = v(G) - 2$, then G is a 2-tree, a contradiction. Hence $t(G) \le t(G) - 3$ and the result holds when $k = 3$.

Assume that $k \ge 4$. By Lemma 10.2.4, there exist two non-adjacent vertices u and v in G such that both $G + uv$ and $G \cdot uv$ are 3-colourable and $N(u) \cap N(v) \ne \emptyset$. By Lemma 10.2.3,

$$2^{k-2} > s_3(G) = s_3(G + uv) + s_3(G \cdot uv).$$

Thus either $s_3(G + uv) < 2^{k-3}$ or $s_3(G \cdot uv) < 2^{k-3}$.

Case 1: $s_3(G + uv) < 2^{k-3}$.

Since $N(u) \cap N(v) \ne \emptyset$, $t(G) \le t(G + uv) - 1$. Observe that

$$e(G + uv) = e(G) + 1 = 2v(G + uv) - (k - 1).$$

If $G + uv$ is non-chordal, then by the induction hypothesis, $t(G + uv) \le v(G + uv) - (k - 1)$, implying that

$$t(G) \le t(G + uv) - 1 \le v(G + uv) - (k - 1) - 1 = v(G) - k.$$

If $G+uv$ is chordal, then $t(G+uv) = v(G+uv)-(k-1)+1$ by Lemma 10.2.5. As G is non-chordal and $G + uv$ is chordal, u and v must be two non-consecutive vertices of a pure cycle C_4 in G. Thus

$$t(G) \le t(G + uv) - 2 = v(G) - k.$$

Case 2: $s_3(G \cdot uv) < 2^{k-3}$.

Since $N(u) \cap N(v) \ne \emptyset$, $e(G \cdot uv) \le e(G) - 1$. Thus

$$e(G \cdot uv) \le e(G) - 1 = 2v(G) - k - 1 = 2v(G \cdot uv) - (k - 1).$$

If $e(G \cdot uv) < 2v(G \cdot uv) - (k-1)$, then $s_3(G \cdot uv) \ge 2^{k-3}$ by Theorem 10.2.1, a contradiction. Hence $e(G \cdot uv) = 2v(G \cdot uv) - (k - 1)$, which implies that $e(G \cdot uv) = e(G) - 1$; i.e., $|N(u) \cap N(v)| = 1$. If $G \cdot uv$ is non-chordal, then by the induction hypothesis,

$$t(G \cdot uv) \le v(G \cdot uv) - (k - 1) = v(G) - k.$$

As $|N(u) \cap N(v)| = 1$, we have $t(G) \le t(G \cdot uv) = v(G) - k$.

Now assume that $G \cdot uv$ is chordal. By Lemma 10.2.5,

$$t(G \cdot uv) = v(G \cdot uv) - (k - 1) + 1 = v(G) - k + 1.$$

Let C be a pure cycle in G. Since $G \cdot uv$ is chordal, we have $u, v \in V(C)$ and $|V(C)| \le 6$. Since $e(G) = e(G \cdot uv) + 1$ and $uv \notin E(G)$, $|V(C)| \ge 5$. Thus $5 \le |V(C)| \le 6$, implying that

$$t(G) \le t(G \cdot uv) - 1 = v(G) - k.$$

□

It is not hard to prove the following result (see Exercise 10.3).

Lemma 10.2.6   *Let G be a 3-colourable 2-connected graph with*

$$e(G) \le 2v(G) - 4.$$

Then G is not a chordal graph.

□

By Lemma 10.2.6 and Theorem 10.2.2, the next result, due to Dong and Koh (1999), follows directly.

Corollary 10.2.1   *Let G be a 3-colourable 2-connected graph with $e(G) = 2v(G) - k$, where $k \ge 4$. If $s_3(G) < 2^{k-2}$, then $t(G) \le v(G) - k$.*

□

10.3   A family of 3-colourable graphs

Let G be a 3-colourable connected graph with $e(G) = 2v(G) - k$, where $k \ge 3$. In Section 10.2, we have showed that $s_3(G) \ge 2^{k-3}$, and further, if $s_3(G) < 2^{k-2}$, then $t(G) \le v(G) - k + 1$, where equality holds if and only if G is chordal. In this section, we shall characterize the structure of G when $s_3(G) < 2^{k-2}$ and $t(G) = v(G) - k$.

We shall first establish some results which are useful towards our goal. Let $\mathcal{C}_3$ be the family of 3-colourable connected chordal graphs. We first have the following (see Exercise 10.5):

Lemma 10.3.1   *For $G \in \mathcal{C}_3$, every block of G is a 2-tree.*

□

Define

$$\mathcal{Y} = \{G \cdot uv : G \in \mathcal{C}_3, d_G(u, v) \ge 4, u, v \in V(G)\}.$$

By definition, every graph in $\mathcal{Y}$ is non-chordal and connected. Observe that $\mathcal{Y}$ can be partitioned into $\mathcal{Y}_1, \mathcal{Y}_2, \cdots$, where for $k \in \mathbb{N}$,

$$\mathcal{Y}_k = \{G \cdot uv : G \in \mathcal{C}_3, b(G) = k, d_G(u, v) \ge 4, u, v \in V(G)\}.$$

The next result gives an equivalent statement of '$d_G(u, v) \ge 4$' (see Exercise 10.6).

Lemma 10.3.2   *Let G be any graph and $u, v \in V(G)$. Then $e(G \cdot uv) = e(G)$ and $t(G \cdot uv) = t(G)$ if and only if $d_G(u, v) \geq 4$.* $\square$

By Lemma 10.3.2, if $G \in \mathcal{C}_3$ and $u, v \in V(G)$ such that $e(G \cdot uv) = e(G)$ and $t(G \cdot uv) = t(G)$, then $G \cdot uv \in \mathcal{Y}$.

Lemma 10.3.3   *Let $H \in \mathcal{Y}_k$, where $k \in \mathbb{N}$. Then $e(H) = 2v(H) - k$ and $t(H) = v(H) - k$.*

Proof.   Let $H = G \cdot uv$, where $G \in \mathcal{C}_3$ and $u, v \in V(G)$ with $d_G(u, v) \geq 4$. Observe that for every 2-tree T, $e(T) = 2v(T) - 3$. Let $G_1, G_2, \cdots, G_k$ be the blocks of G. Then

$$e(G) = \sum_{i=1}^{k} e(G_i) = \sum_{i=1}^{k} (2v(G_i) - 3) = 2(v(G) + k - 1) - 3k = 2v(G) - k - 2.$$

By Lemma 10.2.5, $t(G) = v(G) - k - 1$.
By Lemma 10.3.2, $e(H) = e(G)$ and $t(H) = t(G)$. Thus

$$e(H) = e(G) = 2v(G) - k - 2 = 2v(H) - k$$

and

$$t(H) = t(G) = v(G) - k - 1 = v(H) - k.$$ $\square$

Lemma 10.3.4   *Let H be a graph and $u, v \in V(H)$ such that $uv \notin E(H)$. If $H + uv$ is chordal and $t(H + uv) \leq t(H) + 1$, then H is also chordal.*

Proof.   Suppose that H is not chordal and C is a pure cycle of H. Since $H + uv$ is chordal, both u and v are on C. If $v(C) \geq 5$, then $C + uv$ has a pure cycle in $H + uv$, a contradiction. If $v(C) = 4$, then u, v are not consecutive on C, and thus $t(H + uv) \geq t(H) + 2$, a contradiction. $\square$

Lemma 10.3.5   *Let G be a graph and $u, v \in V(G)$ such that $uv \notin E(G)$. If $G + uv \in \mathcal{Y}$ and $t(G + uv) = t(G) + 1$, then $G \in \mathcal{Y}$.*

Proof.   Since $G + uv$ is connected and $t(G + uv) = t(G) + 1$, G is also connected.

By the definition of $\mathcal{Y}$, $G + uv = H \cdot xy$ for some $H \in \mathcal{C}_3$ and $x, y \in V(H)$ with $d_H(x, y) \geq 4$. Since $d_H(x, y) \geq 4$, $G = H \cdot xy - uv = (H - uv) \cdot xy$. It is clear that $d_{H-uv}(x, y) \geq d_H(x, y) \geq 4$. So $t(H - uv) = t((H - uv) \cdot xy)$ by Lemma 10.3.2. Thus

$$t(H) = t(H \cdot xy) = t(G + uv) = t(G) + 1 = t((H - uv) \cdot xy) + 1 = t(H - uv) + 1.$$

Since H is a chordal graph and $t(H) = t(H - uv) + 1$, by Lemma 10.3.4, $H - uv$ is also chordal. Further, $H \in C_3$ implies that $H - uv$ is 3-colourable. Thus, $H - uv \in C_3$. As $G = (H - uv) \cdot xy$, we have $G \in \mathcal{Y}$. $\qquad\square$

For any connected graph G, the *block-graph* of G is defined to be the graph with vertex set $V = \{B : B \text{ is a block of } G\}$ and edge set

$$E = \{B_1 B_2 : B_1 \text{ and } B_2 \text{ contain a common vertex in } G\}.$$

It is clear that the block-graph of G is a chordal graph. (See Exercise 10.12.)

For $k \in \mathbb{N}$, let $\mathcal{X}_k$ be the family of 2-connected graphs in $\mathcal{Y}_k$. For any $G \in C_3$ and $u, v \in V(G)$ with $d_G(u, v) \geq 4$, if $G \cdot uv \in \mathcal{X}_k$, then the block-graph of G must be the path P_k, and u and v belong to the two blocks corresponding to the end vertices of this path. The structures of graphs in $\mathcal{X}_k$ for $k = 1$ and $k \geq 2$ are shown in Figure 10.2 (a) and (b) respectively.

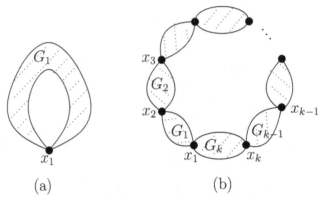

(a)   (b)

Figure 10.2

Let

$$\mathcal{X} = \bigcup_{k \geq 1} \mathcal{X}_k.$$

By definition, every graph of $\mathcal{Y}$ contains only one non-chordal block (all other blocks are 2-trees); and this block, as a graph, is a member in $\mathcal{X}$. For any $G \in \mathcal{Y}$, let $b_c(G)$ be the integer s such that the only non-chordal block of G belongs to $\mathcal{X}_s$. So $b_c(G) = k$ for every graph $G \in \mathcal{X}_k$. Hence we have:

Lemma 10.3.6   *Let $G \in \mathcal{Y}$. If $b_c(G) \geq k$, then there exist k distinct vertices $x_1, x_2, \cdots, x_k$ in G and two components $B_{i,1}$ and $B_{i,2}$ of $[N(x_i)]$ for $i = 1, 2, \cdots, k$ such that for every pure cycle C in G,*

$$|E(C) \cap \{x_i u : u \in V(B_{i,j})\}| = 1$$

for all $i = 1, 2, \cdots, k$ and $j = 1, 2.$ $\qquad\square$

Lemma 10.3.7   *Let G be a connected non-chordal graph with $\omega(G) \leq 3$. Then $G \in \mathcal{Y}$ if and only if there exist a vertex $x \in V(G)$ and a component B of $[N(x)]$ such that*

$$|E(C) \cap \{xu : u \in V(B)\}| = 1$$

for every pure cycle C in G.

Proof. [Necessity]. We need only consider the case that G is 2-connected. Then G is a graph of the form in Figure 10.2. Let $x = x_1$, and B_1 and B_2 be the blocks of $[N(x)]$. It is clear that every pure cycle of G contains precisely one edge xz_1 and one edge xz_2, where $z_i \in V(B_i)$ for $i = 1, 2$. So the necessity holds.

[Sufficiency]. Since G is non-chordal and $|E(C) \cap \{xu : u \in V(B)\}| = 1$ for every pure cycle C of G, we have $B' = N(x) \backslash B \neq \emptyset$. Let H be the graph obtained from $G - x$ by adding two vertices x_1 and x_2 and adding edges joining x_1 to every vertex in B and joining x_2 to every vertex in B'. So $G = H \cdot x_1 x_2$, $e(G) = e(H \cdot x_1 x_2)$ and $t(G) = t(H \cdot x_1 x_2)$, implying that $d_H(x_1, x_2) \geq 4$.

We now show that H is chordal. Suppose that C is a pure cycle of H. Since $d_H(x_1, x_2) \geq 4$, C is also a pure cycle of G. Thus C contains precisely one edge in $\{xu : u \in V(B)\}$. Hence $x_i \in V(C)$ for some $i = 1, 2$. If $x_1 \in V(C)$, then C contains two edges in $\{xu : u \in V(B)\}$, a contradiction; if $x_1 \notin V(C)$, then C contains no edges in $\{xu : u \in V(B)\}$, again a contradiction. Therefore H contains no pure cycle.

Since H is chordal, $G = H \cdot x_1 x_2$ and G is non-chordal, we conclude that H is connected. Thus $H \in \mathcal{C}_3$ by the assumption that $\omega(H) \leq 3$. Further, since $d_H(x_1, x_2) \geq 4$, we have $H \in \mathcal{Y}$. $\qquad\square$

Corollary 10.3.1   *Let G be a connected graph with $\omega(G) \leq 3$. If G contains exactly one pure cycle, then $G \in \mathcal{Y}$.* $\qquad\square$

Lemma 10.3.8   *Let G be a connected graph and $u, v \in V(G)$ such that $uv \notin E(G)$. If $G \cdot uv \in \mathcal{Y}$ with $b_c(G \cdot uv) \geq 2$, $e(G \cdot uv) = e(G) - 1$ and $t(G \cdot uv) = t(G)$, then $G \in \mathcal{Y}$.*

Proof. We first show that G is not chordal. Since $G \cdot uv \in \mathcal{Y}$, by Lemma 10.3.3,

$$v(G \cdot uv) + t(G \cdot uv) = e(G \cdot uv),$$

implying that

$$v(G) - 1 + t(G) = e(G) - 1;$$

i.e., $v(G) + t(G) = e(G)$. Since $G \cdot uv \in \mathcal{Y}$, we have $\omega(G \cdot uv) \leq 3$ and so $\omega(G) \leq 3$. If G is chordal, then $G \in \mathcal{C}_3$, and thus, by Lemma 10.2.5,

$$v(G) + t(G) = e(G) + 1,$$

a contradiction.

Let y be the vertex in $G \cdot uv$ resulting from the identification of u and v in G. Since $G \cdot uv \in \mathcal{Y}$ with $b_c(G) \geq 2$, by Lemma 10.3.6, there exist a vertex $x \in V(G \cdot uv)$ and a component B of $[N_{G \cdot uv}(x)]$ such that $y \notin \{x\} \cup V(B)$ and every pure cycle of $G \cdot uv$ contains precisely one edge in $\{xz : z \in V(B)\}$.

It is clear that $x \in V(G)$ and B is also a component of $[N_G(x)]$. We shall now prove that every pure cycle of G contains precisely one edge in $\{xz : z \in V(B)\}$.

Suppose that C is a pure cycle of G such that C has no edge or two edges in $\{xz : z \in V(B)\}$. If $\{u, v\} \not\subseteq V(C)$, then C is also a pure cycle of $G \cdot uv$, which contradicts the condition that every pure cycle of $G \cdot uv$ contains precisely one edge in $\{xz : z \in V(B)\}$. Thus $\{u, v\} \subseteq V(C)$. Since $e(G \cdot uv) = e(G) - 1$ and $t(G \cdot uv) = t(G)$, we have $v(C) \geq 6$. Hence there exists a pure cycle C' in the subgraph $C \cdot uv$. So C' contains precisely one edge in $\{xz : z \in V(B)\}$, implying that $x \in V(C')$. Therefore C contains two edges xz_1 and xz_2 in $\{xz : z \in V(B)\}$. Since $y \neq x$, we have $u \neq x$ and $v \neq x$, implying that $xz_1, xz_2 \in E(C')$, a contradiction. This shows that every pure cycle of G contains precisely one edge in $\{xz : z \in V(B)\}$. By Lemma 10.3.7, $G \in \mathcal{Y}$. $\qquad\square$

Lemma 10.3.9　　Let $H \in \mathcal{Y}$. If H is 3-colourable and $b_c(H) \leq 2$, then $s_3(H) \geq 2^{2v(H) - e(H) - 2}$.

Proof.　We first consider the case that H has no simplicial vertex. Then H is 2-connected and thus $H \in \mathcal{X}_s$ for some s. So $s = b_c(H)$. By Lemma 10.3.3, $e(H) = 2v(H) - s$. Hence

$$2v(H) - e(H) - 2 = s - 2 \leq 0$$

if $s \leq 2$. Since H is 3-colourable, $s_3(H) \geq 1 \geq 2^{2v(H) - e(H) - 2}$.

Now let x be a simplicial vertex of H. Then $d(x) \in \{1, 2\}$ and

$$P(H, \lambda) = (\lambda - d(x))P(H - x, \lambda).$$

So

$$s_3(H) = (3 - d(x))s_3(H - x).$$

By the induction hypothesis, $s_3(H - x) \geq 2^{2v(H-x)-e(H-x)-2}$. If $d(x) = 1$, then

$$s_3(H) = 2s_3(H - x) \geq 2 \times 2^{2v(H-x)-e(H-x)-2} = 2^{2v(H)-e(H)-2}.$$

If $d(x) = 2$, then

$$s_3(H) = s_3(H - x) \geq 2^{2v(H-x)-e(H-x)-2} = 2^{2v(H)-e(H)-2}.$$

Hence the result holds. □

Theorem 10.3.1 *Let G be a 3-colourable connected graph with $e(G) = 2v(G) - k$, $t(G) = v(G) - k$ and $s_3(G) < 2^{k-2}$, where $k \geq 3$. Then $G \in \mathcal{Y}_k$.*

Proof. We first note that if $G \in \mathcal{Y}$, then as $e(G) = 2v(G) - k$, we have $G \in \mathcal{Y}_k$ by Lemma 10.3.3.

We shall prove that $G \in \mathcal{Y}_k$ by induction on k. For $k = 3$, we have $s_3(G) = 1$. Since $t(G) = v(G) - 3$, $G \ncong K_2$. Hence G is uniquely 3-colourable. By Theorem 7.3.1, $G \in \mathcal{X}_3 \subseteq \mathcal{Y}_3$.

Assume that $k \geq 4$. By Lemma 10.2.4, there exist two non-adjacent vertices u and v in G such that both $G + uv$ and $G \cdot uv$ are 3-colourable and $N(u) \cap N(v) \neq \emptyset$. By Lemma 10.2.3,

$$2^{k-2} > s_3(G) = s_3(G + uv) + s_3(G \cdot uv).$$

Thus either $s_3(G + uv) < 2^{k-3}$ or $s_3(G \cdot uv) < 2^{k-3}$.
Case 1: $s_3(G + uv) < 2^{k-3}$.

Observe that

$$e(G + uv) = e(G) + 1 = 2v(G) - (k - 1) = 2v(G + uv) - (k - 1).$$

Assume that $G + uv$ is chordal. Since G is non-chordal, G contains pure cycles of length 4 only, and every pure cycle of G contains both u and v. By Lemma 10.2.5,

$$t(G + uv) = v(G + uv) - (k - 1) + 1 = v(G) - k + 2 = t(G) + 2.$$

So $|N(u) \cap N(v)| = 2$ and G contains only one pure cycle, implying that $G \in \mathcal{Y}$ by Corollary 10.3.1.

We now consider the case that $G + uv$ is non-chordal. Since $N(u) \cap N(v) \neq \emptyset$,

$$t(G + uv) \geq t(G) + 1 = v(G) - k + 1 = v(G + uv) - (k - 1).$$

By Theorem 10.2.2, $t(G + uv) \leq v(G + uv) - (k - 1)$. Hence $t(G + uv) = v(G + uv) - (k - 1)$, which implies that $t(G + uv) = t(G) + 1$. By the induction hypothesis, $G + uv \in \mathcal{Y}$. Thus $G \in \mathcal{Y}$ by Lemma 10.3.5.

Case 2: $s_3(G \cdot uv) < 2^{k-3}$.

Since $N(u) \cap N(v) \neq \emptyset$,

$$e(G \cdot uv) \leq e(G) - 1 = 2v(G) - k - 1 = 2v(G \cdot uv) - (k - 1).$$

If $e(G \cdot uv) \leq 2v(G \cdot uv) - (k - 1) - 1$, then $s_3(G \cdot uv) \geq 2^{k-3}$ by Theorem 10.2.1, a contradiction. Hence $e(G \cdot uv) = 2v(G \cdot uv) - (k - 1)$, implying that $e(G \cdot uv) = e(G) - 1$ and $|N(u) \cap N(v)| = 1$.

Assume that $G \cdot uv$ is chordal. Since G is non-chordal, G contains no pure $u - v$ paths longer than 3. Since $|N(u) \cap N(v)| = 1$, G contains only one $u - v$ path of length 2. By Lemma 10.2.5,

$$t(G \cdot uv) = v(G \cdot uv) - (k - 1) + 1 = v(G) - k + 1 = t(G) + 1,$$

implying that G contains only one pure $u - v$ path of length 3. Thus G contains only one pure cycle, which is of length 5. Hence $G \in \mathcal{Y}$ by Corollary 10.3.1.

We now assume that $G \cdot uv$ is non-chordal. Since $|N(u) \cap N(v)| = 1$,

$$t(G \cdot uv) \geq t(G) = v(G) - k = v(G \cdot uv) - (k - 1).$$

But, by Theorem 10.2.2, $t(G \cdot uv) \leq v(G \cdot uv) - (k - 1)$. Hence $t(G \cdot uv) = t(G) = v(G \cdot uv) - (k - 1)$. By the induction hypothesis, $G \cdot uv \in \mathcal{Y}$.

Assume that $b_c(G \cdot uv) = s$. Since

$$s_3(G \cdot uv) < 2^{k-3} = 2^{2v(G \cdot uv) - e(G \cdot uv) - 2},$$

by Lemma 10.3.9, $s \geq 3$. Thus, by Lemma 10.3.8, $G \in \mathcal{Y}$. $\qquad\square$

Corollary 10.3.2 *Let G be a 3-colourable 2-connected graph with*

$$e(G) = 2v(G) - k, \ t(G) = v(G) - k \ and \ s_3(G) < 2^{k-2},$$

where $k \geq 3$. Then $G \in \mathcal{X}_k$. $\qquad\square$

In the following, we shall see that Theorem 10.3.1 is no longer true if the condition '$s_3(G) < 2^{k-2}$' is relaxed to '$s_3(G) \leq 2^{k-2}$'.

First, we state the following result due to Dong and Koh (1999).

Lemma 10.3.10 Let $G \in \mathcal{X}_k$, where $k \geq 2$, as shown in Figure 10.2, and let t be the number of G_i's such that both x_i and x_{i+1} are in the same colour class of any 3-colouring of G_i.
(i) If $t = k - 1$, then $\chi(G) \geq 4$.
(ii) If $t \neq k - 1$, then $s_3(G) = (2^{k-1} + (-1)^{k-t}2^t)/3$. □

Corollary 10.3.3 Let $G \in \mathcal{X}_k$, as shown in Figure 10.2, where $k \geq 3$. Then $1 \leq s_3(G) < 2^{k-2}$ if and only if there are at most $k - 3$ G_i's such that both x_i and x_{i+1} are in the same colour class of any 3-colouring of G_i. □

Let $q \in \mathbb{N}$ with $q \geq 3$. By Lemma 10.3.10, there exists $G \in \mathcal{X}_{q+1}$ such that $s_3(G) = 2^{q-2}$ (letting $t = q - 2$). Without loss of generality, assume that x_1 and x_2 are in different colour classes of any 3-colouring of G_1. Let G' be the graph obtained from G by replacing G_1 by a wheel W_s with odd $s \geq 5$ such that x_1 and x_2 are in different colour classes of any 3-colouring of W_s. Observe that G' is 2-connected, 3-colourable and

$$s_3(G') = s_3(G) = 2^{q-2},$$
$$e(G') = 2v(G') - q,$$
$$t(G') = v(G') - q.$$

But $G' \notin \mathcal{X}$.

10.4 Chromaticity of graphs in $\mathcal{X}_k$

In this section, we shall apply Theorem 10.3.1 to solve Problem 10.1.1. We first have:

Theorem 10.4.1 Let $k \in \mathbb{N}$ with $k \geq 3$. Then each of the following families is χ-closed:

$$\{G \in \mathcal{Y}_k : 2^{k-3} \leq s_3(G) < 2^{k-2}\},$$
$$\{G \in \mathcal{X}_k : 2^{k-3} \leq s_3(G) < 2^{k-2}\}.$$

Proof. Since $\{G \in \mathcal{X}_k : 2^{k-3} \leq s_3(G) < 2^{k-2}\}$ is the family of 2-connected graphs in $\{G \in \mathcal{Y}_k : 2^{k-3} \leq s_3(G) < 2^{k-2}\}$, we need only to show that the latter one is χ-closed.

Let $G \in \mathcal{Y}_k$ such that $2^{k-3} \leq s_3(G) < 2^{k-2}$. As $s_3(G) > 0$, G is 3-colourable. Since $G \in \mathcal{Y}_k$, by Lemma 10.3.3,

$$e(G) = 2v(G) - k \text{ and } t(G) = v(G) - k.$$

Now let H be a graph such that $H \sim G$. Observe that

$$v(H) = v(G),$$
$$e(H) = e(G) = 2v(G) - k = 2v(H) - k$$
$$\text{and} \quad t(H) = t(G) = v(G) - k = v(H) - k.$$

As $H \sim G$, H is connected and $s_3(H) = s_3(G)$. By Theorem 10.3.1, $H \in \mathcal{Y}_k$. Hence the family $\{G \in \mathcal{Y}_k : 2^{k-3} \leq s_3(G) < 2^{k-2}\}$ is χ-closed. $\square$

We aim to prove the following result.

Theorem 10.4.2   *For any $n, s \in \mathbb{N}$ with $n \geq 5$ and $2 \leq s \leq n - 3$, the graph $W(n, s)$ is χ-unique.*

For any graph G in $\mathcal{X}$ (see Figure 10.2.), let $g_p(G)$ denote the length of a shortest pure cycle in G and $n'_r(G)$ the number of pure cycles of length r in G. Then for $G \in \mathcal{X}_k$,

$$g_p(G) = \sum_{i=1}^{k} d_{G_i}(x_i, x_{i+1}).$$

We shall show that $g_p(G)$ and $n'_r(G)$, where $r = g_p(G)$, are χ-invariants for graphs in the χ-equivalence class $[G]$, where $G \in \mathcal{X}_k$ with $k \geq 3$ and $1 \leq s_3(G) < 2^{k-2}$.

The chromatic polynomial of each member of $\mathcal{C}_3$ is given below (see Exercise 10.7).

Lemma 10.4.1   *Let $G \in \mathcal{C}_3$. Then*

$$P(G, \lambda) = \lambda(\lambda - 1)^{v(G)-t(G)-1}(\lambda - 2)^{t(G)}. \qquad \square$$

We proceed to prove the following:

Lemma 10.4.2   *Let $G, H \in \mathcal{X}$. If $G \sim H$, then*
(i) $G, H \in \mathcal{X}_k$ for some $k \in \mathbb{N}$;
(ii) $g_p(G) = g_p(H)$;
(iii) $n'_r(G) = n'_r(H)$, where $r = g_p(G)$.

Proof. (i) By Lemma 10.3.3, $G \in \mathcal{X}_k$ for $k = 2v(G) - e(G)$. Since $H \sim G$, we have $v(H) = v(G)$ and $e(H) = e(G)$. Hence $H \in \mathcal{X}_k$ too, by Lemma 10.3.3 again.

(ii) and (iii)   Assume that $G \in \mathcal{X}_k$, where $k \geq 1$. By the definition of $\mathcal{X}$, $G = G' \cdot uv$ for some graph $G' \in \mathcal{C}_3$ and $x, y \in V(G')$. By Lemma 10.3.3, $t(G') = t(G) = n - k$, where $n = v(G)$. Thus, by Lemma 10.4.1,

$$P(G', \lambda) = \lambda(\lambda - 1)^k (\lambda - 2)^{n-k}.$$

Suppose that

$$P(G, \lambda) = \sum_{i=1}^{n} (-1)^{n-i} h_i \lambda^i$$

and

$$P(G', \lambda) = \sum_{i=1}^{n+1} (-1)^{n+1-i} a_i \lambda^i.$$

By Theorem 2.3.1 and by comparing the numbers of broken cycles in G and G', we have

$$h_{n-i} = \begin{cases} a_{n-i} & \text{if } 0 \leq i \leq r - 2, \\ a_{n-i} - n_r'(G) & \text{if } i = r - 1, \end{cases}$$

where $r = g_p(G)$. Thus $g_p(G)$ $(= r)$ and $n_r'(G)$ are completely determined by the coefficients of $P(G, \lambda)$ and $P(G', \lambda)$. But $P(G', \lambda)$ is completely determined by n and k $(= 2v(G) - e(G))$. Thus $g_p(G)$ and $n_r'(G)$ are completely determined by $P(G, \lambda)$.

Since $H \sim G$, we have $g_p(G) = g_p(H)$ and $n_r'(G) = n_r'(H)$, where $r = g_p(G)$. □

It is not hard to verify the following result (see Exercise 1.9(a)).

Lemma 10.4.3   *For $n, s \in \mathbb{N}$ with $n \geq 5$ and $1 \leq s \leq n - 2$,*

$$P(W(n, s), \lambda) = (\lambda - 2)^{s-1} \left((\lambda - 1)^{n-s+1} + (-1)^{n-s} \right) + (-1)^{n-1} \lambda(\lambda - 2).$$

□

By Lemmas 10.3.3 and 10.4.3, we have the following result (see Exercise 10.10).

Lemma 10.4.4   *For $n, s \in \mathbb{N}$ with $n \geq 5$ and $2 \leq s \leq n-3$, $W(n, s) \in \mathcal{X}_k$ and $s_3(W(n, s)) < 2^{k-2}$, where $k = n - s + 1$.*

□

We need one more result (see Exercise 2.2) before proving Theorem 10.4.2.

Lemma 10.4.5   *Let G be a graph, and G' and G'' be two induced subgraphs of G such that $V(G') \cup V(G'') = V(G)$ and $V(G') \cap V(G'') = \{x, y\}$. If $\chi(G') \geq 3$ and $\chi(G'') \geq 3$, then $(\lambda - 2)^2 | P(G, \lambda)$.*   □

We are now ready to establish our main result.

Proof of Theorem 10.4.2:   By Lemma 10.4.4, $W(n, s) \in \mathcal{X}_k$, where $k = n - s + 1 \geq 4$, and $s_3(W(n, s)) < 2^{k-2}$. Let G be a graph such that $G \sim W(n, s)$. Then $G \in \mathcal{X}_k$ by Theorem 10.4.1. Assume G is a graph in Figure 10.2(b).

Suppose that $s = 2$. We have $k = n - s + 1 = n - 1$. As G is of order n, one G_i is a 2-tree of order 3 and all others are the 2-tree of order 2(i.e., K_2). Thus $G \cong W(n, 2)$. Hence the theorem holds for $s = 2$. In the following, we assume that $s \geq 3$.

Since $s \geq 3$, by Lemma 10.4.3, $(\lambda - 2)^2$ is not a factor of $P(W(n, s), \lambda)$, which implies that $(\lambda - 2)^2$ is not a factor of $P(G, \lambda)$. By Lemma 10.4.5, $v(G_i) = 2$ for all G_i but one. Assume that $v(G_1) \geq 3$. Then $G_i \cong K_2$ for $i = 2, 3, \cdots, n$. (See Figure 10.3.)

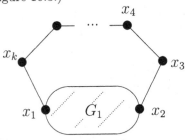

Figure 10.3

By Lemma 10.4.2, $g_p(G) = g_p(W(n, s)) = n - s + 2 = k + 1$, which implies that $d_{G_1}(x_1, x_2) = 2$.

We assert that G contains no simplicial vertices. Suppose that x is a simplicial vertex in G. Since G is 2-connected and 3-colourable, we have $d(x) = 2$. Thus

$$P(G, \lambda) = (\lambda - 2)P(G - x, \lambda).$$

Since

$$t(G - x) = t(G) - 1 = v(G) - k - 1 = n - (n - s + 1) - 1 = s - 2 \geq 1,$$

we have $\chi(G - x) \geq 3$, which implies that $(\lambda - 2)$ is a factor of $P(G - x, \lambda)$. Thus $(\lambda - 2)^2$ is a factor of $P(G, \lambda)$, a contradiction. Therefore G contains no simplicial vertices.

Observe that $v(G_1) = n - (k - 2) = n - (n - s + 1) + 2 = s + 1 \geq 4$. Since G_1 is a 2-tree, G_1 contains at least two simplicial vertices. Observe that each simplicial vertex x of G_1 is a simplicial vertex of G if $x \notin \{x_1, x_2\}$. Since G contains no simplicial vertices, only x_1 and x_2 are simplicial vertices of G_1.

Now G_1 is a 2-tree with only two simplicial vertices x_1 and x_2 and

$$d_{G_1}(x_1, x_2) = 2.$$

It can be proved (see Exercise 10.8) by induction that $G_1 \cong P_s + K_1$ (see Figure 10.4). Hence $G \cong W(n, s)$, as was to be shown. $\qquad\square$

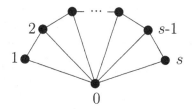

Figure 10.4

Finally, let us give some remarks on determining $[G]$, where $G \in \mathcal{X}_k$.

Remarks

(i) As shown in Figure 10.2, G has no simplicial vertex if and only if x_i and x_{i+1} are the only simplicial vertices of G_i for each $i = 1, 2, \cdots, k$.

(ii) Assume that G has a simplicial vertex x. Then $d(x) = 2$, $G - x \in \mathcal{X}_k$ and $\mathcal{G}[H \cup_2 K_3] \subseteq [G]$ for any $H \in [G - x]$. But it is not known whether every graph in $[G]$ contains a simplicial vertex. If every graph in $[G]$ contains a simplicial vertex, then

$$[G] = \bigcup_{H \in [G-x]} \mathcal{G}[H \cup_2 K_3].$$

(iii) Let G_i be a 2-tree and $x_{i,1}, x_{i,2}$ be any two distinct vertices in G_i for $i = 1, 2, \cdots, k$, where $k \geq 2$. Let G be the graph obtained from the union of $G_1, G_2, \cdots, G_k$ by identifying $x_{i,2}$ and $x_{i+1,1}$ for $i =$

$1, 2, \cdots, k$, where $x_{k+1,1} = x_{1,1}$. Then $G \in \mathcal{X}_k$ if $\sum\limits_{i=1}^{k} d_{G_i}(x_{i,1}, x_{i,2}) \geq 4$. Let $A = \{1, 2, \cdots, k\}$ and $B = \{i \in A : x_{i,1}x_{i,2} \in E(G_i)\}$. Then $P(G, \lambda)$ can be expressed as

$$\prod_{j \in B} \frac{P(G_j, \lambda)}{\lambda(\lambda - 1)} \sum_{I \subseteq A \setminus B} \left(P(C_{|I| + |B|}, \lambda) \left(\prod_{i \in I} \frac{P(G_i + x_{i,1}x_{i,2}, \lambda)}{\lambda(\lambda - 1)} \right) \right.$$
$$\left. \left(\prod_{i \in A \setminus (I \cup B)} \frac{P(G_i \cdot x_{i,1}x_{i,2}, \lambda)}{\lambda} \right) \right). \qquad (10.4)$$

(See Exercise 10.11.) So $P(G, \lambda)$ is completely independent of the order among $G_1, G_2, \cdots, G_k$ and the order between $x_{i,1}$ and $x_{i,2}$ for $i = 1, 2, \cdots, k$ when G is formed. The following problem thus arises: if G has no simplicial vertex and $H \sim G$, can H be obtained from G by applying the operation of interchanging $x_{i,1}$ and $x_{i,2}$ for any i and the operation of interchanging G_i and G_j for any i and j?

(iv) One may study the problem stated in Remark (iii) in the case that each G_i is any 2-tree with only two simplicial vertices $x_{i,1}$ and $x_{i,2}$ such that $d_{G_i}(x_{i,1}, x_{i,2}) = 2$ (i.e. $G_i = P_s + K_1$ for some $s \in \mathbb{N}$ with $s \neq 2$).

(v) As shown in Figure 10.2, let $G_i = K_2$ for $i = 2, 3, \cdots, k$, where $k \geq 4$, and $x_{1,1}$ and $x_{1,2}$ be the only simplicial vertices of G_1. If $d_{G_1}(x_{1,1}, x_{1,2}) = 2$, then G is a broken wheel $W(n, s)$ with $2 \leq s \leq n - 2$; and the problem of chromaticity of this graph has been settled. If $d_{G_1}(x_{1,1}, x_{1,2}) = 3$, it is not hard to show that G is also χ-unique. We wonder whether G is always χ-unique if $d_{G_1}(x_{1,1}, x_{1,2}) \geq 3$.

Exercise 10

10.1 Prove that for any graph G and any $r \in \mathbb{N}$,

$$s_r(G) = \alpha(G, r) + \alpha(G, r - 1) + \sum_{k=\chi(G)}^{r-2} \frac{\alpha(G, k)}{(r - k)!}.$$

10.2 Let G be a 3-colourable connected graph with $e(G) \leq 2v(G) - 4$. Show that G contains two non-adjacent vertices u and v such that both $G + uv$ and $G \cdot uv$ are 3-colourable and $N(u) \cap N(v) \neq \emptyset$.

10.3 Let G be a 3-colourable 2-connected graph with $e(G) \leq 2v(G) - 4$. Show that G is not a chordal graph.

10.4 Let G be a 3-colourable connected chordal graph. Show that

$$e(G) = 2v(G) - 2 - b(G) \quad \text{and} \quad t(G) = v(G) - 1 - b(G).$$

10.5 Let G be a 3-colourable chordal graph. Prove that if G is 2-connected, then G is a 2-tree.

10.6 Let G be any connected graph and $u, v \in V(G)$. Prove that $e(G \cdot uv) = e(G)$ and $t(G \cdot uv) = t(G)$ if and only if $d_G(u, v) \geq 4$.

10.7 Let G be a 3-colourable connected chordal graph. Prove that

$$P(G, \lambda) = \lambda(\lambda - 1)^{v(G)-t(G)-1}(\lambda - 2)^{t(G)}.$$

10.8 Prove that if G is a 2-tree with only two simplicial vertices x and y and $d_G(x, y) = 2$, then $G = P_s + K_1$ for some $s \geq 3$.

10.9 Prove that if G is uniquely 3-colourable, then G is connected and $e(G) \geq 2v(G) - 3$.

10.10 For $n, s \in \mathbb{N}$ with $n \geq 5$ and $2 \leq s \leq n - 3$, prove that $W(n, s) \in \mathcal{X}_k$ and $s_3(W(n, s)) < 2^{k-2}$, where $k = n - s + 1$.

10.11 Let G be the graph defined in Remark (iii) in page 211, $A = \{1, 2, \cdots, k\}$ and $B = \{i \in A : x_{i,1}x_{i,2} \in E(G_i)\}$. Prove that $P(G, \lambda)$ has the expression of (10.4).

10.12 Show that the block-graph of any connected graph is a chordal graph.

Chapter 11

Polynomials Related to Chromatic Polynomials

11.1 Introduction

In Chapter 4, we discuss the chromaticity of multi-partite graphs. We pay particular attention to the χ-uniqueness of bipartite graphs obtained from a complete bipartite graph by removing the edges of some subgraphs with simple structures. In this chapter, we focus our attention on complete graphs with some edges deleted. This topic was first considered by Xu (1983) and Salzberg (1984). By expanding the chromatic polynomials in factorial form (see Theorem 1.4.1), they found all χ-unique graphs determined by deleting five or fewer edges from any complete graph. This was later extended by Gracia and Salzberg (1985) to those with six edges removed.

Let H_1 and H_2 be subgraphs of K_n. Suppose $K_n - E(H_1)$ is χ-equivalent to $K_n - E(H_2)$. Then $K_{n+p} - E(H_1)$ is also χ-equivalent to $K_{n+p} - E(H_2)$, for all $p \in \mathbb{N}$. Consequently, for a subgraph H of K_n, if $K_n - E(H)$ is χ-unique, then so is $\overline{H}$. However, the converse is not true. We will see later that $\overline{P}_5$ is χ-unique while $K_n - P_5$ is not so for all $n > 5$. Note that if $v(H) = n$, then

$$K_{n+p} - E(H) = \overline{pK_1 \cup H} = K_p + \overline{H}.$$

Readers are referred to Section 9.6 for some χ-unique graphs of the form $K_p + G$, where G is χ-unique.

Let us recall Theorem 1.4.1, which says that

$$P(G, \lambda) = \sum_{k=1}^{n} \alpha(G, k)(\lambda)_k,$$

where $v(G) = n$ and $\alpha(G, k)$ is the number of ways of partitioning $V(G)$ into k independent sets. Frucht (1985) noticed that $\alpha(G, k)$ is simply the number of spanning subgraphs of $\overline{G}$ whose connected components are all complete graphs. Thus if $\overline{G}$ has a simpler structure than G, it may be easier to use Frucht's method for computing $\alpha(G, k)$. Indeed, Giudici (1985) had used this method to show that a χ-unique graph always results when the edges of a star $K(1, r)$ are deleted from a complete graph K_n with $n > r$. Later, Han (1986) showed that in most cases an extra edge can be removed to yield a χ-unique graph. The exceptional cases are when $r = 3$ or 4. Moreover, Han determined the χ-equivalence classes for these two exceptional cases. The chromaticity of two more complicated complements was studied by Frucht and Giudici (1983), and Li, Whitehead Jr. and Xu (1987). These include some vertex-gluings of two stars and $K(2, p)$, where $p \geq 2$.

The ξ-polynomial:

$$\xi(G, x) = \sum_{k=1}^{n} \alpha(G, k)x^k,$$

which we introduce in Section 4.3, was first used by Xu (1983) for notational convenience. This is related to the σ-polynomial introduced by Korfhage in 1978 as follows:

$$\sigma(G, x) = \xi(G, x)/x^{\chi(G)}.$$

Liu (1987) introduced the notion of the adjoint polynomial of a graph to study the chromaticity of the complements of some graphs with simple structures. The adjoint polynomial of a graph G is defined by

$$h(G, x) = \sum_{k=1}^{n} \alpha(\overline{G}, k)x^k.$$

We say that G is adjointly equivalent to H, symbolically $G \sim_h H$, if $h(G, x) = h(H, x)$. For each G, let $[G]_h = \{H : H \sim_h G\}$. A graph G is said to be adjointly unique if $[G]_h = \{G\}$. Evidently, we have the following result.

Theorem 11.1.1   *Let G and H be graphs. Then $G \sim H$ if and only if $\overline{G} \sim_h \overline{H}$. In particular, G is χ-unique if and only if $\overline{G}$ is adjointly unique.* $\square$

Consequently, the goal of determining the χ-equivalence class of G can be realized by determining the adjoint equivalence class of $\overline{G}$. Thus, as has been observed by many researchers (see Liu (1997)),if $e(G)$ is very large, it may be advantageous to study the latter. Indeed, for graphs with simple structures (e.g. C_n, P_n), it is possible to compute all the roots of their adjoint polynomials while this may not be possible for the chromatic polynomials of their complements.

Although some researchers, such as Du (1996b), Li, Whitehead Jr. and Xu (1987), have used σ-polynomials to study the chromaticity of some dense graphs, one disadvantage is that $\sigma(G, x)$ does not determine the order of G. This can be seen from the fact that $\sigma(\overline{G}, x) = \sigma(\overline{G \cup tK_1}, x)$ for any $t \in \mathbb{N}$. The adjoint polynomial does not have this fault, and it contains all the information that the σ-polynomial has.

In Section 11.2, we shall provide some basic properties of adjoint polynomials. In Section 11.3, we shall derive a recursive formula for $h(G, x)$. This formula will then be used to compute the adjoint polynomials of some "simple" graphs. In Section 11.4, we shall study the roots of some adjoint polynomials. Two invariants for adjointly equivalent graphs, which will be used to crudely classify graphs by their adjoint polynomials, will be introduced in Section 11.5. After establishing the adjoint uniqueness of some families of graphs in Section 11.6, we shall discuss, in the final section, the irreducibility of a polynomial related to the adjoint polynomial of a graph over the rational field; and report some results on the adjoint uniqueness of certain graphs under this premise.

11.2 Basic properties of adjoint polynomials

Let G be a graph of order n. If H is a spanning subgraph of G whose components are all complete graphs, then we say that H is a *clique cover* of G (Frucht (1985) called it a *special subgraph*, and Liu (1987) called it an *ideal subgraph* of G (see also Farrell (1993))). For $k \in \mathbb{N}$, let $N(G, k)$ be the number of clique covers H in G with $c(H) = k$. Evidently, $N(G, n) = \alpha(\overline{G}, k)$ for all k. Thus

$$h(G, x) = \sum_{k=1}^{n} N(G, k) x^k.$$

Note that

$$h(G, x) = \xi(\overline{G}, x) = x^{\chi(\overline{G})} \sigma(\overline{G}, x).$$

The following result concerning the first four leading coefficients of $h(G, x)$ follows directly from the definition of $N(G, k)$ (see Guo and Li (1989)). Let $n_G(Q)$ denote the number of subgraphs in G which are isomorphic to Q.

Lemma 11.2.1   *For any (n, m)-graph G, we have:*

(i) $N(G, n) = 1$;

(ii) $N(G, n-1) = m$;

(iii) $N(G, n-2) = n_G(2K_2) + n_G(K_3)$;

(iv) $N(G, n-3) = n_G(3K_2) + n_G(K_2 \cup K_3) + n_G(K_4)$.   □

The following alternative expressions for $N(G, n-2)$ and $N(G, n-3)$ were respectively found by Du (1996a) and Dong, Teo, Little and Hendy (2002a).

Corollary 11.2.1   *For any connected (n, m)-graph G, we have:*

(i) $N(G, n-2) = n_G(K_3) + \binom{m}{2} - \sum_{x \in V(G)} \binom{d(x)}{2}$;

(ii) $N(G, n-3) = \binom{m}{3} + n_G(P_4) + 5n_G(K_3) + n_G(K_4) - \sum_{x \in V(G)} d(x)\Delta(x)$

$$+ m \left(n_G(K_3) - \sum_{x \in V(G)} \binom{d(x)}{2} \right) + 2 \sum_{x \in V(G)} \binom{d(x)+1}{3},$$

where $\Delta(x)$ is the number of triangles in G which contain the vertex x.   □

Liu (1990), and Liu and Wang (1992) expressed the above formulas in terms of the degree sequence of the graph.

Corollary 11.2.2   *Let G be an (n, m)-graph with vertex set $\{1, 2, \cdots, n\}$ and $d_i = d_G(i)$ for all $i = 1, 2, \cdots, n$. Then*

(i) $N(G, n-2) = n_G(K_3) + \binom{m+1}{2} - \frac{1}{2} \sum_{i=1}^{n} d_i^2$;

(ii) $N(G, n-3) = \frac{1}{6}m(m^2 + 3m + 4) - \frac{m+2}{2} \sum_{i=1}^{n} d_i^2 + \frac{1}{3} \sum_{i=1}^{n} d_i^3 + \sum_{ij \in E(G)} d_i d_j$

$$- \sum_{i=1}^{n} \Delta(i)d_i + (m+2)n_G(K_3) + n_G(K_4).$$ □

Example 11.2.1

(i) $h(K_1, x) = x$, $h(P_2, x) = x^2 + x$;

(ii) $h(C_3, x) = x^3 + 3x^2 + x$, $h(C_4, x) = x^4 + 4x^3 + 2x^2$;

(iii) $h(P_4, x) = x^4 + 3x^3 + x^2 = h(K_1 \cup C_3, x)$.

Let us now find the explicit expressions for $h(P_n, x)$ and $h(C_n, x)$ in general. Note that if G is a triangle-free graph of order n, then $N(G, n - k)$ is equal to the number of matchings with k edges in G (see Frucht and Giudici (1985)). Farrell (1979) showed that

$$N(P_n, n - k) = \binom{n - k}{k}, \text{ and}$$

$$N(C_n, n - k) = \frac{n}{n - k}\binom{n - k}{k}.$$

Hence we obtain the following result (see also Liu (1987)).

Theorem 11.2.1

(i) $h(P_n, x) = \sum_{k \leq n} \binom{k}{n-k} x^k$ for $n \geq 1$;

(ii) $h(C_n, x) = \sum_{k \leq n} \frac{n}{k}\binom{k}{n-k} x^k$ for $n \geq 4$. $\qquad\square$

R.Y. Liu (1992a) found the following sharp upper bound for $N(G, n-2)$.

Lemma 11.2.2 *Let G be a connected (n, m)-graph. Then*

$$N(G, n - 2) \leq \binom{m - 1}{2},$$

and equality holds if and only $G \cong P_{m+1}$ or $G \cong C_3$. $\qquad\square$

In an unpublished note, L.C. Zhao (1994) showed that the upper bound can be relaxed if G is a connected (n, n)-graph.

Lemma 11.2.3 *If G is a connected (n, n)-graph, then*

$$N(G, n - 2) \leq \binom{n - 1}{2} - 1,$$

and equality holds if and only if $G \cong C_n$ or $G \cong D_n$, where D_n is the third graph shown in Figure 11.2. $\qquad\square$

11.3   Reduction formulas for adjoint polynomials

In this section, we rewrite FRT in terms of adjoint polynomials, and use it to compute the adjoint polynomials of some graphs. Evidently, (1.3) can be rewritten as follows.

Theorem 11.3.1   *Let G be a graph with k components $G_1, G_2, \cdots, G_k$. Then $h(G, x) = \prod_{i=1}^{k} h(G_i, x)$.*   □

In view of Theorem 11.3.1, we say that $h(G, x)$ is *multiplicative* over the components.

Let G be a graph and $e = xy \in E(G)$. We denote by $G * e$ the graph defined as follows:

(i) $V(G * e) = (V(G) - \{x, y\}) \cup \{z\}$, where $z \notin V(G)$;

(ii) $E(G*e) = \{f \ : \ f \in E(G)$ and f is not incident with x or $y\}$
           $\cup \{zv : v \in N(x) \cap N(y)\}$.

We illustrate this graphically as follows:

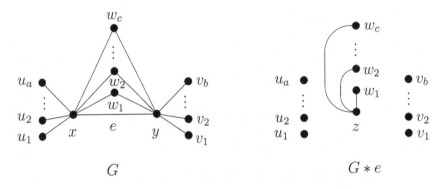

G $G * e$

Figure 11.1

It is easy to see from the definition of $N(G, k)$ that

$$N(G, k) = N(G - e, k) + N(G * e, k),$$

for all $k \in \mathbb{N}$ and $e \in E(G)$. Hence we obtain the following reduction formula due to Liu (1990), which is in fact a rewrite of FRT (see also Du (1996a) and Dong, Teo, Little and Hendy (2002a)).

Theorem 11.3.2   *For any graph G and $e \in E(G)$,*

$$h(G, x) = h(G - e, x) + h(G * e, x).$$ □

By Theorems 11.3.1 and 11.3.2 and the fact that $h(K_1, x) = x$, we obtain the following particular result.

Corollary 11.3.1   *Let $e = uv \in E(G)$. If e is not an edge of any triangle of G, then*

$$h(G, x) = h(G - e, x) + xh(G - \{u, v\}, x).$$ □

Let us illustrate Theorem 11.3.2 by some examples. Let Y_n, U_n, D_n, B_n and J_n be the graphs of order n, shown below:

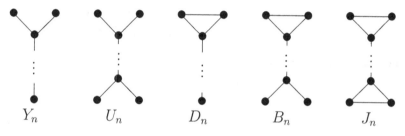

Figure 11.2

It is not hard to verify (see Liu (1997), Liu and Zhao (1997) and Dong, Teo, Little and Hendy(2002a)) that the following recurrence relations hold.

Lemma 11.3.1   *Let $G_n \in \{P_n, C_n, Y_n, U_n, D_n, B_n, J_n\}$. Then*

$$h(G_n, x) = x(h(G_{n-1}, x) + h(G_{n-2}, x))$$

for all n such that G_{n-2} is defined.   □

Let G be a graph and $\partial(G)$ the multiplicity of $x = 0$ as a root of $h(G, x)$. We define $h_1(G, x)$ as the polynomial such that $h(G, x) = x^{\partial(G)} h_1(G, x)$. In view of Lemma 11.3.1, Zhao, Li, Liu and Wang (2002b) established the following useful algebraic property.

Lemma 11.3.2   *Let $\{Q_k(x) : k \in \mathbb{N}\}$ be a sequence of polynomials over $\mathbb{Z}$ satisfying the property*

$$Q_n(x) = x(Q_{n-1}(x) + Q_{n-2}(x)).$$

Then $h_1(P_n, x) | Q_{k(n+1)+i}(x)$ if and only if $h_1(P_n, x) | Q_i(x)$, where $0 \le i \le n$ and $n \ge 2$.   □

As a consequence, the authors showed that

Corollary 11.3.2   (i) For $n \geq 4$, $h_1(P_4, x)|h(D_n, x)$ if and only if $n = 5k + 3$, where $k \geq 1$.

(ii) For $n \geq 6$, $h_1(P_4, x)|h(J_n, x)$ if and only if $n = 5k + 2 \geq 7$.   □

The adjoint polynomial of each graph in Lemma 11.3.1 can be expressed in terms of the adjoint polynomials of paths. For example, we have:

Lemma 11.3.3   (i) $h(C_n, x) = h(P_n, x) + xh(P_{n-2}, x)$ for $n \geq 4$;

(ii) $h(D_n, x) = h(P_n, x) + x(x+1)h(P_{n-3}, x)$ for $n \geq 4$.   □

Note that Theorem 11.2.1 can also be proved by using Lemma 11.3.1 and induction on n. We also have (see R.Y. Liu (1987)):

Theorem 11.3.3   For $n \geq 4$,

$$h(D_n, x) = \sum_{k \leq n} \left(\frac{n}{k} \binom{k}{n-k} + \binom{k-2}{n-k-3} \right) x^k.$$ □

By using Corollary 11.3.1, it is not hard to establish the following results (see Exercise 11.1).

Lemma 11.3.4   (i) $h(P_{2n+1}, x) = h(P_n \cup C_{n+1}, x)$ for any $n \geq 3$;

(ii) $h(Y_n, x) = h(K_1 \cup C_{n-1}, x)$ for any $n \geq 5$;

(iii) $h(U_n, x) = x^3(x+4)h(P_{n-4}, x)$ for any $n \geq 6$;

(iv) $h(U_{2n+1}, x) = h(U_{n+2}, x)h(C_{n-1}, x)$ for any $n \geq 5$.   □

11.4   Roots of adjoint polynomials

The question of when all the roots of a σ-polynomial are real has been studied by Brenti (1992) and Brenti, Royle and Wagner (1994). In particular, it was shown in Brenti, Royle and Wagner (1994) that this is the case for K_3 and any graph with triangle-free complement. This result implies that the adjoint polynomial of any such graph has only real roots. In particular, for $n \geq 4$ and $k \geq 1$, the roots of $h(C_n, x)$ and $h(P_k, x)$ are all real. In this section, we shall find all the roots of these two polynomials, and then study the minimum real roots of some adjoint polynomials. These will be useful in sorting out graphs that are not adjointly equivalent.

The proofs for the next two results were due to Dong, Teo, Little and Hendy (2002b) (see also H.X. Zhao and Liu (2001a) for alternative proofs).

For any $n \in \mathbb{N}$ and $x \in \mathbb{R}$, define

$$g_n(x) = \begin{cases} 1, & \text{if } x = 0 \\ x^n h\left(P_n, \frac{1}{x}\right), & \text{if } x \neq 0. \end{cases}$$

By Lemma 11.3.1, we obtain for all $n \geq 3$,

$$g_n(x) = g_{n-1}(x) + x g_{n-2}(x),$$

where $g_1(x) = 1$ and $g_2(x) = x + 1$. It can be shown by induction that

$$g_n(u^2 + u) = \sum_{i=0}^{n} (1+u)^i (-u)^{n-i}$$

for any $u \in \mathbb{R}$. Thus

$$(2u + 1)g_n(u^2 + u) = (1+u)^{n+1} - (-u)^{n+1}. \tag{11.1}$$

This can be factorized by using the following classical result (see Barnard and Child (1955)).

Lemma 11.4.1 *For any $a, b \in \mathbb{R}$ and $n \in \mathbb{N}$,*

(i) if n is odd, then

$$a^n - b^n = (a - b) \prod_{s=1}^{(n-1)/2} \left(a^2 + b^2 - 2ab \cos \frac{2s\pi}{n}\right);$$

(ii) if n is even, then

$$a^n - b^n = (a - b)(a + b) \prod_{s=1}^{(n-2)/2} \left(a^2 + b^2 - 2ab \cos \frac{2s\pi}{n}\right). \quad \square$$

By letting $x = u^2 + u$ and making use of the definition of $g_n(x)$, we obtain the following result, from which the roots of $h(P_n, x)$ can be found readily.

Theorem 11.4.1 *For any $n \in \mathbb{N}$,*

$$h(P_n, x) = x^{\lceil n/2 \rceil} \prod_{s=1}^{\lfloor n/2 \rfloor} \left(x + 2 + 2\cos \frac{2s\pi}{n+1}\right). \quad \square$$

We now consider $h(C_n, x)$. For any $n \in \mathbb{N}$ with $n \geq 4$, define

$$f_n(x) = \begin{cases} 1, & \text{if } x = 0 \\ x^n h\left(C_n, \frac{1}{x}\right), & \text{if } x \neq 0. \end{cases}$$

By Lemma 11.3.3(i), we have

$$f_n(x) = g_n(x) + x g_{n-2}(x).$$

It follows from (11.1) that

$$f_n(u^2 + u) = (1 + u)^n + (-u)^n.$$

This time we factorize the above by using the following:

Lemma 11.4.2   *For any $a, b \in \mathbb{R}$ and $n \in \mathbb{N}$,*

(i) if n is odd, then

$$a^n + b^n = (a + b) \prod_{s=1}^{(n-1)/2} \left(a^2 + b^2 - 2ab \cos \frac{(2s - 1)\pi}{n}\right);$$

(ii) if n is even, then

$$a^n + b^n = \prod_{s=1}^{n/2} \left(a^2 + b^2 - 2ab \cos \frac{(2s-1)\pi}{n}\right). \qquad \square$$

The following result now follows.

Theorem 11.4.2   *For any $n \in \mathbb{N}$ with $n \geq 4$,*

$$h(C_n, x) = x^{\lceil n/2 \rceil} \prod_{s=1}^{\lfloor n/2 \rfloor} \left(x + 2 + 2 \cos \frac{(2s-1)\pi}{n}\right). \qquad \square$$

Note that the roots of $h(Y_n, x)$ and $h(U_n, x)$ can be found by using Lemma 11.3.4 and the above two theorems.

Du (1996a) found that the knowledge of the minimum real roots of adjoint polynomials is useful in determining adjointly equivalent graphs of the form $\cup C_{n_j}$ or $\cup P_{n_j}$. Subsequently, this idea has been successfully employed by several researchers, such as Wang and Liu (2001) and Zhao, Huo and Liu (2000), to search for more adjointly unique graphs.

For a graph G, let $\beta(G)$ be the minimum real root of $h(G, x)$. Du (1996a) showed that $\beta(P_n) < \beta(P_{n-1})$ and $\beta(C_n) < \beta(C_{n-1})$. Indeed, these inequalities follow from the fact that

$$\beta(P_n) = -2 - 2\cos\frac{2\pi}{n+1} \qquad \text{and} \quad \beta(C_n) = -2 - 2\cos\frac{\pi}{n}.$$

The following inequalities are not hard to establish (see Wang and Liu (2001)).

Lemma 11.4.3

(i) For $n \geq 2$, $-4 < \beta(P_n) < \beta(P_{n-1})$.

(ii) For $n \geq 4$, $-4 < \beta(C_{n+1}) < \beta(C_n) < -3$.

(iii) For $n \geq 4$, $\beta(D_{n+1}) < \beta(D_n) < \beta(C_n) < \beta(P_n)$. $\qquad\square$

Zhao, Liu, Li and Zhang (2002) established the following result (see Exercise 11.10).

Lemma 11.4.4 For $n \geq 6$, $\beta(J_{n-1}) < \beta(J_n) < \beta(D_n)$. $\qquad\square$

The following result of Zhao, Li, Liu and Ye (2004) can be proved by using Theorem 11.3.2 and induction.

Theorem 11.4.3 *Let G be a connected graph and H a proper subgraph of G. Then $\beta(G) < \beta(H)$.* $\qquad\square$

For $k, s, t \in \mathbb{N}$, let $T(k, s, t)$ be the tree with a unique degree 3 vertex v such that

$$T(k, s, t) - \{v\} = P_k \cup P_s \cup P_t.$$

Note that $Y_n = T(1, 1, n - 3)$, where $n \geq 4$. Denote by $\mathcal{T}$ the family of all $T(k, s, t)$, where $1 \leq k \leq s \leq t$.

For $n \geq 3$ and $m \geq 2$, let $C_n(P_m)$ be the graph obtained from C_n and P_m by identifying a degree 1 vertex of P_m with a vertex of C_n, and let $C_n(P_{m_1}, P_{m_2})$ be the graph resulted from identifying a degree 1 vertex of each of P_{m_1} and P_{m_2} with two successive vertices of C_n. Zhao, Li, Zhang and Liu (2004) characterized connected graphs G with $\beta(G) \geq -4$.

Theorem 11.4.4 *Let G be a connected graph. Then*

(i) $\beta(G) = -4$ *if and only if*

$$G \in \{T(1,2,5),\ T(2,2,2),\ T(1,3,3),\ K(1,4),\ C_4(P_2),$$
$$C_3(P_2,P_2),\ K_4 - e, D_8\} \cup \{U_n : n \geq 6\};$$

(ii) $\beta(G) > -4$ *if and only if*

$$G \in \{T(1,2,i) : 2 \leq i \leq 4\} \cup \{D_n : 4 \leq n \leq 7\}$$
$$\cup \{P_n : n \geq 1\} \cup \{C_n : n \geq 3\} \cup \{Y_n : n \geq 4\};$$

(iii) $\beta(G) \geq -3$ *if and only if*

$$G \in \{P_n : 1 \leq n \leq 5\} \cup \{C_3\} \cup \{T(1,1,1)\}. \qquad \square$$

As a consequence of this, Zhao, Li, Zhang and Liu (2004) obtained the following result.

Corollary 11.4.1   *Let G be a connected graph with $\beta(G) \geq -4$. Then all roots of $h(G, x)$ are real.* $\qquad \square$

Problem 11.4.1   *Find all graphs G such that all the roots of $h(G, x)$ are real.*

Note that Zhao, Li and Liu (2003) characterized all connected graphs G with $-(2 + \sqrt{5}) \leq \beta(G) < -4$.

11.5   Invariants for adjointly equivalent graphs

11.5.1   Definitions

To search for adjointly unique graphs, it is very helpful to know as many as possible necessary conditions for two graphs to be adjointly equivalent. A quantity $q(G)$, where G is a graph, is called an *invariant for adjointly equivalent graphs* (or *adj-invariant* in short) if $h(G, x) = h(H, x)$ implies that $q(G) = q(H)$. For example, $v(G)$ and $e(G)$ are adj-invariants (see Lemma 11.2.1). Obviously, $N(G, k)$ is an adj-invariant for all $k \in \mathbb{N}$. Thus any expression in terms of $N(G, 1), N(G, 2), \cdots, N(G, n)$ is an adj-invariant for graphs of order n. However we prefer invariants which have some useful properties such as having constant upper bounds or lower bounds. One invariant that has been used by several researchers, such as R.Y. Liu (1997) and Dong, Teo, Little and Hendy (2002a), to determine adjoint equivalence classes of graphs is $R_1(G)$, which we now define.

Given a polynomial $f(x) = x^n + b_1 x^{n-1} + b_2 x^{n-2} + \cdots + b_{n-1} x + b_n$, let

$$
R_1(f) = \begin{cases} -\binom{b_1-1}{2} + 1, & \text{if } n = 1 \\[2mm] b_2 - \binom{b_1-1}{2} + 1, & \text{if } n \geq 2. \end{cases}
$$

For a given graph G, we define

$$
R_1(G) = R_1(h(G,x)).
$$

In terms of the coefficients of $h(G,x)$, we have

$$
R_1(G) = N(G, n-2) - \binom{m-1}{2} + 1,
$$

for any (n,m)-graph G (see Lemma 11.2.2).

It should be pointed out that the *parameter* $\pi(G) = N(G, n-2) - (m^2 - 3m)/2$, introduced by Du (1996a), is in fact the same as $R_1(G)$.

Dong, Teo, Little and Hendy (2002a) introduced another adj-invariant. Let f be the polynomial as before. Define

$$
R_2(f) = b_3 - \binom{b_1}{3} - (b_1 - 2)\left(b_2 - \binom{b_1}{2}\right) - b_1,
$$

where $b_k = 0$ for $k > n$. For any graph G, define $R_2(G) = R_2(h(G,x))$.

Another two adj-invariants they utilized are defined below:

$$
\begin{aligned}
\overline{\chi}(G) &= \min\{k : N(G,k) > 0\} = \chi(\overline{G}), \\
\tau(G) &= N(G, \overline{\chi}(G)).
\end{aligned}
$$

Remark Note that the number of components $c(G)$ of G, is not an adj-invariant. For example, $h(P_4, x) = h(K_1 \cup C_3, x)$ (see Example 11.2.1(iii)).

In the next two subsections, we provide some basic properties of the adj-invariants $R_1(G)$ and $R_2(G)$. We obtain a recursive formula and a sharp bound for $R_1(G)$. We then obtain alternative formulae for $R_2(G)$ which enable us to compute $R_2(G)$ for some specific graphs.

11.5.2   The adj-invariant $R_1(G)$

A graph function ψ is *additive* over the components of a graph G if $G = G_1 \cup G_2 \cup \cdots \cup G_k$ implies that

$$
\psi(G) = \sum_{i=1}^{k} \psi(G_i).
$$

Evidently, $v(G)$ and $e(G)$ are additive over the components. By definition, $\overline{\chi}(G)$ is also additive over the components. We show that the same is true for $R_1(G)$. Note, however, that $\tau(G)$ is multiplicative over the components. For ease of reference, we state these observations as follows.

Lemma 11.5.1   *Let G be a graph. If $G = G_1 \cup G_2 \cup \cdots \cup G_k$, then*

$$\overline{\chi}(G) = \sum_{i=1}^{k} \overline{\chi}(G_i) \quad and \quad \tau(G) = \prod_{i=1}^{k} \tau(G_i). \qquad \square$$

The following result follows immediately from Corollary 11.2.1 and the definition of $R_1(G)$.

Lemma 11.5.2   *For any graph G,*

$$R_1(G) = n_G(C_3) + e(G) - \sum_{x \in V(G)} \binom{d(x)}{2}. \qquad \square$$

It is now easy to verify the following:

Lemma 11.5.3   *For any graph G, the adj-invariant $R_1(G)$ is additive over the components.* $\qquad \square$

If $e(G) < v(G)$, then the minimum of $R_1(G)$ is attained when

$$G \cong K(1,t) \cup (v(G) - (t+1))K_1$$

for some $t < v(G)$. Thus we obtain Giudici's result (1985) that says $K_n - E(K(1,t))$ is χ-unique for all $n \in \mathbb{N}$ and $t < n$.

If $e(G) = 0$, then from Lemma 11.5.2, we have $R_1(G) = 0$. We shall find a recursive expression for $R_1(G)$ when $e(G) > 0$. For $x, y \in V(G)$, let

$$N(x,y) = (N(x) \cup N(y)) - \{x,y\}.$$

Observe that

$$|N(x,y)| = \begin{cases} d(x) + d(y) - |N(x) \cap N(y)|, & \text{if } xy \notin E(G); \\ d(x) + d(y) - |N(x) \cap N(y)| - 2, & \text{if } xy \in E(G). \end{cases}$$

By using Lemma 11.5.2, we obtain the following recursive formula due to Du (1996a).

Lemma 11.5.4   *For any graph G and $xy \in E(G)$,*

$$R_1(G) = R_1(G - xy) - |N(x,y)| + 1. \qquad \square$$

Our next result says that if two vertices without common neighbours are joined by a path, then this path can be replaced by an edge without changing the value of R_1. Its validity follows directly from Lemma 11.5.4 (see Dong, Teo, Little and Hendy (2002a)).

Lemma 11.5.5 *Let G be a graph and $xy \in E(G)$ with $N(x) \cap N(y) = \emptyset$. Let G' be a graph obtained from G by subdividing the edge xy. Then*

$$R_1(G') = R_1(G).$$ □

Example 11.5.1 *(i) For $n \geq 2$, $R_1(P_n) = R_1(P_2) = 1$.*

(ii) For $n \geq 4$, $R_1(C_n) = R_1(C_4) = 0$.

(iii) For $n \geq 4$, $R_1(D_n) = R_1(D_4) = 0$.

(iv) For $k, s, t \in \mathbb{N}$, $R_1(T(k, s, t)) = R_1(T(1, 1, 1)) = 0$.

Part (i) and (ii) of the following theorem are due to Liu (1992b) and Du (1995) independently, and part (iii) is due to Du (1995), who used a recursive method to construct graphs with $R_1(G) = -i$, where $i \geq 1$.

Theorem 11.5.1 *Let G be a connected graph. Then $R_1(G) \leq 1$ and*

(i) $R_1(G) = 1$ if and only if $G \in \{K_3\} \cup \{P_n : n \geq 2\}$;

(ii) $R_1(G) = 0$ if and only if $G \in \{K_1\} \cup \mathcal{T} \cup \{C_n, D_n : n \geq 4\}$; and

(iii) $R_1(G) = -1$ if and only if G is one of the graphs in Figure 11.3. □

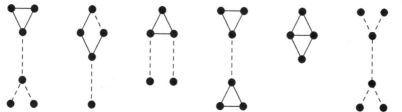

Figure 11.3

Here is Du's algorithm of constructing graphs G with $R_1(G) = -i < 0$:

(1) For each H with $R_1(H) = -t > -i$, construct graphs $H + xy$, where $xy \notin E(H)$ and $x \in V(H)$ (y may or may not be a vertex of H), such that

$$|N_{H+xy}(x, y)| = i + 1 - t.$$

(2) For each graph Q obtained in (1), choose any sequence of pairs of vertices $\{u_j, v_j\}$, where $j = 1, 2, \cdots, k$, such that $u_j v_j \notin E(Q)$ and

$$|N_{Q_j}(u_j, v_j)| = 1, \quad j = 1, 2, \cdots, k,$$

where $Q_1 = Q + u_1 v_1$ and $Q_j = Q_{j-1} + u_j v_j$ for $j = 2, \cdots, k$.

It follows from Lemma 11.5.4 that the graphs G constructed in either step (1) or (2) of the above algorithm have the property that $R_1(G) = 1$. By applying the above algorithm, Du obtained all graphs G with $R_1(G) = -i$ for $i = -1, 0, 1, 2, 3$.

Corollary 11.5.1   *For any graph G,*

$$R_1(G) \le v(G) - e(G) + n_G(K_3),$$

and equality holds if and only if every component of G is a member of the family

$$\{K_4, K_4 - e\} \cup \{P_n, C_{n+1}, D_{n+2}, J_{n+4} : n \ge 2\}. \qquad \square$$

Dong, Teo, Lttle and Hendy (2002a) obtained a sharper bound for $R_1(G)$ when $G \not\cong K_4$. Note that $R_1(K_4) = -2$.

Theorem 11.5.2   *For any connected graph G, if $G \not\cong K_4$, then*

$$R_1(G) \le 2(v(G) - e(G)) + 1. \qquad \square$$

From this the following result follows easily.

Corollary 11.5.2   *Let G be a graph. Suppose $R_1(G) \le -2$ and $G \not\cong K_4$. Then $R_1(G) \le v(G) - e(G) - 1$.* $\qquad \square$

In view of the above results, one may wish to define another adj-invariant:

$$R_3(G) = R_1(G) + e(G) - v(G)$$

and use it to rewrite the results in a neater way. Evidently, $R_3(G)$ is also additive over the components. Zhao, Liu, Li and Zhang (2002) provided the following recursive formula for $R_3(G)$.

Lemma 11.5.6   *Let G be a connected graph and $xy \in E(G)$. Then*

$$R_3(G) = R_3(G - xy) - |N(x, y)| + 2. \qquad \square$$

Corollary 11.5.2 can be written more succinctly as follows (see Zhao, Li and Liu (2003)).

Theorem 11.5.3   *Let G be a connected graph. Then*

(i) $R_3(G) \le 1$, and equality holds if and only if $G \cong K_3$;

(ii) $R_3(G) = 0$ if and only if

$$G \in \{K_4, K_4 - e\} \cup \{P_n, C_{n+1}, D_{n+2}, J_{n+4} : n \ge 2\}. \qquad \square$$

In terms of the adj-invariants $\tau(G)$ and $\overline{\chi}(G)$, Dong, Teo, Little and Hendy (2002a) obtained, by Theorem 11.5.2 and its corollary, the following result.

Lemma 11.5.7   *For any connected graph G, if $\tau(G) = 1$, then*

$$R_1(G) \le 2v(G) - e(G) - 2\overline{\chi}(G),$$

and equality holds if and only if $G \in \{K_1, K_3\} \cup \{P_{2i} : i \ge 1\}$. $\square$

11.5.3   The adj-invariant $R_2(G)$

In terms of $N(G, n - k)$ for $k = 1, 2, 3$, we have

$$R_2(G) = N(G, n-3) - \binom{m}{3} - (m-2)\left(N(G, n-2) - \binom{m}{2}\right) - m.$$

By Corollary 11.2.1, Dong, Teo, Little and Hendy (2002a) established the following formula.

Theorem 11.5.4   *For any graph G,*

$$\begin{aligned} R_2(G) &= 2 \sum_{x \in V(G)} \binom{d(x)}{3} - \sum_{x \in V(G)} d(x) \triangle(x) - e(G) \\ &\quad + n_G(P_4) + 7n_G(K_3) + n_G(K_4). \end{aligned} \qquad \square$$

It is not hard to verify (see Exercise 11.2) that

$$n_G(P_4) = \sum_{xy \in E(G)} (d(x) - 1)(d(y) - 1) - 3n_G(K_3).$$

Hence we can rewrite Theorem 11.5.4 (see Dong, Teo, Little and Hendy (2002a)) as follows.

Theorem 11.5.5   *For any graph G,*

$$R_2(G) = 2 \sum_{x \in V(G)} \binom{d(x)}{3} - \sum_{x \in V(G)} d(x)\triangle(x) - e(G) + 4n_G(K_3)$$

$$+ n_G(K_4) + \sum_{xy \in E(G)} (d(x) - 1)(d(y) - 1). \qquad \square$$

The above expression takes a very simple form if G is triangle-free.

Corollary 11.5.3   *If G is triangle-free, then*

$$R_2(G) = 2 \sum_{x \in V(G)} \binom{d(x)}{3} - e(G) + \sum_{xy \in E(G)} (d(x) - 1)(d(y) - 1). \quad \square$$

Zhao, Li, Liu and Wang (2002b) expressed $R_2(G)$ in terms of the degree sequence of G.

Corollary 11.5.4   *If G is an (n, m)-graph with vertex set $\{1, 2, \cdots, n\}$ and $d_i = d(i)$ for all $i = 1, 2, \cdots, n$, then*

$$R_2(G) = \frac{4m}{3} - 2\sum_{i=1}^{n} d_i^2 + \frac{1}{3}\sum_{i=1}^{n} d_i^3 + \sum_{ij \in E(G)} d_i d_j - \sum_{i=1}^{n} \triangle(i) d_i$$

$$+ 4n_G(K_3) + n_G(K_4). \qquad \square$$

Theorem 11.5.5 also implies that $R_2(G)$ is additive over the components. We state it for reference.

Corollary 11.5.5   *Let G be a graph. If $G = G_1 \cup G_2 \cup \cdots \cup G_k$, then*

$$R_2(G) = \sum_{i=1}^{k} R_2(G_i). \qquad \square$$

We give some simple examples.

Example 11.5.2

(i) $R_2(P_1) = 0$, $R_2(P_2) = -1$ and $R_2(P_n) = -2$ for $n \geq 3$;

(ii) $R_2(C_3) = -2$ and $R_2(C_n) = 0$ for $n \geq 4$;

(iii) $R_2(Y_4) = -1$ and $R_2(Y_n) = 0$ for $n \geq 5$;

(iv) $R_2(D_4) = 0$ and $R_2(D_n) = 1$ for $n \geq 5$;

(v) $R_2(J_6) = 5$ *and* $R_2(J_n) = 4$ *for* $n \geq 7$;

(vi) $R_2(K_4 - e) = 3$ *and* $R_2(K_4) = 7$;

(vii) $R_2(T(1, s, t)) = 1$ *for* $t \geq s \geq 2$ *and* $R_2(T(k, s, t)) = 2$ *for* $t \geq s \geq k \geq 2$. *(Note that* $T(1, 1, n - 3) = Y_n$.)

By using Corollary 11.5.1, Dong, Teo, Little and Hendy (2002c) obtained the following result.

Lemma 11.5.8 *Let G be a graph. If $R_1(G) = v(G) - e(G) + n_G(K_3)$, then*

$$2R_1(G) + R_2(G) \geq 0,$$

and equality holds if and only if every component of G belongs to the family

$$\{D_4\} \cup \{P_n, C_n : n \geq 3\}. \qquad \square$$

11.6 Adjointly equivalent graphs

As pointed out in Section 11.1, the goal of determining the χ-equivalence class of a graph can be realized by figuring out the adjoint equivalence class of its complement. This approach has been found particularly effective for graphs whose complements are of smaller size (see Koh and Teo (1997), Liu and L.C. Zhao (1997) and Liu (1997) for brief surveys).

In this section, we provide some results on the adjoint equivalence classes of some families of graphs of simple structure. We will focus on those results that are in a sense more complete, and will leave other results to the final section.

A family $\mathcal{S}$ of graphs is said to be *adjointly closed* if for any graph H with $H \sim_h G$ for some $G \in \mathcal{S}$, we have $H \in \mathcal{S}$. Evidently, an adjoint equivalence class is adjointly closed, and every adjointly closed family is the union of some adjoint equivalence classes. We will focus our attention on graphs whose components are members of the family:

$$\{P_k, C_{k+2} \; : \; k \geq 1\} \cup \{D_4\}.$$

Note that D_4 is adjointly equivalent to C_4. Indeed,

$$h(D_4, x) = h(C_4, x) = x^4 + 4x^3 + 2x^2.$$

Hence C_4 and D_4 are not adjointly unique. Evidently, $\{C_4, D_4\}$ is an adjoint equivalence class. Let $\mathcal{S}$ be a family of graphs. Write $G \in \langle \mathcal{S} \rangle$ if $G = n_1 G_1 \cup n_2 G_2 \cup \cdots \cup n_k G_k$, for some $n_i \in \mathbb{N}_0$ and $G_i \in \mathcal{S}, i = 1, 2, \cdots, k$.

The following graphs are not adjointly unique (see, for example, Dong, Teo, Little and Hendy (2002c)).

Theorem 11.6.1   *A graph is not adjointly unique if it is a member of the family*

$$\langle \{P_4, C_4, D_4\} \cup \{P_{2n+1}, P_n \cup C_{n+1}, K_1 \cup C_n : n \geq 3\} \rangle.$$

Proof.   The assertion follows from the following equalities:

(i) $h(P_4, x) = h(K_1 \cup C_3, x)$;

(ii) $h(P_{2n+1}, x) = h(P_n \cup C_{n+1}, x)$ for $n \geq 3$;

(iii) $h(Y_n, x) = h(K_1 \cup C_{n-1}, x)$ for $n \geq 5$;

(iv) $h(C_4, x) = h(D_4, x)$.   □

From Theorem 11.6.1, we see that P_n is not adjointly unique if $n = 4$ or $n = 2k + 1$ for $n \geq 3$. Xu (1983) (see also Salzberg (1984)) showed that $\overline{P}_5$ is χ-unique but $K_n - P_5$ is not so for $n \geq 6$. Equivalently, P_5 is adjointly unique but $rK_1 \cup P_5$ is not so for all $r \in \mathbb{N}$. In fact, by comparing the chromatic polynomials of all the graphs of the form $K_5 - E(G)$ with $e(G) = 5$, they obtained a result which is equivalent to the following.

Theorem 11.6.2   *For any $r \in \mathbb{N}$, the adjoint equivalence class of $rK_1 \cup P_5$ is*

$$\{rK_1 \cup P_5, (r-1)K_1 \cup K(1,3) \cup P_2\}.$$

Moreover, P_5 is adjointly unique.   □

Dong, Teo, Little and Hendy (2002c) gave a proof for the above theorem by using the adj-invariant $R_1(G)$. We illustrate their method by establishing the adjoint uniqueness of P_5.

Suppose $H \sim_h P_5$. Then $R_1(H) = R_1(P_5) = 1$. Since $R_1(G)$ is additive over the components and $R_1(G) \leq 1$ for all connected graphs G, H has a component, say H_0, such that $R_1(H_0) = 1$. By Theorem 11.5.1, $H_0 \cong K_3$ or $H_0 \cong P_t$, for some $2 \leq t \leq 5$. Moreover, $h(H_0, x)$ is a factor of $h(H, x)$. Hence $H_0 \cong P_2$ or P_5. Suppose $H_0 \cong P_2$. Write $H = P_2 \cup H_1$. Then $h(H_1, x) = h(P_5, x)/h(P_2, x) = x^3 + 3x^2$. Evidently, H_1 is a graph of order

3 and size 3. Hence $H \cong K_3$. But $h(K_3, x) = x^3 + 3x^2 + x \neq h(H_1, x)$. We conclude that $H_0 \not\cong P_2$, and hence $H \cong P_5$.

Liu (1997) proposed the following conjecture:

Conjecture 11.6.1   *For each even number n with $n \neq 4$, P_n is adjointly unique.*

Note that earlier Du (1996a) had proved a result which implies that P_n is adjointly unique if $n \not\equiv 4 \pmod{10}$, and Liu (1993) (see also Du (1996a)) showed that the conjecture is true if $n + 1$ is a prime number larger than 5.

We will consider a more general problem. We determine the adjointly unique graphs in the family

$$\mathcal{S}_1 = \langle \{K_1, K_3\} \cup \{P_{2i} : i \in \mathbb{N}\} \rangle.$$

The chromaticity of the complements of graphs in $\mathcal{S}_1$ has attracted the attention of several researchers (Liu (1992a), (1993)), B.R. Zhang (1997), X.Y. Zhang (1997), Du (1996a)). Dong, Teo, Little and Hendy (2002c) settled this problem completely. They found necessary and sufficient conditions for a graph to be a member of $\mathcal{S}_1$.

Theorem 11.6.3   *Let G be a graph. Then $G \in \mathcal{S}_1$ if and only if $\tau(G) = 1$ and*

$$R_1(G) = 2v(G) - e(G) - 2\overline{\chi}(G).$$

Proof. It is straightforward to check the necessity. Conversely, suppose $\tau(G) = 1$ and $R_1(G) = 2v(G) - e(G) - 2\overline{\chi}(G)$. Let $G_1, G_2, \cdots, G_k$ be the components of G. By Lemma 11.5.1, $\tau(G_i) = 1$ for all $i = 1, 2, \cdots, k$. Also $\overline{\chi}(G) = \sum_{i=1}^{k} \overline{\chi}(G_i)$. Thus, by Lemma 11.5.7,

$$
\begin{aligned}
R_1(G) &= \sum_{i=1}^{k} R_1(G_i) \\
&\leq \sum_{i=1}^{k} (2v(G_i) - e(G_i) - 2\overline{\chi}(G_i)) = 2v(G) - e(G) - 2\overline{\chi}(G).
\end{aligned}
$$

By hypothesis, we must have $R_1(G_i) = 2v(G_i) - e(G_i) - 2\overline{\chi}(G_i)$ for all $i = 1, 2, \cdots, k$. Again, by Lemma 11.5.7,

$$G_i \in \{K_1, K_3\} \cup \{P_{2j} : j \in \mathbb{N}\}. \qquad \square$$

From this, we have:

Theorem 11.6.4   *The family S_1 is adjointly closed.*   □

We next determine the adjoint equivalence class of each member of S_1. The following result is due to Dong, Teo, Little and Hendy (2002c).

Theorem 11.6.5   *For any $s, r_0, r_1, a_i \in \mathbb{N}_0$, where $i = 1, 2, \cdots, s$, if $G = r_0 K_1 \cup r_1 K_3 \cup \bigcup_{i=1}^{s} a_i P_{2i}$, then*

$$[G]_h = \{(r_0 - a)K_1 \cup (r_1 - a)K_3 \cup (a_2 + a)P_4$$
$$\cup \bigcup_{\substack{1 \le i \le s \\ i \ne 2}} a_i P_{2i} : -a_2 \le a \le \min\{r_0, r_1\}\}.$$

Proof. Let $\mathcal{G}$ denote the family of graphs stated in the theorem. As

$$h(P_4, x) = h(K_1, x)h(K_3, x),$$

$\mathcal{G} \subseteq [G]_h$.

Observe that

$$h(G, x) = x^{r_0 - r_1} (h(P_4, x))^{r_1 + a_2} \prod_{\substack{1 \le i \le s \\ i \ne 2}} (h(P_{2i}, x))^{a_i}.$$

By Theorem 11.4.1, $h(G, x)$ has exactly a_i repeated roots $-2 - 2\cos\left(\frac{2\pi}{2i+1}\right)$ for $1 \le i \le s, i \ne 2$, and has $r_1 + a_2$ repeated roots: $-2 - 2\cos\left(\frac{2\pi}{5}\right)$.

Now suppose $H \in [G]_h$. By Theorem 11.6.4, $H \in S_1$. Thus

$$H \cong r_0' K_1 \cup r_1' K_3 \cup \bigcup_{i \in \mathbb{N}} a_i' P_{2i},$$

for some r_0', r_1' and $a_i' \in \mathbb{N}_0$.

By comparing the roots of $h(H, x)$ and $h(G, x)$, we see that

$a_1' = a_1, a_2' + r_1' = a_2 + r_2, a_i' = a_i$ for $i = 3, 4, \cdots, 5$, and $a_i' = 0$ for $i \ge s+1$.

Let $a = r_1 - r_1'$. Then $a_2' = a_2 + a, r_1' = r_1 - a$. Now $v(G) = v(H)$ yields

$$r_0 + 3r_1 + \sum_{i=1}^{s} 2ia_i = r_0' + 3r_1' + \sum_{i=1}^{s} 2ia_i',$$

which implies that $r_0' = r_0 - a$. Thus $H \in \mathcal{G}$.   □

As a special case, Dong, Teo, Little and Hendy (2002c) obtained all adjointly unique graphs in S_1.

Theorem 11.6.6 *For any $r_0, r_1, a_i \in \mathbb{N}_0$, where $i = 1, 2, \cdots, s$, the graph*

$$r_0 K_1 \cup r_1 K_3 \bigcup_{i=1}^{s} a_i P_{2i}$$

is adjointly unique if and only if $r_0 r_1 = 0$ and $a_2 = 0$. □

As a consequence of Theorem 11.6.6, we now know exactly which paths are adjointly unique; thereby proving that Liu's conjecture(i.e., Conjecture 11.6.1) is true. In fact, Dong, Teo, Little and Hendy (2002c) obtained a more general result.

Theorem 11.6.7 *For any $n, r \in \mathbb{N}_0, n \geq 1$, the graph $rK_1 \cup P_n$ is adjointly unique if and only if one of the following conditions is satisfied:*

(i) $r = 0$, $n = 5$;

(ii) $n \in \{1, 3\} \cup \{2k : k \in \mathbb{N}, k \neq 2\}$. □

By studying the minimum roots of the adjoint polynomials of the components of possible candidates H such that $H \sim_h P_5$, Zhao, Huo and Liu (2000) also proved Liu's conjecture.

We next turn our attention to graphs which contain some cycles as components. From the chromatic polynomials of all the graphs of the form $K_n - E(G)$ with $e(G) \leq 6$ contained in the articles of Salzberg (1984), Xu (1983) and Gracia and Salzberg (1985), we see that the following families are adjoint equivalence classes:

$$\{P_4, K_1 \cup C_3\}, \ \{Y_5, K_1 \cup D_4, K_1 \cup C_4\}, \ \{K_1 \cup C_5, Y_6\}, \ \{K_1 \cup C_6, Y_7\}.$$

From this, we see that C_3, C_5, C_6 are adjointly unique, while C_4 is not so. After investigating by a computer, Farrell and Whitehead Jr (1990) found that $\overline{C}_7$ and $\overline{C}_8$ are χ-unique, or equivalently, C_7 and C_8 are adjointly unique. They made a conjecture which is equivalent to the following:

Conjecture 11.6.2 *For $n \neq 4$, C_n is adjointly unique.*

This was proved by Guo and Li (1989) by using a graph function known as matching polynomial. Du (1996a) (see also Liu and Bao (1993)) established the following more general result.

Theorem 11.6.8 *Let $G = \bigcup_{i=1}^{k} C_{n_i}$, where $n_i \neq 4$ for each $i = 1, \cdots, k$. Then G is adjointly unique.* □

Dong, Teo, Little and Hendy (2002c) considered a more general problem. By studying the roots of the adjoint polynomials of the components and making use of Lemma 11.5.8, they showed that the following family is adjointly closed.

$$\mathcal{S}_2 = \langle \{K_3, D_4\} \cup \{P_n : n \geq 3, n \not\equiv 4 \ (\text{mod } 5)\} \cup \{C_n : n \geq 4\} \rangle.$$

Recall that for any $k \geq 3$, $P_{2k+1} \sim_h P_n \cup C_{n+1}$ and $D_4 \sim_h C_4$. For each $G \in \mathcal{S}_2$, construct a graph $\tilde{G}$ from G by replacing all D_4 with C_4 and all paths of odd length until there is none left. It is not hard to check that if $G \in \mathcal{S}_2$, then so is $\tilde{G}$. Evidently, $\tilde{G} \sim_h G$. Dong, Teo, Little and Hendy (2002c) proved the following:

Theorem 11.6.9   *For each $G \in \mathcal{S}_2$,*

$$[G]_h = \{H \in \mathcal{S}_2 : \tilde{H} \cong \tilde{G}\}. \qquad \square$$

In general, it is not easy to list all graphs in $[G]_h$. They did find a numerical method to decide if two members of $\mathcal{S}_2$ are adjointly equivalent.

For each $a \in \mathbb{N}$, let $\mathcal{S}_2(a) = \{G \in \mathcal{S}_2 : n_G(K_3) = a\}$. Dong, Teo, Little and Hendy (2002c) showed that $\mathcal{S}_2(a)$ is adjointly closed. They determined all adjointly unique graphs in $\mathcal{S}_2$ as follows.

Theorem 11.6.10   *Let $G \in \mathcal{S}_2(a)$, where $a \in \mathbb{N}_0$. Suppose*

$$G = aK_3 \cup bD_4 \cup \bigcup_{1 \leq i \leq s} P_{u_i} \cup \bigcup_{1 \leq j \leq t} C_{v_j},$$

for $s, t, b, u_i, v_j \in \mathbb{N}_0$ with $u_i \geq 3, u_i \not\equiv 4 \ (\text{mod } 5)$ and $v_j \geq 4$. Then G is adjointly unique if and only if $b = 0$, $v_j \geq 5$, u_i is even when $u_i \geq 6$ and

$$\{u_1 + 1, u_2 + 1, \cdots, u_s + 1\} \cap \{v_1, v_2, \cdots, v_t\} = \emptyset.$$

Proof.   The necessity is easy to verify by using Theorem 11.6.1. We prove the sufficiency. Let G be a graph in $\mathcal{S}_2(a)$ that satisfies the condition, and $H \sim_h G$. Then $H \in \mathcal{S}_2(a)$ and $\tilde{H} = \tilde{G} = G$. Hence H does not contain C_4 or D_4. If H contains a component P_{2n+1}, then $\tilde{H}$, and hence G, contains a pair of components P_m and C_{m+1}, contradicting the condition that G satisfies. Hence H does not contain either D_4 or P_{2n+1}, and so $\tilde{H} = H$, which implies that $H \cong G$. $\square$

Theorem 11.6.8 now follows as a consequence. It also implies a result of Du's, which says that if u_i is even and $u_i \not\equiv 4 \ (\text{mod } 10)$, then $aK_3 \cup \bigcup_{1 \leq i \leq s} P_{u_i}$ is adjointly unique.

Let $\mathcal{A} = \{K_3, D_4\} \cup \{P_n : n \geq 3\} \cup \{C_n : n \geq 4\}$. The family $\mathcal{S}_3 = \langle \mathcal{A} \rangle$ is not adjointly closed. For example, $K_1 \cup K_3 \notin \mathcal{S}_3$, but $K_1 \cup K_3 \sim_h P_4$ and $P_4 \in \mathcal{S}_3$.

Let $\mathcal{S}$ be a family of graphs. Define the *adjoint closure* of $\mathcal{S}$ to be the family

$$\mathcal{C}(\mathcal{S}) = \bigcup_{G \in \mathcal{S}} [G]_h.$$

Thus $\mathcal{S}$ is adjointly closed if and only if $\mathcal{C}(\mathcal{S}) = \mathcal{S}$.

It is not hard to check that

$$\mathcal{C}(\mathcal{S}_3) \supseteq \langle \mathcal{A} \cup \{K_1\} \cup \{Y_n : n \geq 5\} \rangle.$$

Dong, Teo, Little and Hendy (2002c) proposed the following two conjectures.

Conjecture 11.6.3 *Let $\mathcal{A} = \{K_3, D_4\} \cup \{P_n : n \geq 3\} \cup \{C_n : n \geq 4\}$ and $\mathcal{S}_3 = \langle \mathcal{A} \rangle$. Then*

$$\mathcal{C}(\mathcal{S}_3) = \langle \mathcal{A} \cup \{K_1\} \cup \{Y_n : n \geq 5\} \rangle.$$

Conjecture 11.6.4 *Let $\mathcal{B} = \{K_1\} \cup \{C_n : n \geq 4\}$ and $\mathcal{S}_4 = \langle \mathcal{B} \rangle$. Then*

$$\mathcal{C}(\mathcal{S}_4) = \langle \mathcal{B} \cup \{D_4\} \cup \{Y_n : n \geq 5\} \rangle.$$

Very recently, Zhao, Li and Liu (2003) showed that the above conjectures are false. They established the following results.

Theorem 11.6.11 *Let $\mathcal{A}$ and $\mathcal{B}$ be families of graphs defined in Conjectures 11.6.3 and 11.6.4. Then*
(i)

$$
\begin{aligned}
\mathcal{C}(\mathcal{S}_3) &= \{G : \beta(G) > -4\} \\
&= \langle \mathcal{A} \cup \{K_1\} \cup \{Y_n : n \geq 5\} \cup \{T(1, 2, i) : 2 \leq i \leq 4\} \\
&\quad \cup \{D_i : 4 \leq i \leq 7\} \rangle ;
\end{aligned}
$$

(ii) $\mathcal{C}(\mathcal{S}_4) = \langle \mathcal{B} \cup \{Y_n : n \geq 5\} \cup \{D_{i+2}, \ T(1, 2, i) : i = 2, 4, 5\} \rangle.$ $\qquad \square$

Zhao, Liu, Li and Zhang (2002) studied the adjoint uniqueness of graphs that satisfy $R_1(G) = v(G) - e(G) + n_G(K_3)$ (see Corollary 11.5.1). That is,

$$G \in \langle \{K_4, K_4 - e\} \cup \{P_n, C_{n+1}, D_{n+2}, J_{n+4} : n \geq 2\} \rangle.$$

They showed a result which includes the following.

Theorem 11.6.12   Let A, B, M, E be finite sets such that $A \subseteq \{i \in \mathbb{N} : i > 2,\ i$ even and $i \not\equiv 4 \pmod{10}\}$, $B \subseteq \{j \in \mathbb{N} : j \geq 5\}$, $M \subseteq \{k \in \mathbb{N} : k \geq 9, k \not\equiv 3 \pmod{5}\}$ and $E \subseteq \{s \in \mathbb{N} : s \geq 6,\ s \not\equiv 2 \pmod{5}\}$. Let

$$G = \bigcup_{i \in A} P_i \cup \bigcup_{j \in B} C_j \cup \bigcup_{k \in M} D_k \cup \bigcup_{s \in E} J_s \cup \ell K_3 \cup t(K_4 - e) \cup r K_4,$$

where $\ell, t, r \in \mathbb{N}_0$. Then G is adjointly unique if and only if $j \neq i + 1$ for all $i \in A$ and $j \in B$.   □

As a special case, the authors obtained the following result.

Corollary 11.6.1   Let $s, t \in \mathbb{N}$, $\{n_1, n_2, \cdots, n_s\}$ and $\{m_1, m_2, \cdots, m_t\} \subseteq \mathbb{N}$. If $n_i \geq 9$, $n_i \not\equiv 3 \pmod{5}$, $m_i \geq 6$ and $m_i \not\equiv 2 \pmod{5}$, then $\bigcup_{i=1}^{s} D_{n_i} \cup \bigcup_{i=1}^{t} J_{m_i}$ is adjointly unique.   □

Next we present a result of Zhao, Li, Zhang and Liu (2004) concerning the adjoint uniqueness of graphs in $\langle\{U_n : n \geq 6\}\rangle$ (see Figure 11.2). By making use of Theorem 11.4.4 and Theorem 11.6.7, they established the following result.

Theorem 11.6.13   For $s \in \mathbb{N}$, let $\{n_1, n_2, \cdots, n_s\} \subseteq \mathbb{N}$ with each $n_i \geq 6$. Then $\bigcup_{i=1}^{s} U_{n_i}$ is adjointly unique if and only if for all $i = 1, 2, \cdots, s$, either $n_i = 7$ or n_i is even with $n_i \geq 10$.   □

Note that $U_6 \sim_h (K_4 - e) \cup 2K_1$, $U_8 \sim_h C_3 \cup K(1,4)$ and $U_9 \sim_h K_1 \cup K(1,3) \cup (K_4 - e)$. Moreover, $U_{2n+1} \sim_h U_{n+2} \cup C_{n-1}$ for $n \geq 5$. Hence Theorem 11.6.13 gives all adjointly unique graphs in $\langle\{U_n : n \geq 6\}\rangle$.

Recall that $C_k(P_2)$ is the graph obtained from C_k by joining a new vertex to one vertex of C_k. We will denote $C_{n-1}(P_2)$ by A_n for convenience. Zhao, Li, Liu and Wang (2002b) proved the following result.

Theorem 11.6.14   Let $s \in \mathbb{N}$ and $\{n_1, n_2, \cdots, n_s\} \subseteq \mathbb{N}$ with $n_i \geq 5$ and $n_i \not\equiv 2 \pmod{4}$ for all $i = 1, 2, \cdots, s$. Let $n \geq 5$ and $n \not\equiv 2 \pmod{3}$. Then $A_n \cup \bigcup_{i=1}^{s} C_{n_i}$ is adjointly unique if and only if $n \neq 7$.   □

In the same paper, Zhao, Li, Liu and Wang (2002b) also proved the following theorem (for notation B_n, see Section 11.3).

Theorem 11.6.15   *Let $s \in \mathbb{N}$ and $\{n_1, n_2, \cdots, n_s\} \subseteq \mathbb{N}$ with $n_i \not\equiv 2$ (mod 4) and $n_i \geq 5$ for all $i = 1, 2, \cdots, s$. Let $n \geq 6$. Then $B_n \cup \bigcup_{i=1}^{s} C_{n_i}$ is adjointly unique if and only if*

$$n \notin \{6, 7, 10\} \quad and \quad (n_i, n) \notin \{(9, 8), (15, 9)\}. \qquad \square$$

Remark   In arriving at the above two results, Zhao, Li, Liu and Wang (2002b) had to make use of the invariant $R_2(G)$ and the ordering of the minimum roots of the adjoint polynomials of the components.

11.7   Further results

Attempts have been made by several researchers to employ the factorization theory of polynomials over the rational field $\mathbb{Q}$ to obtain adjointly unique graphs. However, this has not produced a great many results which we have not covered in the previous section.

We define $h_1(G, x)$ as the polynomial such that

$$h(G, x) = x^{\partial(G)} h_1(G, x).$$

If $h_1(G, x)$ is irreducible over $\mathbb{Q}$, then we say that G is an *irreducible* graph. Liu and others (see Liu (1997)) found adjointly unique graphs whose components belong to $\{P_n : n \geq 6\} \cup \{C_n : n \geq 5\}$, under the assumption that each component is irreducible. As we have seen in the previous section, this problem has been completely settled without the assumption of irreducibility of the components. Note that not all adjointly unique P_n and C_n are irreducible. Indeed, it has been shown by Liu (1994) (see also Du (1996a)) that for $n \geq 4$, C_n is irreducible if and only if $n = 2^k$ or n is a prime. Zhao and Liu (2001b) showed that for $n \geq 4$, P_n is irreducible if and only if $n + 1$ is prime. In general, not a great deal is known as to when a graph G is irreducible. For example, the only J_n known to be irreducible are J_6 and J_{10} (see Liu and Zhao (1997)). Liu (1997) and Z.C. Jin (2001) have found some irreducible D_n and a few $T(k, s, t)$.

We now list some results which we have not mentioned earlier.

1. Wang and Liu (2001) (see also R.Y. Liu (1996), Li, Bao and Liu (1997)): Let

$$
\begin{aligned}
T_1 &= \{T(1, s, t) : 3 \leq s \leq t\}, \\
T_2 &= \{T(k, s, t) : 2 \leq k \leq s \leq t\}.
\end{aligned}
$$

For any $n \in \mathbb{N}$, let $G = \bigcup_{i=1}^{n} G_i$, where $G_i \in \mathcal{T}_1 \cup \mathcal{T}_2$. Then G is adjointly unique if each G_i is irreducible.

Note that $T(1,1,1)$ is adjointly unique and $T(1,1,n)$ and $T(1,2,n)$ are not adjointly unique for all $n \geq 2$. Indeed, $T(1,1,n) = Y_{n+3} \sim_h K_1 \cup C_{n+2}$ and $T(1,2,n) \sim_h K_1 \cup D_{n+3}$. Moreover, $T(1,n,n+3)$, $T(1,n,n)$, $T(2,2,n)$ and $T(1,n,2n+5)$ are not adjointly unique either.

2. Wang (1997): Let A, B be finite sets such that

$$
\begin{aligned}
A &\subseteq \{n \in \mathbb{N} : n \geq 2, n \neq 4\}, \\
B &\subseteq \{(k,s,t) : k,s,t \in \mathbb{N}, 2 \leq k \leq s \leq t\}.
\end{aligned}
$$

Then $G = \bigcup_{i \in A} P_i \cup \bigcup_{(k,s,t) \in B} T(k,s,t)$ is adjointly unique if each component of G is irreducible.

Further results can be found in H.X. Zhao (1997), Jiang and Wang (2001), Jiang (1999), Cui, Zhao and Ma (2000) and Zhao, Li, Liu and Wang (2002b).

We end this chapter by considering some graph polynomials that are related to the chromatic polynomial in the study of chromaticity of graphs. The first of these is the σ-polynomial introduced by Korfhage (1978). This can be written in terms of the adjoint polynomial as follows:

$$
\sigma(G,x) = h(\overline{G},x)/x^{\chi(G)}.
$$

Evidently, $\sigma(G,x) = h_1(\overline{G},x)$, where $h_1(\overline{G},x)$ is the factor of $h(\overline{G},x)$ such that $h_1(\overline{G},0) \neq 0$.

We shall only discuss results relevant to chromaticity of graphs. For general properties of σ-polynomials, the reader is referred to Korfhage (1978), Dhurandhar (1984), Li (1988), Du (1996a) and Wakelin (1993). Whereas Dhurandhar (1984), Xu (1988) and Korfhage (1988) characterized quadratic σ-polynomials by their coefficients, Frucht and Giudici (1983) and Li (1985) characterized them graphically.

Let $a, b, c \in \mathbb{N}_0$, where $b \geq c$ and $a + c > 0$. Consider a complete bipartite graph $K(2,a) = (A, B; E)$, where $A = \{u,v\}$. Let $M(a,b,c)$ be the graph obtained from the above graph by joining b independent vertices to u and c independent vertices to v. Let $N(a,b,c)$ be the graph obtained from $M(a,b,c)$ by joining u to v. The following theorem was obtained by Frucht and Giudici (1983). N.Z. Li (1985) also provided a proof for the result.

Theorem 11.7.1 *A graph G has a quadratic σ-polynomial if and only if $\overline{G}$ is one of the following graphs:*

(i) $C_5 \cup rK_1$,

(ii) $N(a,b,c) \cup rK_1$ or $M(a,b,c) \cup rK_1$, where $(a,b,c) \neq (1,0,0)$, $a+c > 0$ and $r \in \mathbb{N}_0$. □

Li, Whitehead Jr. and Xu (1987) found all χ-unique graphs having quadratic σ-polynomials.

Theorem 11.7.2 *Let G be a graph that has a quadratic σ-polynomial. Then G is χ-unique if and only if $\overline{G}$ is one of the following graphs:*

(i) C_5,

(ii) $N(a,b,0)$, for all $a,b \in \mathbb{N}$,

(iii) $M(a,b,b) \cup rK_1$, where $b \neq 0$ and $ra = 0$,

(iv) $M(a, c+1, c) \cup rK_1$, where $c \neq 0$, $r \in \{0,1\}$ or $a \in \{0,1\}$, and $(a,c) \neq (1,1)$,

(v) $M(a, c+p, c) \cup rK_1$, where $p \geq c+3$, $c \neq 0$, and $a \leq p$ or $r \leq p$. □

For classification of graphs having σ-polynomials with degree higher than 2, the reader is referred to Dhurandhar (1984), N.Z. Li (1992b), and N.Z. Li and Whitehead Jr. (1989).

Some articles using the notion of σ-polynomial could effectively be done using either the properties of chromatic polynomial or adjoint polynomial. For example, Du (1993a) could be presented using $\alpha(G,k)$, and the paper by Du (1996a) as well could be presented more neatly by using the notion of adjoint polynomial.

Du (1996b) introduced the notion of σ-uniqueness in the usual way, studied the relation between σ-uniqueness and χ-uniqueness, and obtained some interesting results in this aspect.

Another graph polynomial connected to the chromatic polynomial is the matching polynomial. Though there is no general agreement on its definition, we will adopt the one used by Farrell (1979). Another definition can be found in Godsil and Gutman (1981). For a graph G, let $\beta(G,k)$

be the number of matchings in G with k edges. The *(simple) matching polynomial* of G is defined to be

$$M(G, x) = \sum_{k=0}^{\lfloor n/2 \rfloor} \beta(G, k)x^{n-k},$$

where $n = v(G)$. Recall that $P(G, \lambda) = \sum_{i=0}^{n} \alpha(G, n - k)(\lambda)_{n-k}$. Frucht and Giudici (1985) showed that if G is triangle-free, then

$$\beta(G, k) = \alpha(\overline{G}, n - k)$$

for all $k = 1, 2, \cdots, \lfloor n/2 \rfloor$. Farrell and Whitehead Jr. (1992) showed that the converse is also true. In fact, they proved a more general result.

Theorem 11.7.3   *Let G be a graph of order n. Then $\beta(G, k) \leq \alpha(\overline{G}, k)$ for all k, and equality holds for all k if and only if G is triangle-free.*   □

From this, we can see that the matching polynomial is a useful tool for studying the chromaticity of triangle-free graphs. Two graphs are *matching equivalent* if they have the same matching polynomial. Farrell and Whitehead Jr. (1992) deduced from Theorem 11.7.3 the following result.

Theorem 11.7.4   *Let G and H be triangle-free graphs. Then $\overline{G}$ and $\overline{H}$ are χ-equivalent (G and H are adjointly equivalent) if and only if G and H are matching equivalent.*   □

A graph G is *matching unique* if $M(G, x) = M(H, x)$ implies that $H \cong G$. An interesting question arises from Theorem 11.7.4:

> Suppose G is triangle-free and matching unique. Does it follow that $\overline{G}$ be χ-unique?"

Farrell and Whitehead Jr. (1992) provided the following example to disprove it.

Example 11.7.1   *The path P_4 is matching unique. But $\overline{P}_4 \cong P_4$ is not χ-unique. Note that $P_4 \sim K(1, 3)$.*

In this connection, Guo and Li (1989) established the following result.

Theorem 11.7.5   *Let G be a 2-regular graph that has no C_3 or C_4 as a subgraph. Then G is matching unique if and only if $\overline{G}$ is χ-unique.*   □

Since C_n is matching unique for all $n \geq 5$ (see Guo and Li (1989)), we conclude that $\overline{C}_n$ is χ-unique for $n \geq 5$. Note that this is a special case of Theorem 11.6.8.

We end this chapter by mentioning that the characteristic polynomial of a tree is closely related to its adjoint polynomial. H.X. Zhao and Liu (2001a) have used this connection to compute the roots of the adjoint polynomials of paths, cycles and certain trees.

Exercise 11

11.1 Show that

(i) $h(P_{2n+1}, x) = h(P_n \cup C_{n+1}, x)$ for any $n \geq 3$;

(ii) $h(Y_n, x) = h(K_1 \cup C_{n-1}, x)$ for any $n \geq 5$;

(iii) $h(U_n, x) = x^3(x+4)h(P_{n-4}, x)$ for any $n \geq 6$;

(iv) $h(U_{2n+1}, x) = h(U_{n+2}, x)h(C_{n-1}, x)$ for any $n \geq 5$.

11.2 For any graph G, show that

$$n_G(P_4) = \sum_{xy \in E(G)} (d(x) - 1)(d(y) - 1) - 3n_G(K_3).$$

(See Dong, Teo, Little and Hendy (2002a).)

11.3 Show that

$$N(G, n-2) = n_G(K_3) + \binom{m+1}{2} - \frac{1}{2}\sum_{v \in V(G)} d(v)^2.$$

(See Liu (1990).)

11.4 Derive a formula for $N(P_n, k)$, and one for $N(C_n, k)$, for all $k \geq n$, $k \in \mathbb{N}$.

(See Farrell (1979).)

11.5 Let G be a connected (n, n)-graph. Show that

$$N(G, n-2) \leq \binom{n-1}{2} - 1.$$

(See Zhao 1994).)

11.6 For $x, y \in V(G)$, let $N(x, y) = (N_G(x) \cup N_G(y)) \backslash \{x, y\}$. Show that

$$R_1(G) = R_1(G - xy) - |N(x, y)| + 1.$$

(See Du (1996a).)

11.7 Let $xy \in E(G)$ with $N_G(x) \cap N_G(y) = \emptyset$. Let G' be a graph obtained from G by subdividing the edge xy. Then $R_1(G') = R_1(G)$.

(See Dong, Teo, Little and Hendy (2002a).)

11.8 Find all connected graphs G such that $R_i(G) = 1$ for all $i \in \{-1, 0, 1\}$.

(See Liu (1992b), Du (1995).)

11.9 Let H be a proper subgraph of a connected graph G. Show that $\beta(G) < \beta(H)$.

(See Zhao, Li, Liu and Ye (2004).)

11.10 For $n \geq 6$, show that $\beta(J_{n-1}) < \beta(J_n) < \beta(D_n)$.

(See Zhao, Liu, Li and Zhang (2002).)

11.11 Let $R_3(G) = R_1(G) + e(G) - v(G)$. Show that if G is connected and $xy \in E(G)$, then

$$R_3(G) = R_3(G - xy) - |N(x, y)| + 2$$

(See Zhao, Liu, Li and Zhang (2002).)

11.12 For any (n, m)-graph, let

$$R_2(G) = N(G, n - 3) - \binom{m}{3} - (m - 2)\left(N(G, n - 2) - \binom{m}{2}\right) - m.$$

Show that if G is triangle-free then

$$R_2(G) = 2 \sum_{x \in V(G)} \binom{d(x)}{3} + \sum_{xy \in E(G)} (d(x) - 1)(d(y) - 1) - m.$$

(See Dong, Teo, Little and Hendy (2002a).)

11.13 Compute $R_2(C_n)$ for all $n \geq 3$, and $R_2(P_n)$ for all $n \geq 1$.

11.14 Show that for all $n \geq 8$, $h(J_n, x) = x(h(J_{n-1}, x) + h(J_{n-2}, x))$.

(See Liu and Zhao (1997).)

Chapter 12

Real Roots of Chromatic Polynomials

12.1 Introduction

For a graph G, a root of $P(G, \lambda)$ is also called a *chromatic root* of G. We shall present in this chapter some results on real chromatic roots of graphs.

An interval is called a *root-free interval* for a chromatic polynomial $P(G, \lambda)$ if G has no chromatic root in this interval. Likewise, an interval is called a *root-free interval* for a family S of graphs if every graph in S has no chromatic root in this interval. It is well-known that $(-\infty, 0)$ and $(0, 1)$ are two maximal root-free intervals for the family of all graphs. Jackson (1993) showed that $(1, 32/27]$ is another maximal root-free interval for the family of all graphs. Thomassen (1997) further proved that for any interval (λ_1, λ_2), where $32/27 \le \lambda_1 < \lambda_2$, there exists a graph having a chromatic root in this interval. The results of Jackson and Thomassen thus completed the search for root-free intervals for the family of all graphs. These results will be introduced in Section 12.2.

In Section 12.3, we shall exhibit some real numbers which can never be chromatic roots of any graph. In Section 12.4, we shall consider the upper root-free intervals for some families of graphs which are closed under minors. In Sections 12.5 and 12.6, we shall focus on planar graphs and plane triangulations respectively. We then proceed to Section 12.7 to introduce Thomassen's result on the relationship between the existence of chromatic roots and that of hamiltonian paths in a 2-connected graph. In Section

12.8, Woodall' result (1977) on the chromatic roots of a bipartite graph will be presented. In Section 12.9, we shall prove some results on root-free intervals for the family of graphs which contain a q-tree as a spanning subgraph. In Section 12.10, we provide a formula on the maximum non-integral chromatic roots for graphs with fixed order n. Finally, we shall state, in Section 12.11, a partial result on Sokal's conjecture about the chromatic roots of a graph.

12.2 Root-free intervals for all chromatic polynomials

By Theorem 2.3.1, for any graph G of order n, we have

$$P(G, \lambda) = \sum_{i=1}^{n} (-1)^{n-i} h_i \lambda^i, \qquad (12.1)$$

where $h_i \in \mathbb{N}_0$ for $i = 1, 2, \cdots, n-1$ and $h_n = 1$. This result can be used to prove that $(-\infty, 0)$ is a root-free interval for all chromatic polynomials (see, for instance, Woodall (1977)).

Theorem 12.2.1   *Let G be a graph of order n. Then*

$$(-1)^n P(G, \lambda) > 0$$

for all real $\lambda < 0$.

Proof.   By (12.1), we have

$$(-1)^n P(G, \lambda) = (-1)^n \sum_{i=1}^{n} (-1)^{n-i} h_i \lambda^i = \sum_{i=1}^{n} h_i (-\lambda)^i > 0,$$

if $\lambda < 0$.   □

Since $P(G, k) = 0$ for all $k \in \mathbb{N}_0$ with $k < \chi(G)$, every root-free interval for $P(G, \lambda)$ does not include any integer in $\{0, 1, 2, \cdots, \chi(G) - 1\}$. In the following, we shall show that $(0, 1)$ is another root-free interval for all chromatic polynomials (see Woodall (1977) also).

Theorem 12.2.2   *Let G be a graph of order n having c components. Then*

$$(-1)^{n-c} P(G, \lambda) > 0,$$

for all real $\lambda \in (0, 1)$.

Proof. If G is a forest, then $P(G, \lambda) = \lambda^c(\lambda - 1)^{n-c}$, and the result is trivial in this case.

Suppose that the result is not true in general and H is a graph for which the result fails. We may choose H such that $n + m$ is minimum, where n and m are its order and size respectively. Thus, H is not a forest, and so H contains cycles. Let xy be an edge on some cycle in H. Observe that $H - xy$ has n vertices, $m - 1$ edges and c components, and $H \cdot xy$ has $n - 1$ vertices, at most $m - 1$ edges and c components. By the assumption of H, the result holds for both $H - xy$ and $H \cdot xy$. Thus for all real $\lambda \in (0, 1)$,

$$(-1)^{n-c}P(H - xy, \lambda) > 0$$

and

$$(-1)^{n-1-c}P(H \cdot xy, \lambda) > 0,$$

and we have, by FRT,

$$
\begin{aligned}
(-1)^{n-c}P(H, \lambda) &= (-1)^{n-c}P(H - xy, \lambda) - (-1)^{n-c}P(H \cdot xy, \lambda) \\
&= (-1)^{n-c}P(H - xy, \lambda) + (-1)^{n-1-c}P(H \cdot xy, \lambda) \\
&> 0.
\end{aligned}
$$

$\square$

The following elegant result, discovered by Jackson (1993), furnishes actually the last maximal root-free interval for all chromatic polynomials.

Theorem 12.2.3 *Let G be a connected graph of order $n \geq 2$ and block number b. Then $P(G, \lambda)$ is non-zero with the sign of $(-1)^{n+b+1}$ for all real $\lambda \in (1, 32/27]$.*

$\square$

The complete proof of Theorem 12.2.3 given by Jackson (1993) is rather lengthy. We sketch here his method only.

A *double subdivision* of a graph G is an operation on G defined below:

> choose an edge uv in G and construct a new graph from $G - uv$ by adding two new vertices and joining each new vertex to both u and v.

A *generalized edge* (resp., *generalized triangle*) is either K_2 (resp., K_3) or any graph that can be obtained from K_2 (resp., K_3) by a sequence of double subdivisions. A connected graph is said to be *separable* if it contains cut-vertices. Jackson first proved the following result (its proof is omitted).

Lemma 12.2.1   *A graph G is a generalized triangle if it satisfies the following properties:*

(i) *G is 2-connected;*
(ii) *for each $e \in E(G)$, $G - e$ has exactly two blocks;*
(iii) *for each 2-element cut $\{u, v\}$, $G - \{u, v\}$ has an odd number of components;*
(iv) *when $\{u, v\}$ is a 2-element cut of G such that $G = G_1 \cup G_2$, $V(G_1) \cap V(G_2) = \{u, v\}$, $v(G_i) \geq 3$ for each $i = 1, 2$, and G_1 is a generalized uv-edge, then G_2 is separable with exactly two blocks.* □

Jackson proved Theorem 12.2.3 by contradiction. When $n = 2$, Theorem 12.2.3 holds, since G is K_2 and we have $P(G, \lambda) = \lambda(\lambda - 1)$. Suppose the theorem is false and let G be a connected graph of order $n \geq 3$ and t be a real number in $(1, 32/27]$ such that $P(G, t)$ does not satisfy the conclusion. We may suppose that G has been chosen such that $v(G) + e(G)$ is as small as possible. The remainder of Jackson's proof is divided into two parts, which establish the required contradiction.

Part A: show that G is a generalized triangle.

Based on the assumptions on G, the following facts can be verified.

Fact 1: G is 2-connected, and hence $b = 1$.

Fact 2: $G \cdot e$ is 2-connected for each $e \in E(G)$.

Fact 3: Let $\{u, v\}$ be a 2-element cut of G. Then $uv \notin E(G)$ and $G - \{u, v\}$ has an odd number of components.

Fact 4: $G - e$ has exactly two blocks for each $e \in E(G)$.

Fact 5: Suppose $\{u, v\}$ is a 2-element cut of G such that $G = G_1 \cup G_2$, $V(G_1) \cap V(G_2) = \{u, v\}$, $v(G_i) \geq 3$ for each $i = 1, 2$, and such that G_1 is a generalized uv-edge. Then G_2 is separable with exactly two blocks.

Part A follows from Lemma 12.2.1 and Facts 1-5. Since the order of each generalized triangle is odd, n is also odd. Thus $n + b + 1$ is odd. Since G and t constitute a counterexample to Theorem 12.2.3, we have $P(G, t) \geq 0$. However, we shall see in what follows that this is impossible.

Part B: show that $P(G, t) < 0$.

To prove Part B, Jackson first showed the following result.

Fact 6: Let G_0 be a 2-connected graph with $v(G_0) + e(G_0) \leq v(G) + e(G)$.

(a) If $G_0 = G_1 \cup G_2$, where G_1 and G_2 are generalized uv-edges such that $V(G_1) \cap V(G_2) = \{u, v\}$ and $v(G_i) \geq 4$ for $i = 1, 2$, then

$$P(G_0, t) \geq \frac{9}{4} t^{-1} P(G_1, t) P(G_2, t).$$

(b) If $G_0 = (G_1 \cup G_2) + uv$, where $V(G_1) \cap V(G_2) = \{w\}$ and G_1 and G_2 are generalized wu-edge and generalized wv-edge respectively, then

$$P(G_0, t) \leq -\frac{5}{9} P(G_0 \cdot uv, t).$$

(c) If G_0 is a generalized uv-edge and $v(G_0) \geq 4$, then

$$P(G_0 + uv, t) \geq \frac{5}{8} P(G_0, t).$$

Since G is a generalized triangle, there exist subgraphs G_1, G_2, G_3 and vertices v_1, v_2, v_3 of G such that $G = G_1 \cup G_2 \cup G_3$, $V(G_i) \cap V(G_{i+1}) = \{v_{i+1}\}$, and G_i is a generalized $v_i v_{i+1}$-edge for all $i \in \{1, 2, 3\}$, where the subscripts are read modulo 3.

If $v(G_i) = 2$ for some i, we may apply Fact 6(b) to deduce that $P(G, t) \leq -(5/9)P(G \cdot v_i v_{i+1}, t)$. Since $v(G \cdot v_i v_{i+1}) + e(G \cdot v_i v_{i+1}) < v(G) + e(G)$, by the choice of G and t, we have $P(G \cdot v_i v_{i+1}, t) > 0$. Thus $P(G, t) < 0$, contradicting the choice of G and t. Hence $v(G_i) \geq 4$ for all $i \in \{1, 2, 3\}$, as $v(G_i)$ must be even.

Let $H = G_2 \cup G_3$. Applying Theorem 1.3.1 to $G = G_1 \cup H$ gives

$$
\begin{aligned}
t(t-1)P(G, t) = \ & P(H', t)\left(tP(G_1', t) - (t-1)P(G_1, t)\right) \\
& + (t-1)P(H, t)\left(P(G_1, t) - P(G_1', t)\right), \quad (12.2)
\end{aligned}
$$

where $G_1' = G_1 + v_1 v_2$ and $H' = H + v_1 v_2$. Applying Fact 6(c) to G_1' gives

$$P(G_1', t) \geq \frac{5}{8} P(G_1, t). \qquad (12.3)$$

Applying Fact 6(b) and 6(a) to H' and $H' \cdot v_1 v_2$ respectively, we obtain

$$
\begin{aligned}
P(H', t) \ & \leq \ -\frac{5}{9} P(H' \cdot uv, t) \\
& \leq \ -\frac{5}{9}\left(\frac{9}{4} t^{-1} P(G_2, t) P(G_3, t)\right) \\
& = \ -\frac{5}{4} t^{-1} P(G_2, t) P(G_3, t). \qquad (12.4)
\end{aligned}
$$

Applying Theorem 1.3.2 to H gives

$$P(H,t) = t^{-1}P(G_2,t)P(G_3,t). \tag{12.5}$$

Now, since $v(G_i) + e(G_i) < v(G) + e(G)$, by the choice of G, we have $P(G_i,t) > 0$ for $i = 1,2,3$. Since $t \le 32/27$, we have

$$tP(G_1',t) - (t-1)P(G_1,t) \ge (1 - \frac{3}{8}t)P(G_1,t) > 0.$$

Substituting (12.3), (12.4) and (12.5) into (12.2), we obtain

$$
\begin{aligned}
t^2(t-1)P(G,t) \quad \le \quad & -\frac{5}{4}\left(1 - \frac{3}{8}t\right)P(G_1,t)P(G_2,t)P(G_3,t) \\
& +(t-1)\frac{3}{8}P(G_1,t)P(G_2,t)P(G_3,t) \\
= \quad & -\frac{52-27t}{32}P(G_1,t)P(G_2,t)P(G_3,t). \tag{12.6}
\end{aligned}
$$

Thus $P(G,t) < 0$, which proves Part B.

In the above proof of Theorem 12.2.3, we note that the only place for which the inequality $t \le 32/27$ is critical is the proof of Fact 6. The method used in this proof can be applied to investigate the root-free interval $(1,\gamma)$, where $\gamma \le 2$, of some special families of graphs. For example, Thomassen (2000) used a similar method to prove Theorem 12.7.1 which determines the largest γ with $1 < \gamma \le 2$ such that $P(G,\lambda)$ is root-free in the interval $(1,\gamma)$ for all graphs G with a hamiltonian path.

Jackson (1993) also proved that the number $32/27$ in Theorem 12.2.3 cannot be replaced by any larger number.

Define a sequence of graphs H_n as follows. Let $H_1 = C_3 : w_1w_2w_3w_1$. For $n \ge 2$, let H_n be the graph obtained from H_{n-1} by applying the double subdivision operation to each edge of H_{n-1} except w_1w_2. Jackson (1993) showed that $P(H_n,\lambda)$ has a root arbitrarily close to $32/27$ as $n \to \infty$.

Theorem 12.2.4   *For any real $\epsilon > 0$, there exists $n \in \mathbb{N}$ such that $P(H_n,\lambda)$ has a root in $(32/27, 32/27 + \epsilon)$.* $\square$

Thomassen (1997) further proved the following result which shows that there is no more root-free interval for all chromatic polynomials beyond what Jackson has found.

Theorem 12.2.5   *For any real numbers λ_1 and λ_2 with $32/27 \leq \lambda_1 < \lambda_2$, there exists a graph G such that $P(G, \lambda) = 0$ for some $\lambda \in (\lambda_1, \lambda_2)$.* □

It can be proved by induction that for any connected graph G of order n, $|P(G, \lambda)| \geq |\lambda(\lambda - 1)^{n-1}|$ for all real $\lambda \in (0, 1)$ (see Exercise 12.1). By Theorem 12.2.3, Thomassen (1997) extended this result to the interval $(0, 32/27]$.

Theorem 12.2.6   *Let G be a graph of order n. If $1 \leq \lambda \leq 32/27$, then*

$$|P(G, \lambda)| \geq \lambda(\lambda - 1)^{n-1}.$$

□

Theorems 12.2.1, 12.2.2, 12.2.3 and 12.2.5 show that there are only three maximal root-free intervals for all chromatic polynomials; namely, $(-\infty, 0)$, $(0, 1)$ and $(1, 32/27]$. In some of the subsequent sections, we shall introduce results on root-free intervals for some families of graphs. Jackson (1993) proposed the following conjecture for the family of 3-connected non-bipartite graphs.

Conjecture 12.2.1   *Let G be a 3-connected non-bipartite graph. Then $P(G, \lambda)$ has no roots in the interval $(1, 2)$.*

Birkhoff and Lewis (1946) showed that for any plane triangulation G, $P(G, \lambda)$ is root-free in $(1, 2)$. Thus Conjecture 12.2.1 is true for all 3-connected triangulations. The conjecture has also been verified for all graphs of order at most nine by G. Royle mentioned in Jackson (1993).

Note that Conjecture 12.2.1 is not necessarily true for bipartite graphs. Jackson (1993) pointed out that the chromatic polynomial of every 2-connected bipartite graph of odd order has a root in $(1, 2)$. A very short proof of this observation is given below.

Theorem 12.2.7   *Let G be a connected bipartite graph with $v(G) \geq 2$. If $v(G) + b(G)$ is even, then G has a chromatic zero in $(1, 2)$.*

Proof. Since G is bipartite, $P(G, 2) > 0$. Since $P(G, \lambda)$ is continuous, there exists a real number δ with $0 < \delta < 1$ such that

$$P(G, \lambda) > 0$$

for all $\lambda \in (2 - \delta, 2)$. As $v(G) + b(G)$ is even,

$$(-1)^{v(G)+b(G)+1} P(G, \lambda) < 0$$

for all $\lambda \in (2 - \delta, 2)$. By Theorem 12.2.3, $(-1)^{v(G)+b(G)+1} P(G, \lambda) > 0$ if $1 < \lambda \leq 32/27$. Since $P(G, \lambda)$ is continuous in the interval $(1, 2)$, the result holds.

□

12.3   Real numbers which are not chromatic roots

In this section, we provide families of real numbers which are not chromatic roots of any graph. It is easy to prove that if $P(x)$ is a monic polynomial with integral coefficients, then every rational root of $P(x) = 0$ is an integer (see Exercise 12.2). Since every chromatic polynomial is a monic polynomial with integral coefficients, the next result follows immediately.

Theorem 12.3.1   *If r is a rational number and $P(G, r) = 0$, then r is a positive integer.* $\square$

It was pointed out in Read and Tutte (1988) that $P(G, \tau + 1) \neq 0$ for any graph G, where $\tau = (1 + \sqrt{5})/2$, called the *golden ratio*. Actually, a more general result can be obtained by Theorems 12.2.1, 12.2.2 and 12.2.3.

Theorem 12.3.2   *Let a and b be rational numbers such that $b > 0$, $\sqrt{b}$ is irrational and $a < 32/27 + \sqrt{b}$. Then $a + \sqrt{b}$ cannot be a root of any chromatic polynomial.*

Proof.   Let G be any graph. Since $P(G, \lambda)$ is a polynomial with integral coefficients and $\sqrt{b}$ is irrational, if $P(G, a + \sqrt{b}) = 0$, then $P(G, a - \sqrt{b}) = 0$. But $a - \sqrt{b}$ is irrational and $a - \sqrt{b} < 32/27$. By Theorems 12.2.1, 12.2.2 and 12.2.3, $P(G, a - \sqrt{b}) \neq 0$. Hence $a + \sqrt{b}$ is not a root of $P(G, \lambda)$. $\square$

Corollary 12.3.1   *Let b be a positive rational number such that $\sqrt{b}$ is irrational. Then $\sqrt{b}$ cannot be a root of any chromatic polynomial.* $\square$

12.4   Upper root-free intervals

Let $\mathcal{F}$ be a family of graphs, and $c(\mathcal{F})$ be the supremum of the roots of the chromatic polynomials of graphs in $\mathcal{F}$. If $c(\mathcal{F}) < \infty$, then $(c(\mathcal{F}), \infty)$ or $[c(\mathcal{F}), \infty)$ is called the *upper root-free interval* for $\mathcal{F}$, depending on whether $c(\mathcal{F})$ is a chromatic root of a graph in $\mathcal{F}$.

It is clear that the family of all graphs has no upper root-free interval. Woodall's result (in Section 12.8) on bipartite graphs also shows that the family of bipartite graphs has no upper root-free interval. More generally, for any integer k with $k \geq 2$, the family of graphs with chromatic number k has no upper root-free interval (see Exercise 12.3). But some families of graphs do have upper root-free intervals. For example, the family of planar graphs has an upper root-free interval, since $[5, \infty)$ is a root-free interval for this family of graphs (see Theorem 12.4.5).

A *minor* of a graph G is any graph obtained from a subgraph of G by contracting edges. A family $\mathcal{F}$ of graphs is said to be *closed under minors* if for every graph $G \in \mathcal{F}$, every minor of G also belongs to $\mathcal{F}$.

Now let $\mathcal{F}$ be a family of graphs closed under minors and assume that $\mathcal{F}$ does not include all graphs. Then $K_k \notin \mathcal{F}$ for some k. The following result of Mader (1967) shows that if K_k is not a minor of G, then $\delta(G) < f(k)$ for some positive integer $f(k)$.

Theorem 12.4.1 *For every $k \in \mathbb{N}$, there exists a (smallest) integer $f(k)$ such that any graph G with $\delta(G) \geq f(k)$ has K_k as a minor.* □

Kostochka (1982) and Thomason (1984) independently proved that

$$c_1 k \sqrt{\ln k} < f(k) < c_0 k \sqrt{\ln k}.$$

By Theorem 12.4.1, there exists a smallest integer $d = d(\mathcal{F})$ such that every graph in $\mathcal{F}$ contains a vertex of degree at most d. By means of this observation, Woodall (1997) and Thomassen (1997) independently proved the following theorem. Their proofs are quite similar; and in the following, we shall rewrite Woodall's proof by using Lemma 2.5.7.

Theorem 12.4.2 *Let $\mathcal{F}$ be a family of graphs closed under minors. If $\mathcal{F}$ does not contain all graphs, then $P(G, \lambda) > 0$ for every graph G in $\mathcal{F}$ and every real $\lambda > d(\mathcal{F})$.* □

Proof. We need only to consider the case that G is connected. Write $d = d(\mathcal{F})$ for simplicity. Also assume $\lambda > d$. The result is true when $v(G) \leq d$. Assume that the result holds if $v(G) \leq k$ for some $k \in \mathbb{N}$. Now let G be a connected graph in $\mathcal{F}$ with $v(G) = k + 1$. Since $\delta(G) \leq d$, there exists $x \in V(G)$ such that $d(x) \leq d$. By Lemma 2.5.7,

$$P(G, \lambda) = (\lambda - d(x))P(G - x, \lambda) + \sum_{G' \in \mathcal{S}} P(G', \lambda), \qquad (12.7)$$

where $\mathcal{S}$ is a family of some minors G' of G with $v(G') = v(G) - 2$. So $G - x \in \mathcal{F}$ and $\mathcal{S} \subseteq \mathcal{F}$. By the induction hypothesis, we have $P(G - x, \lambda) > 0$ and $P(G', \lambda)$ for each $G' \in \mathcal{S}$. Hence $P(G, \lambda) > 0$. □

Let $\mathcal{F}_{max}$ denote the family of maximal graphs in $\mathcal{F}$, i.e., if G is in $\mathcal{F}_{max}$ and x and y are non-adjacent vertices in G, then $G + xy$ is not in $\mathcal{F}$. Thomassen (1997) proved the following result.

Theorem 12.4.3   *If $\mathcal{F}_{max}$ has an upper root-free interval, then so has $\mathcal{F}$, and the two intervals are identical.*   □

Now we consider two examples. For a graph G, its *tree-width* is defined to be the minimum positive integer k such that G is a subgraph of a k-tree. It is clear that the family of all graphs with tree-width at most k is closed under minors (see Exercise 12.4) and that a graph in this family is maximal if and only if it is a k-tree. Since for any k-tree G of order n,

$$P(G, \lambda) = (\lambda)_k (\lambda - k)^{n-k},$$

by Theorem 12.4.3, we have the following result, due to Thomassen (1997) also.

Theorem 12.4.4   *If $k \in \mathbb{N}$ with $k \geq 2$, then the upper root-free interval for the family of graphs with tree-width k is (k, ∞).*   □

Another example concerns planar graphs. It can be shown that the family of planar graphs is closed under minors (see Exercise 12.5) and that every planar graph has a vertex of degree at most 5. By Theorem 12.4.2, we have the following result, due to Birkhoff and Lewis (1946).

Theorem 12.4.5   *For any planar graph G, $P(G, \lambda) > 0$ for all real $\lambda \geq 5$.*   □

12.5   Planar graphs

The most exciting result on chromatic roots of planar graphs is the Four-colour Theorem, which can be expressed in terms of chromatic polynomials. This famous theorem was finally proved by Appel and Haken (1977) with the assistance of extensive computer enumerations.

Theorem 12.5.1 *For any planar graph G, $P(G, 4) > 0$.*   □

Theorem 12.4.5 says that, for any planar graph G, $P(G, \lambda)$ has no roots in the interval $[5, \infty)$. Thus the study on the chromatic roots of planar graphs can be confined to the interval $(32/27, 5)$. Birkhoff and Lewis (1946) proposed the following conjecture.

Conjecture 12.5.1   *For any planar graph G, $P(G, \lambda)$ has no root in $(4, 5)$.*

As we know, this conjecture is still open. No counterexample is found and no interval (a, b) within $(4, 5)$ has been proven to be root-free for $P(G, \lambda)$ for all planar graphs G either.

While Conjecture 12.5.1 remains open, Thomassen (1997) proposed the following.

Conjecture 12.5.2 *The set of chromatic roots of the family of planar graphs consists of $0, 1$ and a dense subset of $(32/27, 4)$.*

The distribution of chromatic roots of planar graphs within $(32/27, 4)$ is really unclear. There might be some root-free intervals within $(32/27, 4)$.

Theorem 12.2.4 shows that for any $\epsilon > 0$, there exists $n \in \mathbb{N}$ such that $P(H_n, \lambda)$ has a root in $(32/27, 32/27 + \epsilon)$. Observe that H_n is planar for all $n \geq 1$. Hence the family of planar graphs has chromatic roots arbitrarily close to $32/27$. But it is unknown whether there exists planar graphs with chromatic roots arbitrarily close to 4.

A *near-triangulation* with k-face F is a 2-connected graph G embedded in the plane in such a way that there are at least two distinct faces, the face F being bounded by a C_k $(k \geq 3)$ and every other face being bounded by a C_3. A near-triangulation with 3-face is a *triangulation*.

Define $B_n = 2 + 2\cos(2\pi/n)$ for any $n \in \mathbb{N}$. The number B_n is referred to as the nth *Beraha number*. Beraha (1975) asked the following question:

Question 12.5.1 *Is it true that for every $\epsilon > 0$, there exists a triangulation G such that $P(G, \lambda)$ has a root in $(B_n - \epsilon, B_n + \epsilon)$?*

The first few Beraha numbers are $4, 0, 1, 2, \tau^2, 3, \cdots$, where $\tau = (1 + \sqrt{5})/2$. Observe that the sequence B_n has a limit 4. If the answer to Question 12.5.1 is positive, then there exist planar graphs whose chromatic roots are arbitrarily close to 4.

It is obvious that the answer to Question 12.5.1 is affirmative for $n = 2, 3, 4, 6$, since, in these cases, $B_n \leq 3$ and there exist 4-chromatic planar graphs. The answer to the question was proved to be positive for $n = 7$ and $n = 10$ by Beraha, Kahane and Reid (1973). For $n = 5$ $(B_5 = \tau^2)$, Beraha, Kahane and Weiss (1980) proved that this answer to the question is conclusive by finding an infinite family of plane triangulations with chromatic roots tending to τ^2. As early as in 1970, Tutte (1970a) proved that

$$0 < |P(G, \tau^2)| \leq \tau^{5-v(G)},$$

for any plane triangulation G. But this result, which shows that the value of $|P(G, \tau^2)|$ is very small, does not give any evidence that $P(G, \lambda)$ has a root arbitrarily near τ^2 for near-triangulations G. This question remains open for $n = 8, 9$ and $n \geq 11$.

In the remaining part of this section, we present some results on upper root-free intervals of planar graphs. Kuratowski's theorem says that the family of planar graphs is the intersection of the two following families:

(i) the family of graphs with no K_5 as a minor; and

(ii) the family of graphs with no $K(3,3)$ as a minor.

Interestingly, Thomassen (1997) showed that all these families have almost identical upper root-free interval, as shown in the following two results.

Theorem 12.5.2   *The following three families of graphs have the same upper root-free interval:*

(i) the family of planar graphs,

(ii) the family of maximal planar graphs,

(iii) the family of graphs having no K_5 as a minor.   □

It is clear that K_5 is a graph having no $K(3,3)$ as a minor and and that 4 is a root of $P(K_5, \lambda)$. Thus the upper root-free interval of the family of graphs having no $K(3,3)$ as a minor does not contain '4'. Thomassen (1997) proved the following:

Theorem 12.5.3   *The upper root-free interval of the family of graphs having no $K(3,3)$ as a minor is $I \backslash \{4\}$, where I is the upper root-free interval of the family of planar graphs.*

Proof.   It was proved in Wagner (1937) that if G is a maximal graph having no $K(3,3)$ as a minor, then $G \in \mathcal{G}[G_0 \cup_{r_1} G_1 \cup_{r_2} \cdots \cup_{r_s} G_s]$, where G_i is either maximal planar or isomorphic to K_5. The result then follows.   □

12.6   Near-triangulations

In this section, we introduce some results on chromatic roots of near-triangulations. Birkhoff and Lewis (1946) first discovered the following result which shows that Conjecture 12.2.1 holds for near-triangulations. The following proof of this result can be found in Woodall (1977).

Theorem 12.6.1   *If G is a near-triangulation and $\lambda \in (1, 2)$, then $P(G, \lambda)$ is non-zero with the sign of $(-1)^n$, where $n = v(G)$.*

Proof. We proceed by induction on $e(G)$. First, the theorem holds for the smallest near-triangulation, i.e., K_3, whose chromatic polynomial is $\lambda(\lambda - 1)(\lambda - 2)$. Now assume that $e(G) \geq 4$.

Clearly, $\delta(G) \geq 2$. If x is a vertex in G with $d(x) = 2$, then x is a simplicial vertex. Thus $G - x$ is also a near-triangulation and

$$P(G, \lambda) = (\lambda - 2)P(G - x, \lambda),$$

and so the result follows by the induction hypothesis.

Now suppose that $\delta(G) \geq 3$. Let F be the external face (or the infinity face) of G. There must be a vertex of G that is not in the border of F but that is adjacent to two consecutive vertices u, v in the border of F. Observe that both $G - uv$ and $G \cdot uv$ are near-triangulations. By the induction hypothesis, $P(G - uv, \lambda)$ and $P(G \cdot uv, \lambda)$ are non-zero with the sign of $(-1)^n$ and $(-1)^{n-1}$ respectively, where $n = v(G)$. By FRT,

$$P(G, \lambda) = P(G - uv, \lambda) - P(G \cdot uv, \lambda),$$

and so $P(G, \lambda)$ is non-zero with the sign of $(-1)^n$. This completes the proof. □

Although we generally consider simple graphs only, in the case of near-triangulations, it would be convenient to allow the existence of multiple edges.

A *separating cycle* in a plane graph is a cycle with at least one vertex inside it and at least one vertex outside it. In a near-triangulation, a *digon* is a separating cycle of length 2, and a *proper digon* with vertices u and v is one that has no other edge uv inside this cycle.

Theorem 12.6.1 is actually a corollary of the following result obtained by Birkhoff and Lewis (1946).

Theorem 12.6.2 *Let G be a near-triangulation with k-face. If $\lambda \leq 2$, then*

$$|P(G, \lambda)| \geq |\lambda(\lambda - 1)(\lambda - 2)^{k+d-2}(\lambda - 3)^{n-k-d}|,$$

where n is the order of G and d is the number of proper digons in G. □

As pointed out in Chapter 2, there are interpretations for the multiplicity of chromatic roots '0' and '1' respectively; but until now there is no interpretation for the multiplicity of chromatic root '2' in general. However, if we confine ourselves to the family of near-triangulations, then we have the following result due to Woodall (1977).

Theorem 12.6.3   *Let G be a near-triangulation with k-face and d be the number of proper digons in G. Then the multiplicity of 2 as a root of $P(G, \lambda)$ is at least $d + 1$, and exactly $d + 1$ if G is a triangulation (so 2 is a simple root of $P(G, \lambda)$ if G is a 3-connected triangulation).*   □

Theorem 12.6.1 has been extended recently by Dong and Koh (2005a). For any loopless plane graph G, let $f'(G)$ be the number of faces which are not bounded by C_3. Thus, if G is a near-triangulation, then $f'(G) \le 1$. They showed that

Theorem 12.6.4   *For any 2-connected loopless plane graph G, if $f'(G) \le 4$ and $c(G - S) \le |S| < f'(G)$ for every independent set S of G, then*

$$(-1)^{v(G)} P(G, \lambda) > 0$$

for all $\lambda \in (1, 2)$.   □

We now consider two special plane triangulations. The chromatic polynomial of the octahedral graph $\overline{K_2} + C_4$ is

$$\lambda(\lambda - 1)(\lambda - 2)(\lambda^3 - 9\lambda^2 + 29\lambda - 32),$$

which has a root at $2.546602 \cdots$. The graph $\overline{K_2} + C_5$ has

$$\lambda(\lambda - 1)(\lambda - 2)(\lambda - 3)(\lambda^3 - 9\lambda^2 + 30\lambda - 35)$$

as its chromatic polynomial which has a root at $2.677814 \cdots$.

Woodall (1977, 1992b) believed that $2.546602 \cdots$ is an important value for studying non-integral chromatic roots of triangulations. Indeed, he proved the following result.

Theorem 12.6.5   *If G is a plane triangulation of order n, then $P(G, \lambda)$ is non-zero with the sign of $(-1)^{n+d+1}$ for all real $\lambda \in (2, 2.546602 \cdots)$, where d is the number of proper digons in G and $2.546602 \cdots$ is the unique real root of $\lambda^3 - 9\lambda^2 + 29\lambda - 32$.*   □

Theorem 12.6.5 was proved in Woodall (1977) for $\lambda \in (2.5, 2.546602 \cdots)$ and in Woodall (1992b) for $\lambda \in (2, 2.5]$. Here we give a sketch of the proof for the first part given by Woodall (1977).

The proof is by induction on n. The result holds for K_3 or K_4. So suppose $n \ge 5$. Then there are four cases to consider.

Case 1: $d \ge 1$.

Case 2: $d = 0$ but G has a separating triangle.

Case 3: $\delta(G) = 4$ but $d = 0$ and G has no separating triangle.

Case 4: $\delta(G) = 5$, but $d = 0$ and G has no separating triangle.

For Cases 1 and 2, by Theorem 1.3.2,

$$P(G, \lambda) = \frac{P(G_1, \lambda) P(G_2, \lambda)}{P(K_i, \lambda)},$$

where $i = 1$ or 2, and both G_1 and G_2 are triangulations. So for these two cases, it is not difficult to prove by induction that the result holds. For Cases 3 and 4, $P(G, \lambda)$ is expressed in terms of the chromatic polynomials of triangulations of order less than n. Then, by the induction hypothesis, each term has the sign $(-1)^{p+1}$, and the result follows.

So far, it is unknown whether $(2.677814\cdots, 3)$ is also root-free for the chromatic polynomials of plane triangulations. The following conjecture was proposed in Woodall (1992b).

Conjecture 12.6.1   *If G is a triangulation, then $P(G, \lambda)$ is non-zero throughout the interval $(2.677814\cdots, 3)$. If G is 4-connected and non-eulerian, then $P(G, \lambda)$ is non-zero with the sign of $(-1)^n$ throughout $(2.677814\cdots, 3)$, and has a unique root in $(2.546602\cdots, 2.677814\cdots)$, where n is the order of G.*

12.7 Graphs with hamiltonian paths

In this section, we consider the family of graphs which contain a hamiltonian path. Thomassen (2000) showed that this family of graphs has no chromatic roots in $(32/27, 32/27 + \epsilon)$ for some $\epsilon > 0$ (in fact, $0 < \epsilon \le t_0 - 32/27$, where t_0 is the number defined in the following result).

Theorem 12.7.1   *Let G be a 2-connected graph of order n having a hamiltonian path. Then*

$$(-1)^n P(G, \lambda) > 0 \qquad when \ 1 < \lambda \le t_0,$$

where t_0 is the unique real root of the polynomial $(t-2)^3 + 4(t-1)^2$; namely,

$$t_0 = \tfrac{2}{3} + \tfrac{1}{3}\sqrt[3]{26 + 6\sqrt{33}} + \tfrac{1}{3}\sqrt[3]{26 - 6\sqrt{33}} = 1.29559\cdots. \qquad \square$$

Thomassen (2000) also showed that Theorem 12.7.1 is no longer true if t_0 is replaced by any larger number. For each $k \in \mathbb{N}$, let J_k be the graph obtained from a path $x_1 x_2 \cdots x_{2k+3}$ by adding the edges $x_1 x_4$, $x_{2k} x_{2k+3}$ and $x_i x_{i+4}$ for $i = 2, 4, 6, \cdots, 2k - 2$. Observe that each J_i has a hamiltonian path, and Thomassen (2000) proved the following:

Theorem 12.7.2   *The union of the sets of chromatic roots of the graphs $J_1, J_2, \cdots$ consists of $0, 1, 2$ and a subset of $(t_0, 2)$ having elements arbitrarily close to t_0.* $\square$

Theorem 12.7.1 motivates the study of the following problem: for each family $\mathcal{S}$ of graphs, determine the maximum real number $\phi(\mathcal{S})$ such that $(1, \phi(\mathcal{S}))$ is a root-free interval for $\mathcal{S}$. It is known that $\phi(\mathcal{S}) = 32/27$ if $\mathcal{S}$ is the family of all graphs or the family of planar graphs. Theorem 12.7.1 shows that $\phi(\mathcal{S}) = t_0 = 1.29559 \cdots$ if $\mathcal{S}$ is the family of graphs having a hamiltonian path. Conjecture 12.2.1 is equivalent to the statement that $\phi(\mathcal{S}) = 2$ if $\mathcal{S}$ is the family of 3-connected non-bipartite graphs.

It was showed in Dong and Koh (2004c) that if $\mathcal{S}$ is the family of bipartite graphs, then $\phi(\mathcal{S}) = 32/27$. They also showed in (2004d) that if $\mathcal{D}_2$ is the family of graphs whose domination numbers do not exceed 2, then

$$\phi(\mathcal{D}_2) = 2 + \frac{1}{6}\sqrt[3]{12\sqrt{93} - 108} - \frac{1}{6}\sqrt[3]{12\sqrt{93} + 108} = 1.317672196\cdots,$$

and if $\mathcal{D}_2'$ is the family of graphs G with $\Delta(G) \geq v(G) - 2$, then

$$\phi(\mathcal{D}_2') = \frac{5}{3} + \frac{1}{6}\sqrt[3]{12\sqrt{69} - 44} - \frac{1}{6}\sqrt[3]{12\sqrt{69} + 44} = 1.430159709\cdots.$$

12.8   Bipartite graphs

For a graph G, let $r_{max}(G)$ denote the largest real root of $P(G, \lambda)$. Is there an upper bound for $r_{max}(G)$ in terms of $\chi(G)$? Such bounds exist for some families of graphs. Any chordal graph G has its chromatic roots all integers, and so $r_{max}(G) = \chi(G) - 1$. The chromatic polynomial of an outerplanar graph G has only roots of the form $1 + u$, where $u^k = 1$ for some k (see Wakelin and Woodall (1992)), and so, again, $r_{max}(G) = \chi(G) - 1$.

However, for the complete bipartite graph $K(m, n)$, Woodall (1977) proved the following result which says that its largest chromatic root is close to $m/2$ if n is sufficiently large.

Theorem 12.8.1   *Let $m, n \in \mathbb{N}$. If m is fixed and n is sufficiently large, then $P(K(m, n), \lambda)$ has real roots arbitrarily close to any integer i with $2 \leq i \leq m/2$.*

Proof. It is known that

$$P(K(m,n),\lambda) = \sum_{k=1}^{m} S(m,k)(\lambda)_k (\lambda - k)^n,$$

where $S(m,k)$ is a Stirling number of the second kind (see Exercise 12.6). When $k = 1$ or $k = m$, we have $S(m,k) = 1$.

For any real λ with $0 < \lambda < m$, the term $S(m,m)(\lambda)_m(\lambda-m)^n$ has the sign of $(-1)^{n+\lfloor m-\lambda \rfloor}$. If $1.5 < \lambda < (m+1)/2$, then

$$|\lambda - m| > |\lambda - k|,$$

for $k = 1, 2, \cdots, m-1$. Thus for any real $\lambda \in (1.5, (m+1)/2)$, there exists an integer $n_0(\lambda)$ such that, when $n \geq n_0(\lambda)$, $P(K(m,n),\lambda)$ has the sign of $(-1)^{n+\lfloor m-\lambda \rfloor}$.

Let $i \in \mathbb{N}$ with $2 \leq i \leq m/2$ and $\lambda_i = i + (-1)^{n+m-i}\epsilon$, where $0 < \epsilon < 1/2$. We now show that $(-1)^{n+\lfloor m-\lambda_i \rfloor} = -1$.

If $n + m - i$ is odd, then $i - 1 < \lambda_i = i - \epsilon < i$, and thus

$$(-1)^{n+\lfloor m-\lambda_i \rfloor} = (-1)^{n+m-i} = -1;$$

if $n + m - i$ is even, then $i < \lambda_i = i + \epsilon < i + 1$, and thus

$$(-1)^{n+\lfloor m-\lambda_i \rfloor} = (-1)^{n+m-i-1} = -1.$$

Hence the term $S(m,m)(\lambda)_m(\lambda-m)^n$ is negative when $\lambda = \lambda_i$ for all i with $2 \leq i \leq m/2$. Let

$$n_0(\epsilon) = \max_{2 \leq i \leq m/2} \{n_0(\lambda_i)\}.$$

Then, when $n \geq n_0(\epsilon)$, we have $P(K(m,n),\lambda_i) < 0$. Since $P(K(m,n),i) > 0$ for any integer $i \geq 2$, $P(K(m,n),\lambda)$ has a root between i and λ_i if $n \geq n_0(\epsilon)$, where $2 \leq i \leq m/2$.

Since ϵ is any real number with $0 < \epsilon < 1/2$, this shows that $P(G,\lambda)$ has a root arbitrarily close to i when n is large enough, where $i \in \mathbb{N}$ with $2 \leq i \leq m/2$. $\square$

Theorem 12.8.1 shows that for any $N > 0$, there exist bipartite graphs which have chromatic roots larger than N.

12.9   Graphs containing spanning q-trees

When G is connected, Theorem 12.2.2 implies that

$$(-1)^{n-1} P(G, \lambda) > 0,$$

for all real $\lambda \in (0, 1)$. This result has been generalized by Dong and Koh
(2004b) to any graph which contains a q-tree as a spanning subgraph.

For $n, q \in \mathbb{N}$ with $n \geq q$, let $\mathcal{G}(n, q)$ denote the family of graphs G of
order n such that G contains a q-tree as a spanning subgraph. Observe
that $\mathcal{G}(n, q) \subset \mathcal{G}(n, q - 1)$ for $q \geq 2$ and $\mathcal{G}(n, 1)$ is the family of connected
graphs of order n.

Lemma 12.9.1   *If $G \in \mathcal{G}(n, q)$ and u and v are non-adjacent vertices in*
G, then $G \cdot uv \in \mathcal{G}(n - 1, q)$.

Proof.   Suppose that the result is false, and n is the minimum number such
that there is a counterexample G. Observe that $n > q$. Let T be a q-tree
which is also a spanning subgraph of G. Then u and v are two vertices in
T.

Since T is a q-tree, T contains at least two simplicial vertices. If T
has a simplicial vertex x with $x \notin \{u, v\}$, then $G - x \in \mathcal{G}(n - 1, q)$. By
assumption, $(G - x) \cdot uv \in \mathcal{G}(n - 2, q)$. Thus $G \cdot uv \in \mathcal{G}(n - 1, q)$.

So both u and v are simplicial vertices of T. But then $T \cdot uv \in \mathcal{G}(n-1, q)$,
since $T - u$ is a q-tree and a spanning subgraph of $G \cdot uv$. This contradicts
the assumption. The result thus follows.   □

Theorem 12.9.1   *Let $G \in \mathcal{G}(n, q)$, where $n, q \in \mathbb{N}$ with $n \geq q$. Then*

$$(-1)^{n-q} P(G, \lambda) > 0$$

for $q - 1 < \lambda < q$.

Proof.   If $n = q$, then G is the complete graph K_q and the result holds.

Now assume that $n > q$. Suppose that the result does not hold. Then
there exists a graph $G \in \mathcal{G}(n, q)$ of minimum size such that

$$(-1)^{n-q} P(G, \lambda) \leq 0$$

for some real λ with $q - 1 < \lambda < q$.

It is clear that G cannot be a q-tree; otherwise,

$$P(G, \lambda) = (\lambda)_q (\lambda - q)^{n-q},$$

and we have

$$(-1)^{n-q}P(G,\lambda) > 0$$

for $q - 1 < \lambda < q$.

Let T be a spanning subgraph of G such that T is isomorphic to a q-tree. Since $G \not\cong T$, there exists $uv \in E(G)$ such that $uv \notin E(T)$. Then

$$P(G,\lambda) = P(G - uv, \lambda) - P(G \cdot uv, \lambda),$$

and we have

$$(-1)^{n-q}P(G,\lambda) = (-1)^{n-q}P(G - uv, \lambda) + (-1)^{n-1-q}P(G \cdot uv, \lambda).$$

It is clear that $G - uv \in \mathcal{G}(n, q)$. By Lemma 12.9.1, $G \cdot uv \in \mathcal{G}(n - 1, q)$. By the minimality of $e(G)$, the theorem holds for both $G - uv$ and $G \cdot uv$. Hence by the above equality,

$$(-1)^{n-q}P(G,\lambda) > 0$$

for $q - 1 < \lambda < q$, a contradiction. $\qquad\square$

Corollary 12.9.1   *Let $G \in \mathcal{G}(n, q)$, where $n, q \in \mathbb{N}$ with $n \geq q$. Then $P(G,\lambda) \neq 0$ for all non-integral real λ in $(0, q)$.* $\qquad\square$

12.10   Largest non-integral chromatic root

For any graph G, define $\xi(G) = 0$ if $P(G,\lambda)$ has no non-integral real roots, and $\xi(G)$ to be the largest non-integral real root of $P(G,\lambda)$ otherwise. Since $P(G,\lambda)$ has no negative roots, we have $\xi(G) \geq 0$ for any graph G.

For $n \in \mathbb{N}$, let $\mathcal{G}_n$ denote the family of graphs of order n, and define

$$\xi(n) = \max_{G \in \mathcal{G}_n} \xi(G). \tag{12.8}$$

The function $\xi(n)$ has been completely determined by Dong (2004).

It is clear that every graph of order less than 4 is chordal, and thus $\xi(n) = 0$ if $n \leq 3$. Since there is only one non-chordal graph of order 4, which is C_4, we also have $\xi(4) = \xi(C_4) = 0$. For $n = 5$, it can be verified that $K(2,3)$ is the only graph having a non-integral real chromatic root (see Exercise 12.7). Thus

$$\xi(5) = \xi(K(2,3)) = \frac{5}{3} - \frac{1}{6}\alpha + \frac{10}{3\alpha} = 1.430159\cdots,$$

where $\alpha = (44 + 12\sqrt{69})^{1/3}$ and $\frac{5}{3} - \frac{1}{6}\alpha + \frac{10}{3\alpha}$ is the only real root of $x^3 - 5x^2 + 10x - 7$. In what follows, we thus assume that $n \geq 6$.

Let us define two graphs $G^1(n)$ and $G^2(n)$ for any $n \geq 6$. The graph $G^1(6)$ is $W(6,4)$, as shown in Figure 12.1(a), and for $n \geq 7$, $G^1(n) = G^1(6) + K_{n-6}$. The graph $G^2(n)$ is the complement of the disjoint union $K_2 \cup K(1, n-3)$, which can also be obtained from K_{n-1} by inserting into an edge a new vertex, as shown in Figure 12.1 (b). Observe that $\chi(G^1(n)) = n - 3$ and $\chi(G^2(n)) = n - 2$.

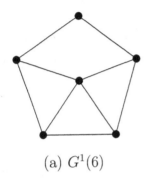

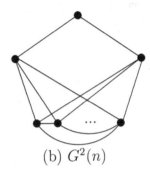

(a) $G^1(6)$ (b) $G^2(n)$

Figure 12.1

Lemma 12.10.1   *For $n \geq 6$,*

(a)

$$\xi(G^1(n)) = n - 4 + \frac{\beta}{6} - \frac{2}{\beta} = n - 3.317672195\cdots, \qquad (12.9)$$

where $\beta = (108 + 12\sqrt{93})^{1/3}$ and $\frac{\beta}{6} - \frac{2}{\beta}$ is the only real root of $x^3 + x - 1$.

(b)

$$\xi(G^2(n)) = \begin{cases} 0, & 6 \leq n \leq 7; \\ \left(n - 1 + \sqrt{(n-3)(n-7)}\right)/2, & n \geq 8. \end{cases}$$

$$(12.10)$$

Proof. It is not hard to show that

$$P(G^1(n), \lambda) = (\lambda)_{n-3} \left((\lambda - n + 4)^3 + \lambda - n + 3\right) \qquad (12.11)$$

and

$$P(G^2(n), \lambda) = (\lambda)_{n-2} \left((\lambda - 2)(\lambda - n + 2) + (\lambda - 1)\right). \qquad (12.12)$$

The lemma thus follows. $\qquad\square$

The following result, due to Dong (2004), shows that $\xi(n)$ is determined by $\xi(G^1(n))$ and $\xi(G^2(n))$.

Theorem 12.10.1   *Let $G \in \mathcal{G}_n$, where $n \geq 6$.*

(i) *If $\chi(G) \geq n - 1$, then G is chordal, and $\xi(G) = 0$.*

(ii) *If $\chi(G) = n - 2$, then $\xi(G) \leq \xi(G^2(n))$; moreover, for $n \geq 8$, equality holds only when $G \cong G^2(n)$.*

(iii) *If $\chi(G) \leq n - 3$, then $\xi(G) \leq \xi(G^1(n))$; moreover, equality holds only when $G \cong G^1(n)$.* $\qquad\square$

The strategy of proving the above result is to determine all possible structures of the graphs with a particular chromatic number. If $\chi(G) = n$, then $G \cong K_n$ and $\xi(G) = 0$. If $\chi(G) = n - 1$, then $G - u \cong K_{n-1}$ for some vertex u, and so $\xi(G) = 0$. If $\chi(G) = n-2$, then either $G-u-v \cong K_{n-2}$ for some vertices u and v, or $G \cong C_5 + K_{n-5}$; in the latter case $\xi(G) = 0$, while in the former case $\xi(G) \leq \xi(G^2(n))$ with equality (for $n \geq 8$) only when $G \cong G^2(n)$. If $\chi(G) \leq n - 3$, the situation is somewhat more complicated.

Observe that $\xi(G^1(n)) > \xi(G^2(n))$ for $6 \leq n \leq 8$ and $\xi(G^1(n)) < \xi(G^2(n))$ for $n \geq 9$. Thus we have:

Theorem 12.10.2   *For $n \in \mathbb{N}$,*

$$
\xi(n) = \begin{cases}
0, & 1 \leq n \leq 4; \\
\frac{5}{3} - \frac{1}{6}\alpha + \frac{10}{3\alpha} = 1.430159\cdots, & n = 5; \\
n - 4 + \frac{\beta}{6} - \frac{2}{\beta} = n - 4 + 0.682327804\cdots, & 6 \leq n \leq 8; \\
\left(n - 1 + \sqrt{(n-3)(n-7)}\right)/2, & n \geq 9,
\end{cases}
$$

where $\alpha = (44 + 12\sqrt{69})^{1/3}$ and $\beta = (108 + 12\sqrt{93})^{1/3}$. $\qquad\square$

12.11   Upper root-free intervals with respect to maximum degrees

Brooks' theorem on chromatic number states that $\Delta(G)$ is an upper bound of $\chi(G)$ for any connected graph G except for complete graphs and odd cycles. Note that if G is a complete graph or an odd cycle, we have $P(G, \lambda) > 0$ for every real $\lambda > \Delta(G)$.

Sokal (2001) proved that all roots (real or complex) of $P(G, \lambda)$ lie in the disc $|z| < 7.963907\Delta(G)$. Hence there exists a constant c with $1 \leq c \leq 7.963907$ such that $P(G, \lambda) > 0$ for all real $\lambda > c\Delta(G)$. In the private communication with one of the authors, Sokal conjectured that $c = 1$ in the above result and proposed also the following stronger conjecture.

Conjecture 12.11.1   *Let G be a graph and $x \in V(G)$. If $\lambda \geq \Delta(G)$, then*
$$P(G, \lambda) \geq (\lambda - d(x))P(G - x, \lambda).$$
In particular, if $\lambda > \Delta(G)$, then $P(G, \lambda) > 0$.

A partial result on Conjecture 12.11.1 was recently obtained by Dong and Koh (2004b) as shown below.

Theorem 12.11.1   *Let G be a graph and $\{A, B\}$ be any partition of $V(G)$, where B can be empty. Let $\Delta_A = \max\{d_G(x) : x \in A\}$. Then for any $u \in A$,*
$$P(G, \lambda) \geq (\lambda - d(u))P(G - u, \lambda),$$
for all real $\lambda \geq \max\{\Delta_A, |B| + \lfloor |A|/3 \rfloor - 1\}$.   □

The above result was proved by induction on $|A|$ in Dong and Koh (2004b). By Lemma 2.5.7,

$$P(G, \lambda) - (\lambda - d(u))P(G - u, \lambda) = \sum_{H \in \mathcal{S}} P(H, \lambda), \qquad (12.13)$$

where $\mathcal{S}$ is a family of graphs of order $v(G) - 2$. It was then shown that $P(H, \lambda) \geq 0$ for all real $\lambda \geq \max\{\Delta_A, |B| + \lfloor |A|/3 \rfloor - 1\}$ and all $H \in \mathcal{S}$, and thus the result follows.

Corollary 12.11.1   *Let G be a graph of order n. Then*

(i) *$P(G, \lambda) \geq (\lambda - d(u))P(G - u, \lambda)$ all real $\lambda \geq \max\{\Delta(G), \lfloor n/3 \rfloor - 1\}$ and for any $u \in V(G)$; and*

(ii) *$P(G, \lambda) > 0$ for all real $\lambda > \max\{\Delta(G), \lfloor n/3 \rfloor - 1\}$.*   □

Remarks

(i) Theorem 12.11.1 verifies Sokal's conjecture only for a special case; namely, the order of G is at most $3\Delta(G) + 5$.

(ii) The method used in the proof of Theorem 12.11.1 is unlikely to be of any use in proving Sokal's conjecture due to the fact that new graphs (the H's) created from the recurrence relation in (12.13) may have their maximum degrees greater than that of the original graph. There might be some hope to establish Sokal's conjecture if one could find a recurrence relation for chromatic polynomials which does not produce any new graph with higher maximum degree even after a finite number of iterations.

(iii) In Sokal's conjecture, the maximum degree is at least 3. It may be more realistic to start off the study by considering the extreme case when $\Delta(G) = 3$.

(iv) The "maximaxflow" of a graph G, denoted by $\Lambda(G)$, is defined as

$$\Lambda(G) = \max_{x \neq y} \lambda(x, y),$$

where

$$
\begin{aligned}
\lambda(x, y) \;=\; & \text{maximum number of edge-disjoint paths from } x \text{ to } y \\
=\; & \text{minimum number of edges separating } x \text{ from } y.
\end{aligned}
$$

(See Brown, Hickman, Sokal and Wagner (2001).) Since $\lambda(x, y) \leq \min\{d(x), d(y)\}$, we have $\Lambda(G) \leq \Delta(G)$. In his private communication, Sokal also conjectured that for any graph G and $\lambda > \Lambda(G)$, $P(G, \lambda) > 0$.

Exercise 12

12.1 Let G be any connected graph of order n. Prove that $|P(G, \lambda)| \geq \lambda(1 - \lambda)^{n-1}$ for all real $\lambda \in (0, 1)$.

12.2 Prove that if $P(x)$ is a monic polynomial with integral coefficients, then every rational root of $P(x)$ is an integer.

12.3 Let $k \in \mathbb{N}$ with $k \geq 3$. Apply Theorem 12.8.1 to prove that for any real r, there exists a graph G with $\chi(G) = k$ such that $P(G, \lambda)$ has a root in (r, ∞).

12.4 Prove that the family of graphs with tree-width at most k is closed under minors.

12.5 Prove that the family of planar graphs is closed under minors.

12.6 Prove that

$$P(K(m, n), \lambda) = \sum_{k=1}^{m} S(m, k)(\lambda)_k (\lambda - k)^n,$$

where $S(m, k)$ is a Stirling number of the second kind.

12.7 Prove that $K(2, 3)$ is the only graph of order 5 whose chromatic polynomial has a non-integral real root.

12.8 For any $p \in \mathbb{N}$ with $p \geq 5$, let H_p be the graph defined below:

$$V(H_p) = \{x_1, x_2, \cdots, x_p\}$$

and

$$E(H_p) = \{x_a x_b : 1 \leq a < b \leq p, (a, b) \notin \{(1, p-1), (2, p), (3, p)\}\}.$$

Show that $p - 3 + i$ and $p - 3 - i$, where $i^2 = -1$, are chromatic roots of H_p.

12.9 Show that $P(K(2, 2k + 1), \lambda)$ has a root in the interval $(1.4, 1.5)$ for all $k \in \mathbb{N}$.

12.10 Show that for any $n \in \mathbb{N}$, there exists $m \in \mathbb{N}$ such that $P(K(m, n), \lambda)$ has a root in the interval $(1, 2)$.

Chapter 13

Integral Roots of Chromatic Polynomials

13.1 Introduction

In Example 1.2.3 (i), we have pointed out that, if G is a chordal graph, then

$$P(G, \lambda) = \lambda^{r_0} (\lambda - 1)^{r_1} \cdots (\lambda - k)^{r_k}, \qquad (13.1)$$

where $k = \chi(G) - 1$ (or $k = \omega(G) - 1$) and $r_i \in \mathbb{N}$ for $i = 0, 1, \cdots, k$.

A polynomial is called an *integral-root* polynomial if all its roots are integers. Let $\mathcal{I}$ be the family of graphs G such that $P(G, \lambda)$ is an integral-root polynomial, and $\mathcal{C}$ the family of chordal graphs. It follows from (13.1) that $\mathcal{C} \subseteq \mathcal{I}$.

It had been conjectured by Braun, Kretz, Walter and Walter (1974) and Vaderlind (1988) that $\mathcal{C} = \mathcal{I}$. This conjecture was, however, disproved by Read (1975) (also, independently, by Dmitriev (1980)), who discovered the non-chordal graph in Figure 13.1 whose chromatic polynomial is given by

$$\lambda(\lambda - 1)(\lambda - 2)(\lambda - 3)^3(\lambda - 4). \qquad (13.2)$$

Hence $\mathcal{C}$ is a proper subfamily of $\mathcal{I}$.

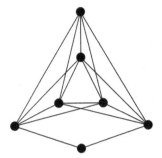

Figure 13.1

For $k, r_0, r_1, \cdots, r_k \in \mathbb{N}$, let $\mathcal{I}(r_0, r_1, \cdots, r_k)$ denote the family of graphs G such that their chromatic polynomials are of the form (13.1). In this chapter, we shall consider the following problems:

Problem 13.1.1   *Given $k, r_0, r_1, \cdots, r_k \in \mathbb{N}$, determine if the family $\mathcal{I}(r_0, r_1, \cdots, r_k)$ is a subfamily of $\mathcal{C}$.*

Problem 13.1.2   *Given $k, r_0, r_1, \cdots, r_k \in \mathbb{N}$, characterize the graphs in the family $\mathcal{I}(r_0, r_1, \cdots, r_k)$.*

Problem 13.1.3   *Construct more graphs in $\mathcal{I} \backslash \mathcal{C}$.*

It is not difficult to prove the following (see Exercise 13.1):

Lemma 13.1.1   *Given $k, r_0, r_1, \cdots, r_k \in \mathbb{N}$, $\mathcal{I}(r_0, r_1, \cdots, r_k) \subseteq \mathcal{C}$ if and only if $\mathcal{I}(1, 1, r_2, \cdots, r_k) \subseteq \mathcal{C}$.*   □

Thus, for Problem 13.1.1, we may confine ourselves to the case that $r_0 = r_1 = 1$. In Section 13.2, we shall introduce some known results on this problem.

It is clear that

$$\mathcal{I}(r_0, r_1, \cdots, r_k) = (\mathcal{I}(r_0, r_1, \cdots, r_k) \backslash \mathcal{C}) \cup (\mathcal{I}(r_0, r_1, \cdots, r_k) \cap \mathcal{C}),$$

and Problem 13.1.1 is equivalent to determining whether $\mathcal{I}(r_0, r_1, \cdots, r_k) \backslash \mathcal{C}$ is empty. Thus Problem 13.1.2 is more difficult than Problem 13.1.1. Also notice that the graphs in the family $\mathcal{I}(r_0, r_1, \cdots, r_k) \cap \mathcal{C}$ can be constructed recursively based on a given graph in the family by adding simplicial vertices. Thus, for Problem 13.1.2, the essential task is to characterize those in $\mathcal{I}(r_0, r_1, \cdots, r_k) \backslash \mathcal{C}$. In Section 13.3, we shall introduce a result, due to

Xu (1997), which solves Problem 13.1.2 for the case where $r_0 = r_1 = 1$ and $r_2 + \cdots + r_k = k + 1$.

Dmitriev (1980) discovered an infinite family of graphs in $\mathcal{I}\backslash\mathcal{C}$. It is, however, very difficult to find out all graphs in $\mathcal{I}\backslash\mathcal{C}$, and we shall focus on searching for some interesting families of graphs in $\mathcal{I}\backslash\mathcal{C}$. The graphs discovered by Dmitriev (1980) have the property that their pure cycles are all of order 4. Let $g^*(G)$ denote the smallest order of pure cycles in G. Dmitriev proposed a problem of searching graphs G in $\mathcal{I}\backslash\mathcal{C}$ such that $g^*(G) = p \geq 5$. In Section 13.4, we shall present some results, due to Dong and Koh (1998), Dong, Teo, Koh and Hendy (2002) and Hernández and Luca (2005), which provide positive answers to Dmitriev's problem for $p = 5, 6, 7, 8, 9, 10, 11, 13$. This problem remains unsolved for $p = 12$ and $p \geq 14$.

13.2 Chromatic polynomials possessed only by chordal graphs

In this section, we introduce some results concerning Problem 13.1.1.

It is clear that if $r_i = 1$ for each $i = 0, 1, \cdots, k$, then $\mathcal{I}(r_0, r_1, \cdots, r_k) = \{K_{k+1}\}$, and so $\mathcal{I}(r_0, r_1, \cdots, r_k) \subseteq \mathcal{C}$. More generally, if $r_k \geq 1$ and $\sum_{i=0}^{k-1} r_i \leq k + 1$, then, by Theorems 6.5.2, 8.3.1 and 8.4.1, $\mathcal{I}(r_0, r_1, \cdots, r_k) \subseteq \mathcal{C}$.

It is also clear that if $k = 1$, then $\mathcal{I}(r_0, r_1)$ is the family of forests of order $r_0 + r_1$ with r_0 components (see Exercise 13.3). When $k = 2$, by applying Theorem 6.5.2, we obtain the following result due to Vaderlind (1988).

Theorem 13.2.1   *Given $r_i \in \mathbb{N}$ for $i = 0, 1, 2$, $\mathcal{I}(r_0, r_1, r_2) \subseteq \mathcal{C}$.* $\square$

When $k = 3$, the following conjecture, proposed by Xu (1997), remains unsettled:

Conjecture 13.2.1   *For all $r_0, r_1, r_2, r_3 \in \mathbb{N}$, $\mathcal{I}(r_0, r_1, r_2, r_3) \subseteq \mathcal{C}$.*

This conjecture is equivalent to the statement that if G is a graph with $\chi(G) \leq 4$ and satisfying (13.1), then G is chordal. When G is a planar graph, this conjecture has been proved to be true by Dong and Koh (1998) (see Exercise 13.2).

Theorem 13.2.2   *For any $r_0, r_1, r_2, r_3 \in \mathbb{N}$, if $G \in \mathcal{I}(r_0, r_1, r_2, r_3)$ and G is a planar graph, then $G \in \mathcal{C}$.*   □

Consider now the case when $k \geq 4$. In our next result, we shall show that if $r_i \geq 3$ for some i with $3 \leq i \leq k - 1$, then $\mathcal{I}(r_0, r_1, \cdots, r_k) \not\subseteq \mathcal{C}$. By Lemma 13.1.1, we assume in what follows that $r_0 = r_1 = 1$. We first prove the following lemma.

Lemma 13.2.1   *Let $r_2, r_3, \cdots, r_k \in \mathbb{N}$ and $r'_2, r'_3, \cdots, r'_s \in \mathbb{N}$, where $s \geq k \geq 4$, such that $r'_{i+t} \geq r_i$ for $i = 2, 3, \cdots, k$ and for some t with $0 \leq t \leq s - k$. If there exists $G \in \mathcal{I}(1, 1, r_2, \cdots, r_k) \backslash \mathcal{C}$ with $\omega(G) = k + 1$, then there exists $H \in \mathcal{I}(1, 1, r'_2, \cdots, r'_s) \backslash \mathcal{C}$ with $\omega(H) = s + 1$.*

Proof.   Let G be a graph in $\mathcal{I}(1, 1, r_2, \cdots, r_k) \backslash \mathcal{C}$ with $\omega(G) = k + 1$, and let $H_1 = G + K_t$. Then

$$P(H_1, \lambda) = (\lambda)_t P(G, \lambda - t) = (\lambda)_{t+2}(\lambda - t - 2)^{r_2} \cdots (\lambda - k - t)^{r_k}.$$

Observe that $\omega(H_1) = \omega(G) + t = k + t + 1$. Let $H_2 \in \mathcal{G}[H_1 \cup_{k+t+1} K_{s+1}]$. Then

$$
\begin{aligned}
&P(H_2, \lambda) \\
=\ & P(H_1, \lambda)(\lambda - k - t - 1) \cdots (\lambda - s) \\
=\ & (\lambda)_{t+2}(\lambda - t - 2)^{r_2} \cdots (\lambda - k - t)^{r_k}(\lambda - k - t - 1) \cdots (\lambda - s).
\end{aligned}
$$

Observe that $\omega(H_2) = s + 1$. Finally, let H_3 be a graph obtained from H_2 by adding $r'_i - 1$ simplicial vertices of degree i for $i = 2, 3, \cdots, t + 1$ and $i = t + k + 1, \cdots, s$ and adding $r'_{i+t} - r_i$ simplicial vertices of degree $i + t$ for $i = 2, 3, \cdots, k$. It can be checked that $H_3 \in \mathcal{I}(1, 1, r'_2, r'_3, \cdots, r'_s)$ but $H_3 \notin \mathcal{C}$.   □

Theorem 13.2.3   *If $k \geq 4$ and $r_i \geq 3$ for some i with $3 \leq i \leq k - 1$, then $\mathcal{I}(1, 1, r_2, \cdots, r_k) \not\subseteq \mathcal{C}$.*

Proof.   Let G be the graph in Figure 13.1. Then

$$P(G, \lambda) = \lambda(\lambda - 1)(\lambda - 2)(\lambda - 3)^3(\lambda - 4).$$

Notice that $\omega(G) = 5$. Since $r_i \geq 3$ for some i with $3 \leq i \leq k - 1$, we have $r_{2+i-3} \geq 1$, $r_{3+i-3} \geq 3$ and $r_{4+i-3} \geq 1$. Thus, by Lemma 13.2.1, $\mathcal{I}(1, 1, r_2, \cdots, r_k) \not\subseteq \mathcal{C}$.   □

13.3   Graphs $G \in \mathcal{I}$ of order $\omega(G) + 2$

For any graph G, we have $1 \le \omega(G) \le v(G)$. If $\omega(G) = v(G)$, then G is complete. If $\omega(G) = v(G) - 1$, then $G - u$ is complete for some vertex u in G, and thus G is chordal (see Exercise 13.4). In this section, we shall introduce a result, due to Xu (1997), which first characterizes graphs $G \in \mathcal{I}$ with $\omega(G) = v(G) - 2$, and then gives a necessary and sufficient condition on $r_0, r_1, \cdots, r_k$, where $\sum_{i=0}^{k} r_i = k + 3$, such that $\mathcal{I}(r_0, r_1, \cdots, r_k) \subseteq \mathcal{C}$.

We first state the following result (see Exercise 13.5), which can be found in Dong (2004).

Lemma 13.3.1   *Let G be a graph of order n and $\chi(G) = n - 2$. Then either $G = C_5 + K_{n-5}$ or $G \backslash \{u, v\} \cong K_{n-2}$ for some vertices u, v in G.* $\square$

It follows from Lemma 13.3.1 that if G is a graph of order n such that $\omega(G) = n - 2$, then either $G \cong C_5 + K_{n-5}$ or $G \backslash \{u, v\} \cong K_{n-2}$ for some vertices u, v in G. Since $C_5 + K_{n-5} \notin \mathcal{I}$, we shall focus on graphs G such that $G \backslash \{u, v\} \cong K_{n-2}$ for some vertices u, v in G. If $uv \notin E(G)$, then G is chordal. So we shall now consider the case that $uv \in E(G)$.

For $a, b, k \in \mathbb{N}$ with $a + b \le k + 1$, let $R(a, b, k)$ be a graph obtained from the complete graph K_{k+1} by adding two adjacent vertices u and v and adding edges joining u to vertices $x_1, \cdots, x_a$ in K_{k+1} and v to vertices $y_1, \cdots, y_b$ in K_{k+1}, where $x_1, \cdots, x_a, y_1, \cdots, y_b$ are distinct, as shown in Figure 13.2. (This family of graphs was constructed in Dmitriev (1980).)

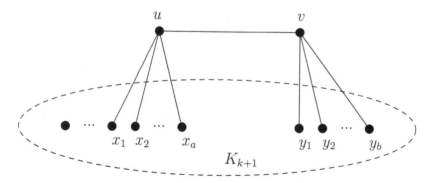

Figure 13.2 $R(a, b, k)$

Lemma 13.3.2   *Let G be a non-chordal graph of order n with $\omega(G) = n - 2$ and $G \not\cong C_5 + K_{n-5}$. Then $G \cong R(a, b, k) + K_r$ for some $r \in \mathbb{N}_0$ and $a, b, k \in \mathbb{N}$ with $a + b \le k + 1$.*

Proof. Let V_0 be the set of vertices in G of degree $n-1$, and $r = |V_0|$. Let $H = G - V_0$. Then $\Delta(H) < n_0 - 1$, where $n_0 = v(H)$. Observe that H is non-chordal and $\omega(H) = n_0 - 2$. Since $G \not\cong C_5 + K_{n-5}$, we have $H \not\cong C_5 + K_{n_0-5}$. By Lemma 13.3.1, $H - \{u, v\} \cong K_{n_0-2}$ for some vertices u and v in H. Since H is non-chordal, we have $uv \in E(H)$, $d_H(u) \geq 2$ and $d_H(v) \geq 2$. Since $\Delta(H) < n_0 - 1$, we have $N_H(u) \cap N_H(v) = \emptyset$. Thus $H \cong R(a, b, k)$ for some $a, b \in \mathbb{N}$ with $a + b \leq k + 1$, where $k = n_0 - 3$. It follows that $G \cong R(a, b, k) + K_r$, as desired. $\square$

Observe that $v(R(a, b, k)) = k + 3$ and $\chi(R(a, b, k)) = k + 1$. It can also be verified that

$$P(R(a, b, k), \lambda) = (\lambda)_{k+1}(\lambda^2 - (a + b + 1)\lambda + ab + a + b). \qquad (13.3)$$

The following result gives a characterization on a and b such that the polynomial in (13.3) is an integral-root polynomial.

Lemma 13.3.3   Let $a, b \in \mathbb{N}$ and $a \leq b$. Then $x^2 - (a+b+1)x + ab + a + b$ is an integral-root polynomial if and only if $a = st$ and $b = (s+1)(t+1)$ for some $s, t \in \mathbb{N}$. Further, if $a = st$ and $b = (s+1)(t+1)$, then the roots of $x^2 - (a+b+1)x + ab + a + b$ are $st + s + 1$ and $st + t + 1$.

Proof. It is easy to see that if $a = st$ and $b = (s+1)(t+1)$, then the roots of $x^2 - (a+b+1)x + ab + a + b$ are $st + s + 1$ and $st + t + 1$. Thus, we need only to prove the necessity.

Assume that $x^2 - (a+b+1)x + ab + a + b$ is an integral-root polynomial. Then its discriminant

$$D = (a + b + 1)^2 - 4(ab + a + b) = (b - a - 1)^2 - 4a$$

is a square of some integer. Let $u_1 = b - a - 1$ and $u_2 = \sqrt{D}$. Then $a = \frac{1}{4}(u_1 - u_2)(u_1 + u_2)$. Since $a, u_1, u_2 \in \mathbb{N}$, both $u_1 - u_2$ and $u_1 + u_2$ are even. Let $s = \frac{u_1 - u_2}{2}$ and $t = \frac{u_1 + u_2}{2}$. Then $a = st$ and

$$b = u_1 + a + 1 = st + 1 + s + t = (s+1)(t+1).$$

This completes the proof. $\square$

Remark   Let $\mathcal{R}$ be the family of graphs whose chromatic polynomials contain real roots only. It is clear that $\mathcal{I} \subseteq \mathcal{R}$. It can be verified that for any $a, b, k \in \mathbb{N}$ with $a \leq b$ and $a + b \leq k + 1$, $R(a, b, k) \in \mathcal{R}$ if and only if

$b \geq a + 1 + 2\sqrt{a}$. All graphs in $\mathcal{R}$ of order at most 9 have been determined by Dántona, Mereghetti and Zamparini (2001).

From Lemma 13.3.3 and (13.3), the next result follows immediately.

Lemma 13.3.4   For $a, b, k \in \mathbb{N}$ with $a + b \leq k + 1$, $R(a, b, k) \in \mathcal{I}$ if and only if $a = st$ and $b = (s+1)(t+1)$ for some $s, t \in \mathbb{N}$. $\square$

Theorem 13.3.1   Let G be a non-chordal graph of order $k+3$ and chromatic number $k+1$. Then $G \in \mathcal{I}$ if and only if $G \cong R(st, (s+1)(t+1), k-r) + K_r$ for some $s, t \in \mathbb{N}$ and $r \in \mathbb{N}_0$ with $st + (s+1)(t+1) \leq k - r + 1$.

Proof. Let $G \cong R(st, (s+1)(t+1), k-r) + K_r$ for some $s, t \in \mathbb{N}$ and $r \in \mathbb{N}_0$ with $st + (s+1)(t+1) \leq k - r + 1$. Clearly, G is non-chordal, $v(G) = k + 3$ and $\chi(G) = k+1$. By Lemma 13.3.4, we have $R(st, (s+1)(t+1), k-r) \in \mathcal{I}$, implying $G \in \mathcal{I}$. This proves the sufficiency.

Now let $G \in \mathcal{I}$. Observe that

$$P(C_5 + K_{k-3}, \lambda) = (\lambda)_{k-2} P(C_5, \lambda - (k-2)),$$

which is not an integral-root polynomial. Thus $G \not\cong C_5 + K_{k-3}$, and by Lemma 13.3.2, $G \cong R(a, b, w) + K_r$ for some $a, b, w \in \mathbb{N}$ with $a + b \leq w + 1$ and some $r \in \mathbb{N}_0$. Since $\chi(G) = k + 1$, we have $w + 1 + r = k + 1$, implying that $w = k - r$. Note that $R(a, b, k-r) + K_r \in \mathcal{I}$ if and only if $R(a, b, k-r) \in \mathcal{I}$. Hence, by Lemma 13.3.4, $a = st$ and $b = (s+1)(t+1)$ for some $s, t \in \mathbb{N}$. This proves the necessity. $\square$

In the following, we apply Theorem 13.3.1 to determine non-chordal graphs in the family $\mathcal{I}(1, 1, r_2, \cdots, r_{k-1}, 1)$, where $r_i \in \mathbb{N}$ and $r_2 + r_3 + \cdots + r_{k-1} = k$.

Theorem 13.3.2   Let $k, p, q \in \mathbb{N}$ with $k \geq 3$ and $2 \leq p \leq q \leq k - 1$. Then G is a non-chordal graph having

$$P(G, \lambda) = \lambda(\lambda - 1) \cdots (\lambda - k)(\lambda - p)(\lambda - q) \qquad (13.4)$$

if and only if $G \cong R(st, (s+1)(t+1), k-r+1) + K_{r-1}$, for some $s, t, r \in \mathbb{N}$ such that $r \geq p + q - k - 1$, $p = st + s + r$ and $q = st + t + r$.

Proof. [Sufficiency] Let $G = R(st, (s+1)(t+1), k - (r-1)) + K_{r-1}$. Since $r + k + 1 \geq p + q$, we have

$$st + (s+1)(t+1) \leq k + 1 - (r-1).$$

It is clear that G is non-chordal. By (13.3) and Lemma 13.3.3, we have

$$
\begin{aligned}
P(G, \lambda) &= (\lambda)_{r-1} P(R(st, (s+1)(t+1), k - (r-1)), \lambda - (r-1)) \\
&= (\lambda)_{r-1}(\lambda - (r-1))_{k+1-(r-1)} \\
&\quad (\lambda - st - s - 1 - (r-1))(\lambda - st - t - 1 - (r-1)) \\
&= (\lambda)_{k+1}(\lambda - p)(\lambda - q). \tag{13.5}
\end{aligned}
$$

[Necessity] Let G be a non-chordal graph such that

$$
P(G, \lambda) = \lambda(\lambda - 1) \cdots (\lambda - k)(\lambda - p)(\lambda - q),
$$

where $p \le q \le k - 1$. Then $v(G) = k + 3$ and $\chi(G) = k + 1$. By Theorem 13.3.1, $G \cong R(st, (s+1)(t+1), k - r + 1) + K_{r-1}$ for some $s, t, r \in \mathbb{N}$, where $s \le t$. By the definition of $R(st, (s+1)(t+1), k - r + 1)$, we have $st + (s+1)(t+1) \le k - r + 1$. By (13.5), we have $p = st + s + r$ and $q = st + t + r$, which imply that $r \ge p + q - k - 1$. $\qquad \square$

Remark The structure of graphs G satisfying (13.4) when $p = q$ (resp. $q = p + 1$) has been characterized by Tang (1999).

We are now in a position to characterize $r_2, r_3, \cdots, r_k$, where $\sum_{i=2}^{k} r_i = k + 1$, such that $\mathcal{I}(1, 1, r_2, \cdots, r_k) \subseteq \mathcal{C}$. This result was first established by Xu (1997). In the following, we give a new proof.

Theorem 13.3.3 *Let $k, p, q \in \mathbb{N}$ with $k \ge 3$ and $2 \le p \le q \le k - 1$. Then*

$$
P(G, \lambda) = \lambda(\lambda - 1) \cdots (\lambda - k)(\lambda - p)(\lambda - q) \tag{13.6}
$$

holds only for chordal graphs G if and only if either $q \ge 2p - 2$ or $2q \ge k + p$.

Proof. [Sufficiency] Assume that (13.6) holds for some non-chordal graph. Then, by Theorem 13.3.2, $p = st + s + r$, $q = st + t + r$ for some $s, t, r \in \mathbb{N}$ with $r + k + 1 \ge p + q$. If $q \ge 2p - 2$, then

$$
st + t + r \ge 2(st + s + r) - 2.
$$

As $r \ge 1$, we have $(s - 1)(t + 2) \le -1$, a contradiction. If $2q \ge k + p$, then as $p + q \le k + r + 1$, we have

$$
q - p \ge k - q \ge p - r - 1.
$$

Thus $t - s + 1 \ge st + s$, which implies that $(s - 1)(t + 2) \le -1$, a contradiction.

[Necessity] Assume that $p \leq q \leq 2p - 3$ and $2q \leq k + p - 1$. Now let $s = 1$, $t = q - p + 1$ and $r = 2p - q - 2$. As $q \leq 2p - 3$, $r \geq 1$. Since $2q \leq k + p - 1$, we have

$$p + q = 2q + p - q \leq (k + p - 1) + p - q = k + (2p - q - 2) + 1 = k + r + 1.$$

Finally, it can be verified that

$$p = st + s + r \quad \text{and} \quad q = st + t + r.$$

By Theorem 13.3.2, there exists a non-chordal graph G such that (13.6) holds.

□

It follows from Theorem 13.2.3 that $\mathcal{I}(r_0, r_1, \cdots, r_k) \not\subseteq \mathcal{C}$ if $r_i \geq 3$ for some i with $3 \leq i \leq k - 1$. In the following, we shall apply Theorems 13.3.2 and 13.3.3 to establish a more genertal result.

Theorem 13.3.4 Let $k \geq 4$ and $r_i \in \mathbb{N}$ for $i = 0, 1, \cdots, k$. If $r_i \geq 3$ for some i with $3 \leq i \leq k - 1$, or $r_i = r_j = 2$ for some i, j with

$$2j - k + 1 \leq i < j \leq 2i - 3,$$

then $\mathcal{I}(r_0, r_1, \cdots, r_k) \not\subseteq \mathcal{C}$.

Proof. Assume that $r_i = 3$ for some i with $3 \leq i \leq k - 1$. By Theorem 13.3.3, there exists a non-chordal graph G_1 with $\omega(G_1) = k + 1$ such that

$$P(G_1, \lambda) = \lambda(\lambda - 1) \cdots (\lambda - k)(\lambda - i)^2.$$

Thus, by Lemma 13.2.1, $\mathcal{I}(r_0, r_1, \cdots, r_k) \backslash \mathcal{C} \neq \emptyset$.

Now assume that $r_i = r_j = 2$ for some i, j with $i < j \leq 2i - 3$ and $2j \leq k + i - 1$. By Theorem 13.3.3, there exists a non-chordal graph G_2 with $\omega(G_2) = k + 1$ such that

$$P(G_2, \lambda) = \lambda(\lambda - 1) \cdots (\lambda - k)(\lambda - i)(\lambda - j).$$

Since $r_i = 2$ and $r_j = 2$, by Lemma 13.2.1, $\mathcal{I}(r_0, r_1, \cdots, r_k) \backslash \mathcal{C} \neq \emptyset$. □

Remark The condition that $2j - k + 1 \leq i < j \leq 2i - 3$ in Theorem 13.3.4 implies that $4 \leq i < j \leq k - 2$. Thus, Theorem 13.3.4 applies only when $k \geq 7$.

13.4   Dmitriev's Problem

In this section, we shall present some results on Problem 13.1.3. Recall that
the graph in Figure 13.1 is actually the graph $R(1, 4, 4)$. All pure cycles of
this graph are of order 4. Dmitriev (1980) discovered a family of graphs in
$\mathcal{I} \backslash \mathcal{C}$ (i.e., $R(st, (s+1)(t+1), k)$, where $k = st + (s+1)(t+1) - 1$). All
pure cycles in each graph of this family are of order 4. Recall that $g^*(G)$
denotes the order of the smallest pure cycles in a graph G. Dmitriev (1980)
asked the following:

Problem 13.4.1   *For any integer $p \geq 5$, does there exist a graph $G \in \mathcal{I}$
such that $g^*(G) = p$?*

Dong and Koh (1998) gave a positive answer to Dmitriev's problem for
$p = 5$. Dong, Teo, Koh and Hendy (2002) further answered this problem
affirmatively for $p = 6$ and 7. They first constructed a family of graphs.
For $k_i \in \mathbb{N}$, $i = 1, 2, \cdots, n$, where $n \geq 2$, let $H_{k_1, k_2, \cdots, k_n}$ denote the graph
obtained from the disjoint union of n complete graphs $K_{k_1}, K_{k_2}, \cdots, K_{k_n}$
and a vertex w by adding edges joining each vertex in K_{k_i} to each vertex
in $K_{k_{i+1}}$ for $i = 1, 2, \cdots, n-1$, and edges joining w to each vertex in K_{k_1}
and K_{k_n}, as shown in Figure 13.3. Clearly, when $n \geq 3$, all pure cycles in
$H_{k_1, k_2, \cdots, k_n}$ are of order $n + 1$.

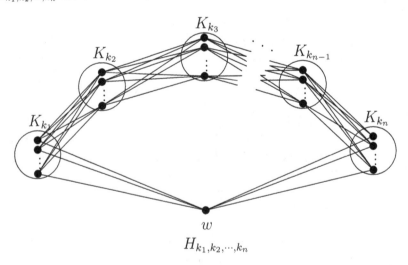

Figure 13.3

Lemma 13.4.1   *For any* $n, k_1, k_2, \cdots, k_n \in \mathbb{N}$, *where* $n \geq 2$,

$$P(H_{k_1,k_2,\cdots,k_n}, \lambda) = \frac{\prod\limits_{i=1}^{n-1} (\lambda)_{k_i+k_{i+1}}}{\lambda \prod\limits_{i=2}^{n-1} (\lambda)_{k_i+1}} \left(\prod_{i=1}^{n} (\lambda - k_i) + (-1)^{n+1} \prod_{i=1}^{n} k_i \right).$$

$$\tag{13.7}$$

Proof. It is easy to verify the result for $n = 2, 3$. Now let $n \geq 4$. Consider the graph H' obtained from $H_{k_1,k_2,\cdots,k_n}$ by adding edges joining w to all the vertices in the clique $K_{k_{n-1}}$ (see Figure 13.3). The chromatic polynomial of H' is

$$P(H', \lambda) = \frac{(\lambda)_{k_{n-1}+k_n+1}}{(\lambda)_{k_{n-1}+1}} P(H_{k_1,k_2,\cdots,k_{n-1}}, \lambda).$$

Thus, by applying FRT repeatedly, we have

$$
\begin{aligned}
P(H_{k_1,k_2,\cdots,k_n}, \lambda) &= \frac{(\lambda)_{k_{n-1}+k_n+1}}{(\lambda)_{k_{n-1}+1}} P(H_{k_1,k_2,\cdots,k_{n-1}}, \lambda) \\
&+ \frac{k_{n-1}(\lambda)_{k_{n-1}+k_n}(\lambda)_{k_{n-2}+k_{n-1}}}{(\lambda)_{k_{n-1}}(\lambda)_{k_{n-2}+1}} P(H_{k_1,k_2,\cdots,k_{n-2}}, \lambda).
\end{aligned}
$$

The result then follows by induction from the above recursive expression.

$\square$

For $n, k_1, k_2, \cdots, k_n \in \mathbb{N}$ and real number x, define

$$f(k_1, k_2, \cdots, k_n, x) = \prod_{i=1}^{n} (x - k_i) + (-1)^{n+1} \prod_{i=1}^{n} k_i. \tag{13.8}$$

For $k_1, k_2, \cdots, k_n \in \mathbb{N}$, let $\mathcal{H}(k_1, k_2, \cdots, k_n)$ denote the family of graphs $H_{k_{t(1)},k_{t(2)},\cdots,k_{t(n)}}$ for all permutations t of $1, 2, \cdots, n$. By Lemma 13.4.1 and (13.8), we have:

Lemma 13.4.2   *Let* $n \in \mathbb{N}$ *with* $n \geq 2$ *and* $k_1, k_2, \cdots, k_n \in \mathbb{N}$. *Then*

$$\mathcal{H}(k_1, k_2, \cdots, k_n) \subseteq \mathcal{I}$$

if and only if $f(k_1, k_2, \cdots, k_n, x)$ *is an integral-root polynomial of* x. $\square$

Now we are ready to present the following main result in this section, which was established by Dong, Teo, Koh and Hendy (2002).

Theorem 13.4.1 *Let* $n, k_1, k_2, \cdots, k_n \in \mathbb{N}$, *where* $n \geq 3$. *Under each of the following conditions,* $\mathcal{H}(k_1, k_2, \cdots, k_n) \subseteq \mathcal{I}$:

(i) $n = 3$, $k_1 = a_1 b_1$, $k_2 = a_2 b_2$ *and* $k_3 = (a_1 + a_2)(b_1 + b_2)$ *for any* $a_1, a_2, b_1, b_2 \in \mathbb{N}$;

(ii) $n = 4$, $k_1 = a_1 b_1$, $k_2 = a_2 b_2$, $k_3 = (a_1 + a_2)(2b_1 + b_2)$ *and* $k_4 = (a_1 + 2a_2)(b_1 + b_2)$ *for any* $a_1, a_2, b_1, b_2 \in \mathbb{N}$;

(iii) $n = 5$, $k_1 = 1$, $k_2 = \frac{a(a-3)}{2}$, $k_3 = \frac{a(a+3)}{2}$, $k_4 = (a - 1)(a + 2)$ *and* $k_5 = (a - 2)(a + 1)$ *for any* $a \in \mathbb{N}$ *with* $a \geq 4$;

(iv) $n = 6$, $k_1 = a_1 b_1$, $k_2 = a_2 b_2$, $k_3 = (a_1 + a_2)(b_1 + 3b_2)$, $k_4 = (3a_1 + a_2)(b_1 + b_2)$, $k_5 = (2a_1 + a_2)(2b_1 + 3b_2)$ *and* $k_6 = (3a_1 + 2a_1)(b_1 + 2b_2)$ *for any* $a_1, a_2, b_1, b_2 \in \mathbb{N}$.

Proof. By Lemma 13.4.2, we need only to show that $f(k_1, k_2, \cdots, k_n, x)$ is an integral-root polynomial under each condition.

(i) If $n = 3$, $k_1 = a_1 b_1$, $k_2 = a_2 b_2$ and $k_3 = (a_1 + a_2)(b_1 + b_2)$, then

$$f(k_1, k_2, k_3, x) = x(x - a_1 b_1 - a_2 b_2 - a_1 b_2)(x - a_1 b_1 - a_2 b_2 - a_2 b_1).$$

(ii) If $n = 4$, $k_1 = a_1 b_1$, $k_2 = a_2 b_2$, $k_3 = (a_1 + a_2)(2b_1 + b_2)$ and $k_4 = (a_1 + 2a_2)(b_1 + b_2)$, then

$$\begin{aligned}f(k_1, k_2, k_3, k_4, x) &= x(x - a_2 b_2 - 2a_2 b_1 - a_1 b_1)(x - a_2 b_2 - a_1 b_2 - a_1 b_1) \\ &\quad (x - 2a_2 b_2 - 2a_2 b_1 - a_1 b_2 - 2a_1 b_1).\end{aligned}$$

(iii) If $n = 5$, $k_1 = 1$, $k_2 = \frac{a(a-3)}{2}$, $k_3 = \frac{a(a+3)}{2}$, $k_4 = (a - 1)(a + 2)$ and $k_5 = (a - 2)(a + 1)$, then

$$\begin{aligned}f(k_1, k_2, k_3, k_4, k_5, x) &= x(x - a^2)(x - a^2 - 1) \\ &\quad \left(x - \frac{a(a - 1)}{2} + 2\right)\left(x - \frac{a(a + 1)}{2} + 2\right).\end{aligned}$$

(iv) If $n = 6$, $k_1 = a_1 b_1$, $k_2 = a_2 b_2$, $k_3 = (a_1 + a_2)(b_1 + 3b_2)$, $k_4 = (3a_1 + a_2)(b_1 + b_2)$, $k_5 = (2a_1 + a_2)(2b_1 + 3b_2)$ and $k_6 = (3a_1 + 2a_2)(b_1 + 2b_2)$, then

$$\begin{aligned}&f(k_1, k_2, k_3, k_4, k_5, k_6, x) \\ &= x(x - 3a_1 b_2 - 3a_1 b_1 - 3a_2 b_2 - 2a_2 b_1) \\ &\quad (x - 3a_1 b_2 - a_1 b_1 - a_2 b_2)(x - 6a_1 b_2 - 4a_1 b_1 - 4a_2 b_2 - 2a_2 b_1) \\ &\quad (x - a_1 b_1 - a_2 b_2 - a_2 b_1)(x - 3a_1 b_1 - 3a_2 b_2 - a_2 b_1 - 6a_1 b_2). \quad \square\end{aligned}$$

Notes

(1) In Theorem 13.4.1(i), if $a_1 = a_2 = b_1 = b_2 = 1$, then $k_1 = k_2 = 1$ and $k_3 = 4$, and $\mathcal{H}(1, 1, 4)$ contains only one graph, i.e., the graph in Figure 13.1.

(2) The family of graphs discovered by Dmitriev (1980) is a subfamily of the family of graphs in Theorem 13.4.1 (i).

(3) Theorem 13.4.1 provides a positive answer to Dmitriev's problem for $p \leq 7$.

(4) Very recently, Hernández and Luca (2005) showed that for any integer $n \geq 3$, there exist positive integers $k_1, k_2, \cdots, k_n$ such that the polynomial $f(k_1, k_2, \cdots, k_n, x)$ has only integral roots if and only if the Prouhet-Tarry-Escott problem has an ideal solution of degree $n - 1$, i.e., there exist two distinct sets of integers $\{\alpha_1, \alpha_2, \cdots, \alpha_n\}$ and $\{\beta_1, \beta_2, \cdots, \beta_n\}$ such that

$$\sum_{i=1}^{n} \alpha_i^j = \sum_{i=1}^{n} \beta_i^j$$

holds for all $j = 1, 2, \cdots, n - 1$. It can be found in Browein (2002) that the Prouhet-Tarry-Escott problem admits ideal solutions for all n with $2 \leq n \leq 12$ and $n \neq 11$. Thus, by Lemma 13.4.2, Dmitriev's problem has an affirmative answer for $3 \leq p \leq 13$ and $p \neq 12$.

(5) For $p = 12$ or $p \geq 14$, Dmitriev's problem remains unsolved.

To end this chapter, we would like to propose the following:

Conjecture 13.4.1 *For any integer $n \geq 3$, there exist $k_1, k_2, \cdots, k_n \in \mathbb{N}$ such that the polynomial $f(k_1, k_2, \cdots, k_n, x)$ is an integral-root polynomial of x.*

By Lemma 13.4.2, if this conjecture is true, then the answer to Dmitriev's problem is in the affirmative for all $p \geq 5$.

Exercise 13

13.1 Let $k, r_0, r_1, \cdots, r_k \in \mathbb{N}$. Prove that there exists a graph $G \in \mathcal{I} \backslash \mathcal{C}$ such that

$$P(G, \lambda) = \lambda^{r_0}(\lambda - 1)^{r_1}(\lambda - 2)^{r_2} \cdots (\lambda - k)^{r_k}$$

if and only if there exists a graph $H \in \mathcal{I} \backslash \mathcal{C}$ such that

$$P(H, \lambda) = \lambda(\lambda - 1)(\lambda - 2)^{r_2} \cdots (\lambda - k)^{r_k}.$$

13.2 Let G be a planar graph with

$$P(G, \lambda) = \lambda^{r_0}(\lambda - 1)^{r_1}(\lambda - 2)^{r_2}(\lambda - 3)^{r_3}.$$

Prove that G is a chordal graph.

(See Dong and Koh (1998).)

13.3 Let G be a graph with $P(G, \lambda) = \lambda^{r_0}(\lambda - 1)^{r_1}$. Prove that G is a forest of order $r_0 + r_1$ having r_0 components. (See Exercise 1.1.)

13.4 Show that if G is a graph such that $w(G) = v(G) - 1$, then $G - u$ is complete for some $u \in V(G)$, and thus G is chordal.

13.5 Let G be a graph of order n and $\chi(G) = n - 2$. Prove that either $G = C_5 + K_{n-5}$ or $G \backslash \{u, v\} \cong K_{n-2}$ for some vertices u, v in G.

(See Dong (2004).)

13.6 Given the polynomial P of (13.6), let $\alpha(P)$ denote the number of non-isomorphic graphs G with $P(G, \lambda) = P$. Show that

$$\alpha(P) = \begin{cases} 2, & q = k \neq p + 1; \\ 3, & q = k = p + 1; \\ \min\{p, k - q + 1\} + 2, & q \neq k, q \neq p + 1; \\ \min\{p, k - q\} + 3, & q \neq k, q = p + 1. \end{cases}$$

(See Xu(1997).)

13.7 Let $p, q, r \in \mathbb{N}$. Find the chromatic polynomials of the following graphs.

(See Dong and Koh (1998).)

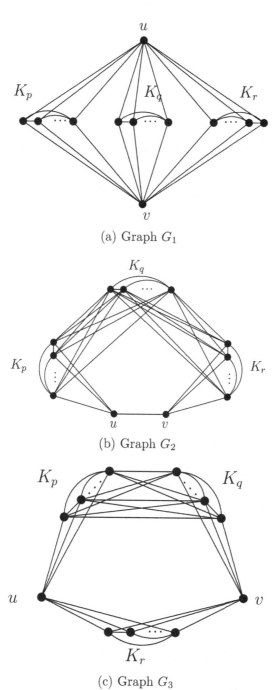

(a) Graph G_1

(b) Graph G_2

(c) Graph G_3

Chapter 14

Complex Roots of Chromatic Polynomials

14.1 Introduction

In Chapters 12 and 13, we have viewed all chromatic polynomials as polynomials over the real number field, and discussed their real roots. In this chapter, we shall consider them as polynomials over the complex number field.

We know that for any graph G of order n, all real roots of $P(G, \lambda)$ are within the interval $[0, n-1]$. So for the complex chromatic roots z, the analogous target is to find a bounded region for $|z|$ in terms of the order and size of G. The first such result was obtained by Thier (1983), and it was then substantially improved by Brown (1998a): for any connected (n, m)-graph, all roots of $P(G, \lambda)$ are within $\{z : |z - 1| \leq m - n + 1\}$. Brown (1998b) further proved that the chromatic polynomial of any connected (n, m)-graph has a root with modulus at least $(m-1)/(n-2)$. In Section 14.2, we shall introduce these results.

Biggs, Damerell and Sands (1972) conjectured that the chromatic roots of k-regular graphs must be within the disc $|z| < f(k)$, where $f(k)$ is independent of the orders of these graphs. This conjecture was further generalized by Brenti, Royle and Wagner (1994) that 'k-regular graphs' can be replaced by 'graphs with maximum degree k'. This more general conjecture has recently been proved by Sokal (2001). In Section 14.3, we shall introduce Sokal's result. In fact, Sokal showed the existence of such a function $f(k)$ with $f(k) < 8k$.

In Section 14.4, we shall present Brown's result (1999) which says that if each edge of a graph G is subdivided into a sufficiently long path, then the chromatic roots of the resulting graph lie "close" to the disk $|z-1| \le 1$.

In Section 14.5, we shall focus on the search for chromatic roots with negative real part. Based on the fact that chromatic polynomials have no negative real roots, Farrell (1980a) conjectured that no chromatic polynomial has a complex root with negative real part. This conjecture was first disproved by Read and Royle (1991). Shrock and Tsai (1998) then proved that there exist graphs which have chromatic roots z with $Re(z) < 0$ and $|Re(z)|$ arbitrarily large, where $Re(z)$ is the real part of z. Sokal (2004) further showed that chromatic roots are dense in the whole complex plane, i.e., for any complex z_0 and real $\epsilon > 0$, there exists a graph G such that $P(G, \lambda)$ has a root in $|z - z_0| < \epsilon$.

In Section 14.6, we shall introduce the partition function of the q-state Potts model of a graph, which is a polynomial with multiple variables q and w_e for all edges e in the graph, where w_e is considered as the weight of e. The chromatic polynomial of a graph can be obtained from the partition function of the Potts model by taking $w_e = -1$ for all edges e. Some researchers in the area of statistical mechanics, such as Sokal, Shrock and Tsai, have obtained some nice results on the Potts model and some of these results actually solved some open problems on chromatic polynomials. For example, Shrock and Tsai (1998) proved that for each real a, there exists a chromatic root whose real part is less than a, and Sokal (2004) proved the chromatic roots are dense in the whole plane. We shall show clearly how the partition function of the Potts model arises in statistical mechanics and how it is related to some other famous polynomials, such as the dichromatic polynomial and the Tutte polynomial. We shall also reveal a property on the partition function of the Potts model, which cannot be transformed into an analogous property of chromatic polynomials. By applying this property, we prove a result, mentioned by Sokal, which is more general than an existing result, namely, Theorem 14.4.3.

14.2 Location of chromatic roots

Locating chromatic roots in the complex plane is a difficult problem. The first general result on the location of chromatic roots, as stated below, was obtained by Thier (1983).

Theorem 14.2.1 *Let G be an (n, m)-graph. Then the roots of $P(G, \lambda)$ lie in $U_1 \cap U_2$, where*

$$U_1 = \{z : |z| \leq m - 1 \text{ or } |z - m| \leq m\}$$

and

$$U_2 = \{z : |z - 1| \leq m - 1 \text{ or } |z - m + n - 2| \cdot |z - 1| \leq m(m - 1)\}. \quad \square$$

Brenti, Royle and Wagner (1994) pointed out that the regions in Theorem 14.2.1 appear to be much larger than necessary. Brown (1998a) substantially improved Theorem 14.2.1 by establishing the following result.

Theorem 14.2.2 *Let G be a connected (n, m)-graph that is not a tree. Then the roots of $P(G, \lambda)$ lie in the disc $\{z : |z - 1| \leq m - n + 1\}$, with equality if and only if G is unicyclic.*
$\square$

Note that if G is an (n, m)-graph with c components, then $m - n + c$ is called the *cyclomatic number* or the *nullity* of G, denoted by $r_c(G)$.

For a complex number z, let $Im(z)$ denote the imaginary part of z. We have another result due to Brown (1998b).

Theorem 14.2.3 *Let G be a connected (n, m)-graph, where $n \geq 3$, with t triangles, and let*

$$\begin{cases} D = (n - 3)(m - 1)(n - m - 1) + 2t(n - 2)(n - 3), \\ B = (m - 1)/(n - 2), \\ W = \begin{cases} B, & \text{if } n = 3, \\ B + \sqrt{D}/(n - 2)(n - 3), & \text{if } n \geq 4. \end{cases} \end{cases} \quad (14.1)$$

If $D \geq 0$, then $P(G, \lambda)$ has a root z with $Re(z) \geq W$; and if $D < 0$, then $P(G, \lambda)$ has roots z_1 and z_2 (not necessarily distinct) such that $Re(z_1) \geq B$ and $Im(z_2) \geq \sqrt{-D}/(n - 2)(n - 3)$.

Proof. If G is a tree or K_3, then $D = 0$, $\chi(G) = B + 1$ and $W(= B)$ is a root of $P(G, \lambda)$. Thus the theorem holds for $n = 3$. We now assume that $n \geq 4$.

By Theorem 2.3.1,

$$P(G, \lambda) = \sum_{i=1}^{n} (-1)^{n-i} h_i \lambda^i,$$

where $h_n = 1$, $h_{n-1} = m$ and $h_{n-2} = \binom{m}{2} - t$. Since $\chi(G) \geq 2$, $\lambda(\lambda - 1)$ is a factor of $P(G, \lambda)$. Thus $g(\lambda) = P(G, \lambda)/(\lambda(\lambda - 1))$ is a polynomial. Indeed,

$$g(\lambda) = \lambda^{n-2} - (m-1)\lambda^{n-3} + \left(\binom{m-1}{2} - t \right) \lambda^{n-4} - \cdots . \quad (14.2)$$

The $(n-4)$th derivative of $g(\lambda)$ is given by

$$g^{(n-4)}(\lambda) = \frac{(n-2)!}{2}\lambda^2 - (m-1)(n-3)!\lambda + \left(\binom{m-1}{2} - t \right)(n-4)!. \quad (14.3)$$

Observe that W is a root of (14.3). Applying the following result on polynomials (due to Lucas, see Marden (1966)):

> if f is a non-constant polynomial, then the roots of the derivative f' of f lie in the convex hull of the roots of f,

it follows that the roots of $g^{(n-4)}$ must lie in the convex hull of the roots of g, and hence of $P(G, \lambda)$. Thus $P(G, \lambda)$ must have roots z_1 and z_2 such that $Re(z_1) \geq Re(W)$ and $Im(z_2) \geq Im(W)$. The result then follows. $\square$

In an unpublished manuscript, Sokal proved a stronger result than Theorem 14.2.3.

Theorem 14.2.4   *Let G be a connected (n, m)-graph, where $n \geq 3$. Then*

(i) $P(G, \lambda)$ has a root z such that $|z| \geq (m-1)/(n-2)$;

(ii) the moduli of all roots of $P(G, \lambda)$ are at most $(m-1)/(n-2)$ if and only if G is a tree or a 2-tree. $\square$

It was mentioned in Brown (1998b) that, by an argument similar to the proof of Theorem 14.2.3, one can show that if $\chi(G) \geq k \geq 2$, then $P(G, \lambda)$ has a root z such that $|z| \geq (m - \binom{k}{k})/(n - k)$ (a proof can be found in Bielak (2001)). This bound is better than $B = (m-1)/(n-2)$ whenever $B > k - 1 \geq 2$.

We know that chordal graphs which are not trees possess plenty triangles and the chromatic polynomials of chordal graphs have only real roots (in fact, only integral roots). In the following, we shall see that if G contains very few triangles, then $P(G, \lambda)$ has, quite possibly, a non-real root.

Corollary 14.2.1   *If G is a K_3-free graph that is not a forest, then $P(G, \lambda)$ has a non-real root.*

Proof. Assume that G is a connected (n,m)-graph with $m \geq n \geq 4$. As $t = 0$, we have

$$D = (m-1)(n-3)(n-m-1) < 0.$$

By Theorem 14.2.3, $P(G,\lambda)$ has a root z such that

$$Im(z) \geq \frac{\sqrt{-D}}{(n-2)(n-3)} > 0. \qquad \square$$

One can weaken the condition in Corollary14.2.1 by considering Sturm sequence (see Brown (1998b) for details): if G is a connected (n,m)-graph with t triangles, where $n \geq 4$, then $P(G,\lambda)$ has a non-real root if

$$t < \frac{m(m-n)+n-1}{2(n-2)}.$$

Let $Im(\mathcal{G}_n)$ denote the largest imaginary part of a chromatic root among all graphs of order n.

Corollary 14.2.2 *For all $n \geq 4$, $Im(\mathcal{G}_n) \geq \sqrt{n}/4$.*

Proof. Let $G = K(\lfloor n/2 \rfloor, \lceil n/2 \rceil)$. Then G has $m = \lfloor n^2/4 \rfloor$ edges and contains no triangles. If n is even, then by (14.1),

$$-D = \frac{1}{16}(n^2-4)(n-3)(n-2)^2 > \frac{1}{16}n(n-3)^2(n-2)^2,$$

and if n is odd, then

$$-D = \frac{1}{16}(n^2-5)(n-3)^2(n-1) > \frac{1}{16}n(n-3)^2(n-2)^2.$$

By Theorem 14.2.3, $P(G,\lambda)$ has a root z with

$$Im(z) \geq \sqrt{-D}/(n-2)(n-3) > \sqrt{n}/4.$$

Thus $Im(\mathcal{G}_n) > \sqrt{n}/4$. $\qquad \square$

14.3 Chromatic roots within $|z| \leq 8\Delta$

Biggs, Damerell and Sands (1972) proposed the following conjecture.

Conjecture 14.3.1 *There is a function $f : \mathbb{N} \to \mathbb{R}$ such that if G is a k-regular graph and $P(G,z) = 0$, then $|z| \leq f(k)$. Furthermore, $f(3) = 3$.*

Biggs, Damerell and Sands (1972) had verified the above conjecture for all 3-regular graphs with at most 10 vertices (see Brenti, Royle and Wagner (1994)). Read and Royle (1991) extended this verification to all 3-regular graphs with at most 16 vertices, and some larger graphs.

Brenti, Royle and Wagner (1994) strengthened the above conjecture as follows.

Problem 14.3.1   *Is there a function $f : \mathbb{N} \to \mathbb{R}$ such that if G is a graph with $\Delta(G) = k$ and $P(G, z) = 0$, then $|z| \leq f(k)$?*

Sokal (2001) completely solved the above problem.

Theorem 14.3.1   *Let C be the smallest number for which*

$$\inf_{\alpha > 0} \alpha^{-1} \sum_{n=2}^{\infty} \exp(\alpha n) C^{-(n-1)} \frac{n^{n-1}}{n!} \leq 1. \tag{14.4}$$

(It can be shown that $C \leq 7.963907$.) If G is a graph and $P(G, z) = 0$, then $|z| < C\Delta(G)$.   □

Sokal (2001) further showed that the "maximum degree" in Theorem 14.3.1 can be replaced by the "second-largest degree" provided that one pays a small price.

Theorem 14.3.2   *Let C be the constant defined in Theorem 14.3.1. If G is a graph in which all but one of the vertices have degrees not exceeding k, and $P(G, z) = 0$, then either $|z| < Ck$ or $|z - 1| < Ck$ (and in particular $|z| < Ck + 1$).*   □

Motivated by an idea due to Shrock and Tsai (1999), Brown, Hickman, Sokal and Wagner (2001) conjectured that the "second-largest degree" can be further weakened to the "maximaxflow", defined as

$$\Lambda(G) = \max_{\substack{x \neq y \\ x, y \in V(G)}} \lambda(x, y),$$

where $\lambda(x, y)$ is the maximum number of edge-disjoint $x - y$ paths, which is equal to the minimum number of edges separating x from y. (See Remark (iv) after Corollary 12.11.1.) Since $\lambda(x, y) \leq \min\{d(x), d(y)\}$, $\Lambda(G)$ cannot exceed the second-largest degree of G.

Conjecture 14.3.2   *For any $k \in \mathbb{N}$, there exists a constant $C(k)$ such that if G is a graph with $\Lambda(G) = k$ and $P(G, z) = 0$, then $|z - 1| \leq C(k)$.*

14.4 Subdivisions

In this section, we shall study the effect of taking subdivisions on the locations of chromatic roots of a graph. Recall that the cyclomatic number of an (n,m)-graph G with component number c is $r_c(G) = m - n + c$. Observe that $r_c(G)$ is invariant under subdivision. Thus, by Theorem 14.2.2, for any subdivision G' of G, the chromatic roots of G' still lie in the disk: $|z-1| \le m-n+1$. Brown (1999) showed that every graph has a subdivision whose chromatic roots lie "close" to the disk $|z-1| \le 1$.

Theorem 14.4.1 *Let G be an (n,m)-graph and let $\epsilon > 0$. Then there is a subdivision G' of G such that $P(G', \lambda)$ has all its roots lie in the disk $|z-1| < 1 + \epsilon$.*

Proof. We may assume that G is connected. We shall prove the result by induction on the cyclomatic number $r_c(G) = m - n + 1$ of G. If $r_c(G) = 0$, then G is a tree and every subdivision of G is still a tree. The result thus holds in this case.

Now assume that $r_c(G) \ge 1$. Then G is not a tree. Let $e = uv$ be an edge on some cycle of G. Then $G - e$ is connected and $r_c(G - e) = r_c(G) - 1$. By the induction hypothesis, $G - e$ has a subdivision G'' such that all chromatic roots of G'' lie in $|z - 1| < 1 + \epsilon$. Consider the graph G_t which is formed from G'' by adding a new $u - v$ path of length $t + 1$. It is clear that G_t is a subdivision of G for all $t \in \mathbb{N}$. By Theorem 1.3.3,

$$P(G_t, z) = (-1)^{t+1} \left(\frac{(1-z)^{t+1} - 1}{z} P(G'', z) + P(G'' \cdot uv, z) \right). \quad (14.5)$$

Let s be the minimum value of $|P(G'', z)|$ on the annulus

$$A = \{ z : 1 + \epsilon \le |z - 1| \le m - n + 1 \}.$$

Since $P(G'', z)$ has no roots in $|z - 1| \ge 1 + \epsilon$, we have $s > 0$. Let S be the maximum value of $|P(G'' \cdot uv, z)|$ on A. Then it follows from (14.5) that for $z \in A$,

$$
\begin{aligned}
|P(G_t, z)| &\ge \left| \frac{(1-z)^{t+1} - 1}{z} P(G'', z) \right| - |P(G'' \cdot uv, z)| \\
&\ge \frac{(1+\epsilon)^{t+1} - 1}{m - n + 1} s - S. \quad (14.6)
\end{aligned}
$$

It is clear that the right hand side of (14.6) is positive when t is large enough. Thus, for such a t, $P(G_t, z)$ has no roots in the annulus A. By

Theorem 14.2.2, all roots of $P(G_t, z)$ satisfy $|z - 1| \leq m - n + 1$. Hence $P(G_t, z)$ has no roots z with $|z - 1| \geq 1 + \epsilon$. □

Theorem 14.4.1 was soon strengthened by Brown and Hickman (2002b) as shown below.

Theorem 14.4.2 *Let G be any graph and let $\epsilon > 0$. Then there is a constant $L = L(G, \epsilon)$ such that, if we divide each edge of G into a path of length at least L, then all chromatic roots of the resulting graph G' lie in the disk $|z - 1| < 1 + \epsilon$.* □

One may also prove Theorem 14.4.2 by modifying the above proof of Theorem 14.4.1.

Brown and Hickman (2002b) obtained also the following interesting result that subdividing each edge of a graph into a path of even length is enough to guarantee that the resulting graph has no real chromatic roots in the interval $[2, \infty)$.

Theorem 14.4.3 *Let G be a graph and G' be a graph obtained by subdividing each edge of G into a path of even length. Then $P(G', \lambda) > 0$ for all real $\lambda \in [2, \infty)$.* □

Note that the graphs G' in Theorem 14.4.3 are always bipartite. We shall present, in what follows, a stronger result. For any graph G, a path $v_0 v_1 \cdots v_r$ in G is called an *odd-subdivided path* if r is odd, $d(v_0) \geq 3$, $d(v_r) \geq 3$ and $d(v_i) = 2$ for $i = 1, \cdots, r - 1$. In particular, every edge uv with $d(u) \geq 3$ and $d(v) \geq 3$ is an odd-subdivided path. Let $\mathcal{G}_{os}$ denote the family of graphs in which every cycle contains no more than two odd-subdivided paths. It is clear that $\mathcal{G}_{os}$ includes not only the graphs G' in Theorem 14.4.3, but also some 3-chromatic graphs such as odd cycles and generalized θ-graphs.

Theorem 14.4.4 *For every $G \in \mathcal{G}_{os}$, $P(G, \lambda) > 0$ for all real $\lambda \in (2, \infty)$.*

Proof. Let $G \in \mathcal{G}_{os}$ and let λ be a real number in $(2, \infty)$. If $e(G) \leq 1$, then it is obvious that $P(G, \lambda) > 0$. Assume that $e(G) > 1$.

Let H be a non-trivial component of G. Observe that $\delta(H) \leq 2$; otherwise, H is not a tree and every edge in H is an odd-subdivided path, implying that every cycle of H contains at least 3 odd-subdivided paths, a contradiction.

Let u be a vertex of H such that $1 \leq d(u) \leq 2$. Now we consider two cases.

Case 1: u is a simplicial vertex.

Then

$$P(G, \lambda) = (\lambda - d(u))P(G - u, \lambda) > 0,$$

since $d(u) \leq 2$ and $P(G - u, \lambda) > 0$ by the induction hypothesis.

Case 2: u is not simplicial.

Then $d(u) = 2$. Let $N(u) = \{v, w\}$. Observe that both $G - u$ and $(G - u) \cdot vw$ belong to $\mathcal{G}_{os}$. By the induction hypothesis, $P(G - u, \lambda) > 0$ and $P((G - u) \cdot vw, \lambda) > 0$. Hence, by FRT,

$$P(G, \lambda) = (\lambda - 2)P(G - u, \lambda) + P((G - u) \cdot vw, \lambda) > 0. \qquad \square$$

Theorem 14.4.4 implies that $\chi(G) \leq 3$ for every graph $G \in \mathcal{G}_{os}$. If $\chi(G) \leq 2$ and $G \in \mathcal{G}_{os}$, then by Theorem 14.4.4, $P(G, \lambda) > 0$ for every real $\lambda \geq 2$. Thus Theorem 14.4.3 follows immediately from Theorem 14.4.4.

14.5 Chromatic roots in the whole plane

From the fact that the coefficients of chromatic polynomials alternate in signs, it can be shown that chromatic polynomials have no negative real roots (see Theorem 12.2.1). Based on this observation, Farrell (1980a) conjectured that there are no chromatic roots with negative real part.

This conjecture was first disproved by Read and Royle (1991) by numerical computation; they found the following result:

Theorem 14.5.1 *There exist 3-regular graphs of order 18 and girth at least 5 which have a chromatic root with negative real part.* $\qquad \square$

As was introduced in Section 5.2, given $k, f \in \mathbb{N}$, the k-bridge graph $\theta_k(f)$ is the graph obtained from two vertices by joining them with k internally disjoint paths of length f. By applying FRT, we have

$$\begin{aligned} \lambda^{k-1}P(\theta_k(f), \lambda) &= (\lambda - 1)((\lambda - 1)^f + (-1)^{f+1})^k \\ &\quad + ((\lambda - 1)^f + (-1)^f(\lambda - 1))^k. \end{aligned} \qquad (14.7)$$

While the graphs found by Read and Royle in Theorem 14.5.1 have chromatic roots z with $Re(z) < 0$ and $|Re(z)|$ is very small, Shrock and Tsai (1998) showed that $P(\theta_k(f), \lambda)$ has roots z with negative $Re(z)$ such that $|Re(z)|$ is arbitrarily large.

Theorem 14.5.2   *For any $m \in \mathbb{N}$, there exist f and k such that $\theta_k(f)$ has a chromatic root z with $Re(z) < -m$.* $\square$

By a different approach, Brown and Hickman (2002a) proved the following related result.

Theorem 14.5.3   *The graph $\theta_3(f)$ has a chromatic root with negative real part if and only if $f \geq 8$.* $\square$

Sokal (2004) obtained the following result which is more general than Theorem 14.5.2.

Theorem 14.5.4   *Let z_0 be any complex number with $|z_0 - 1| \geq 1$. For any real $\epsilon > 0$, there exist constants $f_0(z_0)$ and $k_0(f_0, z_0)$ such that for all $f > f_0(z_0)$ and $k > k_0(f_0, z_0)$, $P(\theta_k(f), \lambda)$ has a root in $|z - z_0| < \epsilon$.* $\square$

Theorem 14.5.4 shows that the chromatic roots of the graphs $\theta_k(f)$'s are dense in the region: $\{z : |z - 1| \geq 1\}$. The chromatic polynomial of the join $K_s + \theta_k(f)$ is given by

$$P(K_s + \theta_k(f), \lambda) = (\lambda)_s P(\theta_k(f), \lambda - s). \tag{14.8}$$

Thus, by Theorem 14.5.4 and (14.8), the chromatic roots of the graphs $K_s + \theta_k(f)$'s over all k, f with $k \geq 3$ and $f \geq 3$ are dense in the region $\{z : |z - 1| \leq 1\}$, where $s \geq 2$.

Corollary 14.5.1   *For any complex number z_0 and real $\epsilon > 0$, there exists a graph G such that $P(G, \lambda)$ has a root in $|z - z_0| < \epsilon$.* $\square$

Corollary 14.5.1 can be stated as: the closure of the set of chromatic roots of all chromatic polynomials is the whole complex plane!

It follows from Theorem 14.5.2 or Theorem 14.5.4 that Theorem 14.3.2 is no longer true if 'all but one' is replaced by 'all but two'; for in this case, the moduli of the chromatic roots of the generalized θ-graphs can be arbitrarily large.

14.6   Remarks

Although the results in Section 14.5 show that the chromatic roots are dense in the whole complex plane, it is interesting to study root-free regions for chromatic polynomials in terms of some invariants of graphs, such as order, size, maximum degree, etc.

So far, we know only that there are three root-free intervals in the line $Im(z) = 0$, i.e., $(-\infty, 0)$, $(0, 1)$ and $(1, 32/27]$. We guess that chromatic polynomials are also root-free on the line $Re(z) = 0$ except the origin.

Conjecture 14.6.1   *There is no chromatic root z with $Re(z) = 0$ and $Im(z) \neq 0$.*

The following conjecture was recently proposed by Sokal in his private communication with one of the authors.

Conjecture 14.6.2   *For any graph G, if $P(G, z) = 0$, then $Re(z) \leq \Delta(G)$.*

We are quite confident in the correctness of the following conjecture which is weaker than Conjecture 14.6.2.

Conjecture 14.6.3   *For any graph G of order n, if $P(G, z) = 0$, then $Re(z) \leq n - 1$.*

Recall that the tree-width of a graph G is the minimum integer k such that G is a subgraph of a k-tree. It is clear that Conjecture 14.6.3 is also weaker than the following conjecture.

Conjecture 14.6.4   *For any graph G of tree-width k, if $P(G, z) = 0$, then $Re(z) \leq k$.*

14.7   Potts model

The Potts model partition function was introduced by Potts in 1952. Roughly speaking, the Potts model partition function is a function of graphs with multiple variables, and the chromatic polynomial of a graph can be obtained from the Potts model by assigning special values to some variables. So the study of roots of chromatic polynomials is extended to that of the Potts model partition function, and the study of roots of the Potts model partition function also pushes the development of the study of roots of chromatic polynomials. Sokal (2001) obtained Theorems 14.3.1 and 14.5.4 through the study of the Potts model. Shrock and Tsai also have done a lot on the study of roots of the Potts model partition function. Hence it is worth to introduce the Potts model in this monograph.

Let G be a graph (in this section, G can be a multigraph) and $q \in \mathbb{N}$. Suppose an 'atom' (or 'spin') at each $x \in V(G)$ can exist in any one state in $\{1, 2, \cdots, q\}$, i.e., each vertex $x \in V(G)$ is associated with a number σ_x in $\{1, 2, \cdots, q\}$. A *spin configuration* of G is a mapping $\sigma : V(G) \to \{1, 2, \cdots, q\}$. Let J_e be the interaction of e for all $e \in E(G)$. The *Hamiltonian* $H(\sigma)$ (or *energy*) of a configuration σ is

$$H(\sigma) = \sum_{e \in E(G)} (-J_e)\delta(\sigma_{x_1(e)}, \sigma_{x_2(e)}), \qquad (14.9)$$

where δ is the Kronecker delta (i.e., $\delta(i, j) = 1$ if $i = j$ and $\delta(i, j) = 0$ otherwise) and $x_1(e)$ and $x_2(e)$ are the two ends of the edge e. The *Boltzmann weight* of a configuration σ is then $\exp(-\beta H(\sigma))$, where $\beta \geq 0$ is the inverse temperature. The *partition function* is the sum of, over all configurations, their Boltzmann weights, i.e.,

$$\sum_{\sigma \in \Omega(q)} \exp(-\beta H(\sigma)) = \sum_{\sigma \in \Omega(q)} \exp\left(-\beta \sum_{e \in E(G)} (-J_e)\delta(\sigma_{x_1(e)}, \sigma_{x_2(e)})\right),$$
$$(14.10)$$

where $\Omega(q)$ is the set of all configurations, i.e., all mappings $\sigma : V(G) \to \{1, 2, \cdots, q\}$.

Lemma 14.7.1   *For any graph G and $q \in \mathbb{N}$,*

$$\sum_{\sigma \in \Omega(q)} \exp(-\beta H(\sigma)) = \sum_{\sigma \in \Omega(q)} \prod_{e \in E(G)} \left(1 + w_e\delta(\sigma_{x_1(e)}, \sigma_{x_2(e)})\right), \qquad (14.11)$$

where $w_e = \exp(\beta J_e) - 1$ for all $e \in E(G)$.

Proof.   Observe that

$$\exp\left(-\beta \sum_{e \in E(G)} (-J_e)\delta(\sigma_{x_1(e)}, \sigma_{x_2(e)})\right) = \prod_{e \in E(G)} \exp\left(\beta J_e\delta(\sigma_{x_1(e)}, \sigma_{x_2(e)})\right).$$
$$(14.12)$$

For any numbers a, i and j, we always have

$$\exp(a\delta(i, j)) = 1 + (e^a - 1)\delta(i, j). \qquad (14.13)$$

The result then follows from (14.10). $\square$

In statistical mechanics, the partition function of the q-*state Potts model* is defined as:

$$Z_G(q, \{w_e\}) = \sum_{\sigma \in \Omega(q)} \prod_{e \in E(G)} \left(1 + w_e \delta(\sigma_{x_1(e)}, \sigma_{x_2(e)})\right), \qquad (14.14)$$

where w_e is the weight of e for all $e \in E(G)$. It is clear that $Z_G(q, \{w_e\})$ is a polynomial in q and w_e's. Actually, we have the following result (its proof is left to the reader).

Lemma 14.7.2 *For any graph G and $q \in \mathbb{N}$,*

$$Z_G(q, \{w_e\}) = \sum_{F \subseteq E(G)} q^{c(F)} \prod_{e \in F} w_e, \qquad (14.15)$$

where $c(F)$ is the number of components of the spanning subgraph of G with the edge set F. □

The partition function of the q-state Potts model is associated with some other polynomials. If $w_e = -1$ for all $e \in E(G)$, then $Z_G(q, \{w_e\})$ is actually the chromatic polynomial of G, i.e.,

$$Z_G(q, \{w_e\}) = P(G, q). \qquad (14.16)$$

If $w_e = w$ for all $e \in E(G)$, then $Z_G(q, \{w_e\})$ is actually the *dichromatic polynomial* $Z_G(q, w)$:

$$Z_G(q, w) = \sum_{F \subseteq E(G)} q^{c(F)} w^{|F|}, \qquad (14.17)$$

which can be verified by Lemma 14.7.2. The dichromatic polynomial $Z_G(q, w)$ can be directly reduced from the *Tutte polynomial* $T_G(x, y)$:

$$Z_G(q, w) = q^{c(G)} w^{r(G)} T_G((q + w)/w, 1 + w), \qquad (14.18)$$

where $c(G)$ is the number of components of G and $r(G) = v(G) - c(G)$ is the *rank* of G. The Tutte polynomial of G is a function of graphs defined below:

$$T_G(x, y) = \sum_{F \subseteq E(G)} (x - 1)^{r(G) - r(F)} (y - 1)^{r_c(F)}, \qquad (14.19)$$

where $r(F)$ and $r_c(F)$ are the rank and the cyclomatic number of the spanning subgraph of G with edge set F, respectively. More details on the

Tutte polynomial and dichromatic polynomial can be found in Brylawski and Oxley (1992) and Welsh (1993).

By (14.16), a chromatic polynomial $P(G, q)$ can be obtained from the partition function $Z_G(q, \{w_e\})$ of the Potts model by taking w_e to be -1 for all $e \in E(G)$. But it does not mean that every property of the partition function of the Potts model is also a property of chromatic polynomial. We shall now introduce a result on the partition function of the Potts model, which has no analogue on chromatic polynomials. This result can be found in Sokal (2004), but no proof is provided there.

Let G be any graph and x, y be distinct vertices of G. For any $F \subseteq E(G)$, 'F connects x and y' means that x and y are in the same component of the spanning subgraph of G with edge set F, and 'F does not connect x and y' means that x and y are in different components of the spanning subgraph of G with edge set F. Define

$$Z_G^{(x \leftrightarrow y)}(q, \{w_e\}) = \sum_{\substack{F \subseteq E(G) \\ F \text{ connects } x \text{ and } y}} q^{c(F)} \prod_{e \in F} w_e, \tag{14.20}$$

and

$$Z_G^{(x \nleftrightarrow y)}(q, \{w_e\}) = \sum_{\substack{F \subseteq E(G) \\ F \text{ does not connect } x \text{ and } y}} q^{c(F)} \prod_{e \in F} w_e. \tag{14.21}$$

By Lemma 14.7.2,

$$Z_G(q, \{w_e\}) = Z_G^{(x \leftrightarrow y)}(q, \{w_e\}) + Z_G^{(x \nleftrightarrow y)}(q, \{w_e\}). \tag{14.22}$$

Lemma 14.7.3   *Let G be any multigraph and x, y be distinct vertices of G. Let H be the multigraph obtained from G by adding an edge $e' = xy$ with weight $w_{e'}$. Then*

$$\begin{aligned} Z_H(q, \{w_e\}) &= Z_G(q, \{w_e\}) + w_{e'} Z_G^{(x \leftrightarrow y)}(q, \{w_e\}) \\ &\quad + \frac{w_{e'}}{q} Z_G^{(x \nleftrightarrow y)}(q, \{w_e\}). \end{aligned} \tag{14.23}$$

Proof. For any $F \subseteq E(H)$, let $c'(F)$ be the number of components of the spanning subgraph of H with edge set F, and let $c(F)$ be the number of components of the spanning subgraph of G with edge set F, where $F \subseteq E(G)$ for the latter case. By Lemma 14.7.2,

$$Z_H(q, \{w_e\}) = \sum_{F \subseteq E(H)} q^{c'(F)} \prod_{e \in F} w_e$$

$$
\begin{aligned}
= \quad & \sum_{F \subseteq E(G)} \left(q^{c'(F)} \prod_{e \in F} w_e + q^{c'(F \cup \{e'\})} \prod_{e \in F \cup \{e'\}} w_e \right) \\
= \quad & Z_G(q, \{w_e\}) + w_{e'} \sum_{F \subseteq E(G)} q^{c'(F \cup \{e'\})} \prod_{e \in F} w_e \\
= \quad & Z_G(q, \{w_e\}) + w_{e'} \sum_{\substack{F \subseteq E(G) \\ F \text{ connects } x \text{ and } y}} q^{c'(F \cup \{e'\})} \prod_{e \in F} w_e \\
& + w_{e'} \sum_{\substack{F \subseteq E(G) \\ F \text{ does not connect } x \text{ and } y}} q^{c'(F \cup \{e'\})} \prod_{e \in F} w_e \\
= \quad & Z_G(q, \{w_e\}) + w_{e'} \sum_{\substack{F \subseteq E(G) \\ F \text{ connects } x \text{ and } y}} q^{c(F)} \prod_{e \in F} w_e \\
& + w_{e'} \sum_{\substack{F \subseteq E(G) \\ F \text{ does not connect } x \text{ and } y}} q^{c(F)-1} \prod_{e \in F} w_e \\
= \quad & Z_G(q, \{w_e\}) + w_{e'} Z_G^{(x \leftrightarrow y)}(q, \{w_e\}) \\
& + \frac{w_{e'}}{q} Z_G^{(x \not\leftrightarrow y)}(q, \{w_e\}).
\end{aligned}
$$

$\square$

Lemma 14.7.4   *Let G be any multigraph and x, y be distinct vertices of G. Suppose G contains two submultigraphs G_1 and G_2 such that $V(G_1) \cup V(G_2) = V(G)$, $V(G_1) \cap V(G_2) = \{x, y\}$, $E(G_1) \cup E(G_2) = E(G)$ and $E(G_1) \cap E(G_2) = \emptyset$. Then*

$$
\begin{aligned}
Z_G(q, \{w_e\}) \quad = \quad & \frac{1}{q^2} Z_{G_1}(q, \{w_e\}) Z_{G_2}^{(x \not\leftrightarrow y)}(q, \{w_e\}) \\
& + \frac{1}{q^2} Z_{G_1}^{(x \not\leftrightarrow y)}(q, \{w_e\}) Z_{G_2}^{(x \leftrightarrow y)}(q, \{w_e\}) \\
& + \frac{1}{q} Z_{G_1}^{(x \leftrightarrow y)}(q, \{w_e\}) Z_{G_2}^{(x \leftrightarrow y)}(q, \{w_e\}). \quad (14.24)
\end{aligned}
$$

Proof.   For simplicity, let Z_{G_i}, $Z_{G_i}^{(x \leftrightarrow y)}$ and $Z_{G_i}^{(x \not\leftrightarrow y)}$ denote

$$
Z_{G_i}(q, \{w_e\}), \quad Z_{G_i}^{(x \leftrightarrow y)}(q, \{w_e\}) \text{ and } Z_{G_i}^{(x \not\leftrightarrow y)}(q, \{w_e\})
$$

respectively. For $i = 1, 2$ and any $F_i \subseteq E(G_i)$, let $c_i(F_i)$ be the number of components of the spanning subgraph of G_i with edge set F_i. Let $c(F_1 \cup F_2)$ be the number of components of the spanning subgraph of G with edge set

$F_1 \cup F_2$. Then

$$c(F_1 \cup F_2) = \begin{cases} c_1(F_1) + c_2(F_2) - 1, & \text{if each } F_i \text{ connects } x \text{ and } y \\ & \text{in } G_i \text{ for } i = 1, 2; \\ c_1(F_1) + c_2(F_2) - 2, & \text{otherwise.} \end{cases}$$

$$(14.25)$$

By Lemma 14.7.2,

$$Z_G(q, \{w_e\})$$

$$= \sum_{F_1 \subseteq E(G_1),\ F_2 \subseteq E(G_2)} q^{c(F_1 \cup F_2)} \left(\prod_{e \in F_1} w_e \right) \left(\prod_{e \in F_2} w_e \right)$$

$$= \sum_{\substack{F_1 \subseteq E(G_1),\ F_2 \subseteq E(G_2) \\ \text{each } F_i \text{ connects } x \text{ and } y \text{ in } G_i}} q^{c(F_1 \cup F_2)} \left(\prod_{e \in F_1} w_e \right) \left(\prod_{e \in F_2} w_e \right)$$

$$+ \sum_{\substack{F_1 \subseteq E(G_1),\ F_2 \subseteq E(G_2) \\ \text{some } F_i \text{ does not connect} \\ x \text{ and } y \text{ in } G_i}} q^{c(F_1 \cup F_2)} \left(\prod_{e \in F_1} w_e \right) \left(\prod_{e \in F_2} w_e \right)$$

$$= \sum_{\substack{F_1 \subseteq E(G_1),\ F_2 \subseteq E(G_2) \\ \text{each } F_i \text{ connects } x \text{ and } y \text{ in } G_i}} q^{c_1(F_1) + c_2(F_2) - 1} \left(\prod_{e \in F_1} w_e \right) \left(\prod_{e \in F_2} w_e \right)$$

$$+ \sum_{\substack{F_1 \subseteq E(G_1),\ F_2 \subseteq E(G_2) \\ \text{some } F_i \text{ does not connect} \\ x \text{ and } y \text{ in } G_i}} q^{c_1(F_1) + c_2(F_2) - 2} \left(\prod_{e \in F_1} w_e \right) \left(\prod_{e \in F_2} w_e \right)$$

$$= \frac{1}{q} Z_{G_1}^{(x \leftrightarrow y)} Z_{G_2}^{(x \leftrightarrow y)} + \frac{1}{q^2} Z_{G_1}^{(x \leftrightarrow y)} Z_{G_2}^{(x \not\leftrightarrow y)} + \frac{1}{q^2} Z_{G_1}^{(x \not\leftrightarrow y)} Z_{G_2}^{(x \leftrightarrow y)}$$

$$+ \frac{1}{q^2} Z_{G_1}^{(x \not\leftrightarrow y)} Z_{G_2}^{(x \not\leftrightarrow y)}$$

$$= \frac{1}{q} Z_{G_1}^{(x \leftrightarrow y)} Z_{G_2}^{(x \leftrightarrow y)} + \frac{1}{q^2} Z_{G_1}^{(x \not\leftrightarrow y)} Z_{G_2}^{(x \leftrightarrow y)} + \frac{1}{q^2} Z_{G_1} Z_{G_2}^{(x \not\leftrightarrow y)}.$$

$$\square$$

Theorem 14.7.1 *Let G be any multigraph and x, y be distinct vertices of G. Suppose that G contains two submultigraphs G_1 and G_2 such that $V(G_1) \cup V(G_2) = V(G)$, $V(G_1) \cap V(G_2) = \{x, y\}$, $E(G_1) \cup E(G_2) = E(G)$ and $E(G_1) \cap E(G_2) = \emptyset$. Let H be the multigraph obtained from G_1 by adding one new edge e' joining x and y. If each edge e of G_1 has the same* *weight w_e in G_1 as in H and the new edge e' of H has its weight*

$$w_{e'} = \frac{q Z_{G_2}^{(x \leftrightarrow y)}(q, \{w_e\})}{Z_{G_2}^{(x \not\leftrightarrow y)}(q, \{w_e\})}, \qquad (14.26)$$

then

$$Z_G(q, \{w_e\}) = Z_{G_2}^{(x \not\leftrightarrow y)}(q, \{w_e\}) \times Z_H(q, \{w_e\})/q^2. \qquad (14.27)$$

Proof. By Lemma 14.7.4,

$$
\begin{aligned}
q^2 Z_G(q, \{w_e\}) &= Z_{G_1}(q, \{w_e\}) Z_{G_2}^{(x \not\leftrightarrow y)}(q, \{w_e\}) \\
&\quad + q Z_{G_1}^{(x \leftrightarrow y)}(q, \{w_e\}) Z_{G_2}^{(x \leftrightarrow y)}(q, \{w_e\}) \\
&\quad + Z_{G_1}^{(x \not\leftrightarrow y)}(q, \{w_e\}) Z_{G_2}^{(x \leftrightarrow y)}(q, \{w_e\}) \\
&= Z_{G_2}^{(x \not\leftrightarrow y)}(q, \{w_e\}) \Big(Z_{G_1}(q, \{w_e\}) + w_{e'} Z_{G_1}^{(x \leftrightarrow y)}(q, \{w_e\}) \\
&\quad + w_{e'} Z_{G_1}^{(x \not\leftrightarrow y)}(q, \{w_e\})/q \Big) \\
&= Z_{G_2}^{(x \not\leftrightarrow y)}(q, \{w_e\}) \times Z_H(q, \{w_e\}),
\end{aligned}
$$

where the last equality follows from Lemma 14.7.3. $\qquad \square$

Let H be any multigraph and for each edge $e \in E(H)$, let G_e be a multigraph with two selected distinct vertices x_e, y_e. We now apply Theorem 14.7.1 repeatedly to obtain the following result in Sokal (2004).

Theorem 14.7.2 *Let H be any multigraph and G_e be as above for all $e \in E(H)$. Let H' be a multigraph obtained from H by replacing each $e \in E(H)$ with a copy of G_e, attaching x_e and y_e to the two ends of e in H. Then $E(H') = \bigcup_{e \in E(H)} E(G_e)$. For each $e' \in E(G_e)$, let $w_{e'}$ be the weight of e' in G_e and also in H'. For any $e \in E(H)$, define the weight of e in H to be*

$$w_e = q Z_{G_e}^{(x \leftrightarrow y)}(q, \{w_{e'}\})/Z_{G_e}^{(x \not\leftrightarrow y)}(q, \{w_{e'}\}). \qquad (14.28)$$

Then

$$Z_{H'}(q, \{w_{e'}\}) = Z_H(q, \{w_e\}) \times \prod_{e \in E(H)} Z_{G_e}^{(x \not\leftrightarrow y)}(q, \{w_{e'}\})/q^2. \qquad (14.29)$$
$\qquad \square$

We shall end this section by applying Theorem 14.7.2 to establish the following result, which was mentioned by Sokal in his communication with one of the authors. This result is an extension of Theorem 14.4.3.

Theorem 14.7.3   *Let G be any multigraph and G' be a multigraph obtained from G by replacing each edge of e by an even path G_e. Let $w_{e'}$ be the weight of e' for each $e' \in E(G')$. If $q > 2$ and $-1 \le w_{e'} \le 0$ for all $e' \in E(G')$, then*

$$Z_{G'}(q, \{w_{e'}\}) > 0. \tag{14.30}$$

Proof. For each $e' \in E(G_e)$, let $w_{e'}$ be also the weight of e' in G_e. It is easy to show that

$$Z_{G_e}^{(x \leftrightarrow y)}(q, \{w_{e'}\}) = q \prod_{e' \in E(G_e)} w_{e'} \tag{14.31}$$

and

$$Z_{G_e}(q, \{w_{e'}\}) = q \prod_{e' \in E(G_e)} (q + w_{e'}). \tag{14.32}$$

(The latter can be proved by the result in Exercise 14.3.) By (14.22),

$$Z_{G_e}^{(x \not\leftrightarrow y)}(q, \{w_{e'}\}) = q \prod_{e' \in E(G_e)} (q + w_{e'}) - q \prod_{e' \in E(G_e)} w_{e'}. \tag{14.33}$$

Since $e(G_e)$ is even, if $q > 2$ and $-1 \le w_{e'} \le 0$, then

$$Z_{G_e}^{(x \leftrightarrow y)}(q, \{w_{e'}\}) \ge 0$$

and

$$Z_{G_e}^{(x \not\leftrightarrow y)}(q, \{w_{e'}\}) > 0,$$

implying that for each $e \in E(G)$,

$$w_e = \frac{q Z_{G_e}^{(x \leftrightarrow y)}(q, \{w_{e'}\})}{Z_{G_e}^{(x \not\leftrightarrow y)}(q, \{w_{e'}\})} \ge 0.$$

Thus, by Theorem 14.7.2, $Z_{G'}(q, \{w_{e'}\}) > 0$, as required. $\square$

Exercise 14

14.1 Prove Lemma 14.7.2.

14.2 Let G_i be a multigraph for $i = 1, 2$. Prove that the partition function of the q-state Potts model of the disjoint union of G_1 and G_2 is

$$Z_{G_1 \cup G_2}(q, \{w_e\}) = Z_{G_1}(q, \{w_e\}) \times Z_{G_2}(q, \{w_e\}).$$

14.3 Let G be a multigraph with two submultigraphs G_1 and G_2 such that $V(G_1) \cup V(G_2) = V(G)$, $V(G_1) \cap V(G_2) = \{x\}$, $E(G_1) \cup E(G_2) = E(G)$ and $E(G_1) \cap E(G_2) = \emptyset$. Prove that

$$Z_G(q, \{w_e\}) = Z_{G_1}(q, \{w_e\}) \times Z_{G_2}(q, \{w_e\})/q.$$

14.4 Let G be a multigraph and e' be a loop of G. Prove that

$$Z_G(q, \{w_e\}) = (1 + w_{e'}) \times Z_{G-e'}(q, \{w_e\}).$$

14.5 Let G be a multigraph and e' be a bridge of G. Prove that

$$Z_G(q, \{w_e\}) = (1 + w_{e'}/q) \times Z_{G-e'}(q, \{w_e\}).$$

14.6 Let G be a multigraph and e' be an edge of G which is neither a loop nor a bridge. Prove that

$$Z_G(q, \{w_e\}) = Z_{G-e'}(q, \{w_e\}) + w_{e'} \times Z_{G \cdot e'}(q, \{w_e\}).$$

14.7 For any multigraph G and two distinct vertices x, y in G, let $G * xy$ be the multigraph obtained from G by contracting x and y to a single vertex, and all edges of G with the ends x, y become loops in $G * xy$. Prove that

$$Z_{G*xy}(q, \{w_e\}) = Z_G^{(x \leftrightarrow y)}(q, \{w_e\}) + Z_G^{(x \not\leftrightarrow y)}(q, \{w_e\})/q.$$

Chapter 15

Inequalities on Chromatic Polynomials

15.1 Introduction

As we know, it is very difficult to find a simple or explicit expression for the chromatic polynomial of a graph, except for some special graphs such as chordal graphs, cycles, wheels, etc. Thus it is quite meaningful to study bounds for the chromatic polynomials.

In Section 15.2, we shall present some bounds for the chromatic polynomials of graphs in terms of their order, size, girth and the number of minimum cycles, obtained by Lazebnik (1990) and Dohmen (1993b, 1996, 1998, 1999).

Given a family of graphs and $\lambda \in \mathbb{N}_0$, what is the maximum value of $P(G, \lambda)$ over all graphs G in this family? In Section 15.3, we shall introduce Tomescu's results on this problem, respectively, for the family of connected graphs; non-bipartite connected graphs; and 2-connected graphs.

In Section 15.4, we shall focus on Welsh's conjecture that

$$(P(G, \lambda))^2 \geq P(G, \lambda + 1)P(G, \lambda - 1)$$

holds for any graph G and $\lambda \in \mathbb{N}$. This conjecture was disproved by Seymour (1997), but we shall show that it holds if λ is sufficiently large.

The *mean colour number* of a graph G of order n, denoted by $\mu(G)$, is defined to be the average of numbers of colours used over all n-colourings of G. It was conjectured by Bartels and Welsh (1995) that $\mu(G) \geq \mu(H)$

if H is a spanning subgraph of G. This conjecture is equivalent to proving the following inequality that

$$P(G,n)P(H,n-1) \geq P(G,n-1)P(H,n),$$

where $n = v(G)$ and H is a spanning subgraph of G. Counterexamples to this conjecture were found by Mosca (1998). In Section 15.5, we shall present some results which show that the conjecture holds under certain conditions.

15.2   Bounds of chromatic polynomials

In this section, we introduce some known results on the bounds of $P(G, \lambda)$.

Lazebnik (1990) showed that if G is an (n, m)-graph and $\lambda \in \mathbb{N}$, then

$$P(G, \lambda) \leq \frac{(\lambda - 1)\lambda^n}{\lambda - 1 + m} \tag{15.1}$$

and

$$P(G, \lambda) \leq \lambda^n - m\lambda^{n-1} + \binom{m}{2}\lambda^{n-2}. \tag{15.2}$$

Inequalities (15.1) and (15.2) were then strengthened to the following results by Dohmen (1993b).

Theorem 15.2.1   *Let G be an (n, m)-graph with girth g and $\lambda, k \in \mathbb{N}$ with $k \leq g - 1$. Then*

(i) for $\lambda \geq 2$, $P(G, \lambda) \leq \lambda^n \div \sum\limits_{s=0}^{\lfloor (k-1)/2 \rfloor} \binom{m}{s}(1/(\lambda - 1))^s$;

(ii) $(-1)^k P(G, \lambda) \leq (-1)^k \sum\limits_{s=0}^{k}(-1)^s \binom{m}{s}\lambda^{n-s}$;

(iii) for $r = n_G(C_g)$,

$$(-1)^g P(G, \lambda) \leq (-1)^g \left\{ \sum_{s=0}^{g-2}(-1)^s \binom{m}{s}\lambda^{n-s} - \left[\binom{m}{g-1} - r \right]\lambda^{n-g+1} \right.$$
$$\left. + \left[\binom{m}{g} - r \right]\lambda^{n-g} \right\}.$$

□

The inequalities of Theorem 15.2.1 (ii) and (iii) were further improved by Dohmen (1999). Write

$$P(G, \lambda) = \sum_{s=1}^{n} a_s(G)\lambda^s, \tag{15.3}$$

where $n = v(G)$. Note that $a_n(G) = 1$. Dohmen (1999) obtained the following result.

Theorem 15.2.2   *Let G be a graph of order n and $\lambda \in \mathbb{N}$. Then for any $k = 0, 1, \cdots, n - 1$,*

$$(-1)^k P(G, \lambda) \le (-1)^k \sum_{i=0}^{k} a_{n-i}(G) \lambda^{n-i}. \tag{15.4}$$

Proof. If G is empty, then $P(G, \lambda) = \lambda^n$ and the result holds. We shall prove the theorem by induction on $e(G)$. Assume that the theorem holds for all graphs with less than m edges, where $m \ge 1$. Let G be an (n, m)-graph and e be an edge in G. By definition, $P(G, \lambda) \le \lambda^n$. Hence the theorem holds for $k = 0$. Assume that

$$P(G - e, \lambda) = \sum_{s=1}^{n} a_s(G - e) \lambda^s$$

and

$$P(G \cdot e, \lambda) = \sum_{s=1}^{n-1} a_s(G \cdot e) \lambda^s.$$

By FRT, for $s = 1, 2, \cdots, n$,

$$a_s(G) = a_s(G - e) - a_s(G \cdot e),$$

where $a_n(G \cdot e) = 0$. Since both graphs $G - e$ and $G \cdot e$ have at most $m - 1$ edges, by the induction hypothesis, for $k = 1, 2, \cdots, n - 1$,

$$(-1)^k P(G - e, \lambda) \le (-1)^k \sum_{i=0}^{k} a_{n-i}(G - e) \lambda^{n-i}$$

and

$$(-1)^{k-1} P(G \cdot e, \lambda) \le (-1)^{k-1} \sum_{i=0}^{k-1} a_{n-1-i}(G \cdot e) \lambda^{n-1-i}.$$

Thus by FRT,

$$\begin{aligned}
&(-1)^k P(G, \lambda) \\
= \ &(-1)^k P(G - e, \lambda) + (-1)^{k-1} P(G \cdot e, \lambda) \\
\le \ &(-1)^k \sum_{i=0}^{k} a_{n-i}(G - e) \lambda^{n-i} + (-1)^{k-1} \sum_{i=0}^{k-1} a_{n-1-i}(G \cdot e) \lambda^{n-1-i}
\end{aligned}$$

$$= (-1)^k \sum_{i=0}^{k} (a_{n-i}(G-e) - a_{n-i}(G \cdot e))\lambda^{n-i}$$

$$= (-1)^k \sum_{i=0}^{k} a_{n-i}(G)\lambda^{n-i},$$

which completes the proof. $\qquad\square$

Let G be an (n,m)-graph with girth g. Then, by Corollary 2.3.1,

$$P(G,\lambda) = \sum_{i=1}^{n} (-1)^{n-i} h_i \lambda^i,$$

where $h_i = \binom{m}{n-i}$ for $i = n, n-1, \cdots, n-g+2$, and

$$h_{n-g+1} = \binom{m}{g-1} - n(C_g). \tag{15.5}$$

Thus Theorem 15.2.1(ii) follows immediately from Theorem 15.2.2. When $g = 3$, by Theorem 2.3.2(i),

$$h_{n-3} = \binom{m}{3} - (m-2)n(C_3) - i(C_4) + 2n(K_4), \tag{15.6}$$

and when $g \geq 4$, by Theorem 2.3.4,

$$h_{n-g} = \binom{m}{g} - (m-g+1)n(C_g) - n(C_{g+1}). \tag{15.7}$$

It is easy to show by induction on m that

$$(m-3)n(C_3) \geq 2n(K_4).$$

(See Exercise 15.1.) Thus, by (15.6),

$$h_{n-3} \leq \binom{m}{3} - n(C_3) - i(C_4) \leq \binom{m}{3} - n(C_3)$$

and by (15.7), we have

$$h_{n-g} \leq \binom{m}{g} - n(C_g), \tag{15.8}$$

for any g with $g \geq 3$. Hence Theorem 15.2.1(iii) follows immediately from Theorem 15.2.2 for $k = g$.

It can be proved that $P(G, \lambda) \leq \lambda^n$ for $\lambda \in \mathbb{R}$ with $\lambda \geq n - 1$. Thus, by modifying the proof of Theorem 15.2.2, one can actually show more generally that Theorem 15.2.2 holds for $\lambda \in \mathbb{R}$ with $\lambda \geq n - 1$. (See Exercise 15.2.)

Theorem 15.2.1 was further strengthened by Dohmen (1996) as shown below.

Theorem 15.2.3 *Let G be an (n, m)-graph with girth g, where $m \geq 1$, and let $\lambda \in \mathbb{N}$. Then for $k = 0, 1, \cdots, g - 1$,*

$$(-1)^k P(G, \lambda) \leq (-1)^k \left(\sum_{s=0}^{k} (-1)^s \binom{m}{s} \lambda^{n-s} - \binom{m-1}{k} \lambda^{n-k-1} \right).$$

$\square$

By letting $k = 1$ and 2, we have:

Corollary 15.2.1 *Suppose that G is an (n, m)-graph with $m \geq 1$. Then for any $\lambda \in \mathbb{N}$,*

$$P(G, \lambda) \geq \lambda^n - m\lambda^{n-1} + (m - 1)\lambda^{n-2},$$

$$P(G, \lambda) \leq \lambda^n - m\lambda^{n-1} + \binom{m}{2}\lambda^{n-2} - \binom{m-1}{2}\lambda^{n-3}.$$

15.3 Maximum chromatic polynomials

In the section, we shall consider the following problem: given a family $\mathcal{S}$ of graphs, determine

$$f_{\mathcal{S}}(\lambda) = \max_{G \in \mathcal{S}} P(G, \lambda),$$

where $\lambda \in \mathbb{N}_0$. For any graph G and any edge e in G, by FRT, we have

$$P(G, \lambda) = P(G - e, \lambda) - P(G \cdot e, \lambda) \leq P(G - e, \lambda) \quad (15.9)$$

for all $\lambda \in \mathbb{N}_0$, where equality holds if and only if $\lambda \leq \chi(G \cdot e) - 1 \leq \chi(G)$. Thus we have the following result.

Lemma 15.3.1 *Let G be a non-empty graph, and H be a proper spanning subgraph of G. Then*

$$P(G, \lambda) \leq P(H, \lambda), \quad (15.10)$$

for all $\lambda \in \mathbb{N}_0$, where inequality is strict if $\lambda \geq \min_{uv \in E(G) \backslash E(H)} \chi(H \cdot uv).$

Proof. By (15.9), $P(G, \lambda) \leq P(H, \lambda)$ for all $\lambda \in \mathbb{N}_0$. Now assume that $\lambda \in \mathbb{N}$ with $\lambda \geq \chi(H \cdot uv)$ for some $uv \in E(G) \backslash E(H)$. Then $H + uv$ is a spanning subgraph of G. By (15.9), $P(G, \lambda) \leq P(H + uv, \lambda)$. Since $\lambda \geq \chi(H \cdot uv)$, we have $P(H \cdot uv, \lambda) > 0$. Thus, by FRT,

$$P(H + uv, \lambda) = P(H, \lambda) - P(H \cdot uv, \lambda) < P(H, \lambda).$$

Therefore $P(G, \lambda) < P(H, \lambda)$. This proves the lemma. $\square$

We shall now apply Lemma 15.3.1 to study the above problem for some families of graphs. First, we consider the family of graphs of order n, where $n \in \mathbb{N}$. It is easy to obtain the following result by Lemma 15.3.1 (see Exercise 15.3).

Theorem 15.3.1   *Let G be a graph of order n. Then for all $\lambda \in \mathbb{N}_0$,*

$$P(G, \lambda) \leq \lambda^n,$$

where equality holds for $\lambda \geq 1$ if and only if G is empty.

We next consider the family of connected graphs of order n.

Theorem 15.3.2   *Let G be a connected graph of order n. Then for all $\lambda \in \mathbb{N}_0$,*

$$P(G, \lambda) \leq \lambda(\lambda - 1)^{n-1},$$

where equality holds for $\lambda \geq 3$ if and only if G is a tree.

Proof. If G is a tree, then $P(G, \lambda) = \lambda(\lambda - 1)^{n-1}$. Assume that G is not a tree. Let T be a spanning tree of G. It is clear that T is a proper subgraph of G, $\chi(T) = 2$ and $\chi(T \cdot uv) \leq 3$ for any $uv \in E(G) \backslash E(T)$. The result now follows directly from Lemma 15.3.1. $\square$

In the next result, we confine ourselves to the family of graphs which are non-bipartite. This result was first obtained by Tomescu (1990). We give here an alternative proof by applying Lemma 15.3.1.

Theorem 15.3.3   *Let G be a connected graph of order n and $\chi(G) \geq 3$.*

(i) If n is odd, then for all $\lambda \in \mathbb{N}_0$,

$$P(G, \lambda) \leq (\lambda - 1)^n - (\lambda - 1),$$

where equality holds for $\lambda \geq 3$ if and only if $G \cong C_n$; and

(ii) if n is even, then for all $\lambda \in \mathbb{N}_0$,

$$P(G, \lambda) \leq (\lambda - 1)^n - (\lambda - 1)^2,$$

where equality holds for $\lambda \geq 3$ if and only if $G - x \cong C_{n-1}$ for some $x \in V(G)$ with $d(x) = 1$.

Proof. Since $\chi(G) \geq 3$, G contains an odd cycle C_k, where $3 \leq k \leq n$ and k is odd. Let H be a connected spanning subgraph of size n which contains C_k. Clearly, C_k is the only cycle in H. Observe that

$$P(H, \lambda) = (\lambda - 1)^{n-k} \left((\lambda - 1)^k - (\lambda - 1) \right) = (\lambda - 1)^n - (\lambda - 1)^{n-k+1}.$$

By Lemma 15.3.1, for all $n \in \mathbb{N}_0$,

$$P(G, \lambda) \leq P(H, \lambda) = (\lambda - 1)^n - (\lambda - 1)^{n-k+1}.$$

If $k \leq n - 2$ for odd n, or $k \leq n - 3$ for even n, then for $\lambda \geq 3$,

$$P(G, \lambda) < \begin{cases} (\lambda - 1)^n - (\lambda - 1), & \text{if } n \text{ is odd} \\ (\lambda - 1)^n - (\lambda - 1)^2, & \text{if } n \text{ is even.} \end{cases}$$

It remains to consider the case that $k = n$ when n is odd, and $k = n - 1$ when n is even. So $H \cong C_n$ when n is odd, or $H - x \cong C_{n-1}$ for some x with degree 1 when n is even. Observe that $\chi(H \cdot uv) \leq 3$ for any vertices u and v in H. The result now follows from Lemma 15.3.1. □

We have considered respectively in Theorems 15.3.2 and 15.3.3 the families of connected graphs with chromatic number at least 2 and 3. In general, Tomescu (1990) proposed the following conjecture on the family of connected graphs with chromatic number at least k, where $k \geq 4$.

Conjecture 15.3.1 *Let $k \in \mathbb{N}$ with $k \geq 4$ and G be a connected graph of order n with $\chi(G) \geq k$. Then*

$$P(G, \lambda) \leq (\lambda)_k (\lambda - 1)^{n-k}$$

for all $\lambda \in \mathbb{N}_0$, where equality holds for $\lambda \geq k$ if and only if G is a connected graph of order n and size $\binom{k}{2} + (n - k)$ with $\omega(G) = k$.

Tomescu (1994b) also obtained the following result on the family of 2-connected graphs. We present here a simpler proof by applying Lemma 15.3.1 again.

Theorem 15.3.4   *Let G be a 2-connected graph of order n, where $n \geq 3$. Then for any $\lambda \in \mathbb{N}$ with $\lambda \geq 3$,*

$$P(G, \lambda) \leq (\lambda - 1)^n + (-1)^n(\lambda - 1),$$

where equality holds if and only if $G \cong C_n$ (or $G \cong K(2,3)$ for the case that $n = 5$ and $\lambda = 3$).

Proof.   It is easy to verify that the result holds for $n \leq 4$. Now let $n \geq 5$ and G be a 2-connected (n, m)-graph with $m \geq n + 1$. We may assume that the theorem holds for all 2-connected graphs of order n and size less than m.

For convenience, let

$$f_n(\lambda) = (\lambda - 1)^n + (-1)^n(\lambda - 1).$$

We first show that the theorem holds if $G - uv$ is still 2-connected for some edge uv.

Case 1: $G - uv$ is 2-connected for some edge uv.

By Lemma 15.3.1 and the induction hypothesis,

$$P(G, \lambda) \leq P(G - uv, \lambda) \leq f_n(\lambda),$$

where the first inequality is strict if $\lambda \geq \chi(G \cdot uv)$ and the second inequality is strict if $G - uv \notin \{C_n, K(2,3)\}$. But if $G - uv \in \{C_n, K(2,3)\}$, then $\lambda \geq 3 \geq \chi(G \cdot uv)$. Hence $P(G, \lambda) < f_n(\lambda)$ in this case.

We now assume that $G - uv$ is not 2-connected for every $uv \in E(G)$. So G has a cut $\{x, y\}$ with $xy \notin E(G)$. If every component of $G - x - y$ is an isolated vertex, then $G \cong K(2, n - 2)$. We now consider this case.

Case 2: $G \cong K(2, n - 2)$.

We have

$$P(G, \lambda) = \lambda(\lambda - 1)(\lambda - 2)^{n-2} + \lambda(\lambda - 1)^{n-2}.$$

It can be verified that if $\lambda = 3$, then $P(G, \lambda) \leq f_n(\lambda)$, where equality holds if and only if $n = 5$. Now assume that $n \geq 5$ and $\lambda \geq 4$. Then

$$
\begin{aligned}
&\frac{f_n(\lambda) - P(G, \lambda)}{\lambda - 1} \\
=\ & (\lambda - 1)^{n-1} + (-1)^n - \lambda(\lambda - 1)^{n-3} - \lambda(\lambda - 2)^{n-2} \\
=\ & (\lambda^2 - 3\lambda + 1)(\lambda - 1)^{n-3} - \lambda(\lambda - 2)^{n-2} + (-1)^n \\
>\ & (\lambda - 2)^2(\lambda - 1)^{n-3} + (\lambda - 3)(\lambda - 1)^{n-3} - (\lambda - 1)^2(\lambda - 2)^{n-3} - 1 \\
>\ & (\lambda - 1)^2(\lambda - 2)^2 \left((\lambda - 1)^{n-5} - (\lambda - 2)^{n-5}\right) \\
>\ & 0.
\end{aligned}
$$

So the theorem holds for case 2.

We now establish the following claim, which will be applied in proof of the last case.

Let $\{x, y\}$ be a cut of G with $xy \notin E(G)$. Let S be the vertex set of a component of $G - x - y$, $G_1 = [S \cup \{x, y\}]$ and $G_2 = G - S$. Let G' be the graph obtained from G_2 by adding an $x - y$ path of length $|S| + 1$, i.e., G' is the graph obtained from G by replacing G_1 by an $x - y$ path of the same order.

Claim: If G_1 is not a path, then $P(G, \lambda) \leq P(G', \lambda)$ for $\lambda \geq 3$, where inequality is strict if $\lambda \geq \max\{\chi(G_2 + xy), \chi(G_2 \cdot xy)\}$.

By FRT, we have

$$
\begin{aligned}
P(G, \lambda) &= P(G + xy, \lambda) + P(G \cdot xy, \lambda) \\
&= \frac{P(G_1 + xy, \lambda)P(G_2 + xy, \lambda)}{\lambda(\lambda - 1)} + \frac{P(G_1 \cdot xy, \lambda)P(G_2 \cdot xy, \lambda)}{\lambda}.
\end{aligned}
$$

Since both $G_1 + xy$ and $G_1 \cdot xy$ are 2-connected and of order less than n, by the induction hypothesis,

$$
P(G_1 + xy, \lambda) \leq f_{2+|S|}(\lambda) \quad \text{and} \quad P(G_1 \cdot xy, \lambda) \leq f_{1+|S|}(\lambda)
$$

for $\lambda \geq 3$, and if G_1 is not an $x - y$ path, then both $G_1 \cdot xy$ and $G_1 + xy$ are not cycles and either $G_1 \cdot xy$ or $G_1 + xy$ is not $K(2, 3)$, implying that at least one of the above two inequalities is strict. Hence

$$
\begin{aligned}
P(G, \lambda) &\leq \frac{f_{2+|S|}(\lambda)P(G_2 + xy, \lambda)}{\lambda(\lambda - 1)} + \frac{f_{1+|S|}(\lambda)P(G_2 \cdot xy, \lambda)}{\lambda} \\
&= P(G' + xy, \lambda) + P(G' \cdot xy, \lambda) \\
&= P(G', \lambda),
\end{aligned}
$$

where inequality is strict if $\lambda \geq \max\{\chi(G_2 + xy), \chi(G_2 \cdot xy)\}$.

We now prove the remaining case by applying the above claim.

Case 3: $G \not\cong K(2, n - 2)$ and G has a cut $\{x, y\}$ with $xy \notin E(G)$.

We first assume that G contains two adjacent vertices u and v such that $d(u) = d(v) = 2$. Then G has a path $u_0 u_1 u_2 u_3$ of length 4. Let $G_1 = G - u_1 - u_2$ and $G_2 = u_0 u_1 u_2 u_3$. Since $G \not\cong C_n$, G_1 is not a path. As $\chi(G_2 + u_0 u_3) = 2$ and $\chi(G_2 \cdot u_0 u_3) = 3$, by the above claim, we have

$$
P(G, \lambda) < P(G', \lambda) = P(C_n, \lambda) = f_n(\lambda)
$$

for $\lambda \geq 3$. This result will be applied later.

We now assume that G has no two adjacent vertices u and v such that $d(u) = d(v) = 2$.

Since $G \not\cong K(2, n-2)$, $G - x - y$ has a component with vertex set S such that $|S| \geq 2$. Since G has no two adjacent vertices u and v such that $d(u) = d(v) = 2$, $G[S \cup \{x, y\}]$ is not a path. By the above claim, we have

$$P(G, \lambda) \leq P(H, \lambda) \leq f_n(\lambda),$$

where H is the graph obtained from $G - S$ by adding an $x - y$ path of length $|S| + 1$, the first inequality is strict if $G - S$ is a path and the second inequality follows from the induction hypothesis (since $e(H) < e(G)$).

Observe that H contains two adjacent vertices u and v with $d_H(u) = d_H(v) = 2$. If $G - S$ is not a path, then $H \not\cong C_n$ and thus the inequality

$$P(H, \lambda) < f_n(\lambda)$$

follows from the result in the first paragraph of this case.

Therefore the inequality $P(G, \lambda) < f_n(\lambda)$ holds in case 3.   □

15.4   An open problem

It was mentioned in Seymour (1997) that the following conjecture was first proposed by Welsh in the early 1970's and later, independently, by Brenti (1992).

Conjecture 15.4.1   *For all $\lambda \in \mathbb{N}$ and all graphs G,*

$$(P(G, \lambda))^2 \geq P(G, \lambda + 1)P(G, \lambda - 1).$$

Let H be the graph of Figure 15.1 and G be the graph with vertex set $\bigcup_{1 \leq i \leq 6} A_i$, where $A_1, A_2, \cdots, A_6$ are parwise disjoint sets with $|A_i| = n$ for $i = 1, 2, \cdots, 6$, and any two vertices $u \in A_i$ and $v \in A_j$ are adjacent in G if and only if $i \neq j$ and the two vertices i and j of H are adjacent.

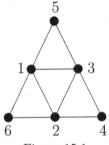

Figure 15.1

After constructing the above graph G, Seymour (1997) proved the following:

Theorem 15.4.1 *Let G be the graph constructed above. Then*

$$P(G,5) \geq 27^n, \quad P(G,7) \geq 216^n$$

and

$$P(G,6) \leq 1080 \cdot 72^n + 210 \cdot 64^n + 360 \cdot 48^n + 360 \cdot 36^n + 90 \cdot 16^n. \quad \square$$

By Theorem 15.4.1, we have

$$\lim_{n \to \infty} \frac{(P(G,6))^2}{P(G,5)P(G,7)} = 0. \tag{15.11}$$

Thus Conjecture 15.4.1 is not true in general. However, it is true for some families of graphs such as graphs in $\mathcal{I}$. In the following, we shall show that for any connected (n,m)-graph, the conjecture holds if $\lambda \geq \max\{n - 1, \sqrt{2}(m - n + 2.5)\}$ and $\lambda \in \mathbb{R}$.

Lemma 15.4.1 *Let $Q(\lambda)$ be a monic polynomial with real coefficients. Then*

$$(Q(\lambda))^2 \geq Q(\lambda + 1)Q(\lambda - 1)$$

if $\lambda \geq a + \sqrt{b^2 + 0.5}$ for every root $a + ib$ of $Q(\lambda)$.

Proof. Assume that the real roots of $Q(\lambda)$ are $d_1, \cdots, d_k$, and the non-real roots are $a_j + ib_j, a_j - ib_j$ for $j = 1, 2, \cdots, r$. Then

$$Q(\lambda) = \prod_{t=1}^{k}(\lambda - d_t) \prod_{j=1}^{r}((\lambda - a_j)^2 + b_j^2).$$

It is clear that

$$(\lambda - d_t)^2 = (\lambda + 1 - d_t)(\lambda - 1 - d_t) + 1.$$

Thus

$$(\lambda - d_t)^2 \geq |(\lambda + 1 - d_t)(\lambda - 1 - d_t)|$$

if $\lambda \geq d_t + \sqrt{0.5}$. Also,

$$\begin{aligned}
&((\lambda - a_j)^2 + b_j^2)^2 - ((\lambda + 1 - a_j)^2 + b_j^2)((\lambda - 1 - a_j)^2 + b_j^2) \\
=\ & 2(\lambda - a_j)^2 - 2b_j^2 - 1 \\
\geq\ & 0
\end{aligned}$$

if $\lambda \geq a_j + \sqrt{b_j^2 + 1/2}$. Hence the lemma holds. $\qquad \square$

Theorem 15.4.2   *Let G be a connected (n, m)-graph. If $\lambda \in \mathbb{R}$ and $\lambda \geq \max\{n - 1, \sqrt{2}(m - n + 2.5)\}$, then*

$$(P(G, \lambda))^2 \geq P(G, \lambda + 1)P(G, \lambda - 1).$$

Proof.   It is easy to check that the theorem holds if G is complete and $\lambda \geq n - 1$. Now assume that G is not complete. It can also be shown that $P(G, \lambda) > 0$ for real $\lambda > n - 2$ (see Exercise 15.6). So every real root of $P(G, \lambda)$ is in the interval $[0, n - 2]$.

Assume that $a + ib$ is a complex root of $P(G, \lambda)$. By Theorem 14.2.2,

$$|a + ib - 1| \leq m - n + 1,$$

i.e.,

$$(a - 1)^2 + b^2 \leq (m - n + 1)^2.$$

Thus

$$
\begin{aligned}
a + |b| &\leq \sqrt{2}\sqrt{a^2 + b^2} \\
&\leq \sqrt{2}\left(1 + \sqrt{(a - 1)^2 + b^2}\right) \\
&\leq \sqrt{2}(m - n + 2).
\end{aligned}
$$

If $\lambda \geq \sqrt{2}(m - n + 2.5)$, then

$$\lambda \geq a + |b| + \sqrt{2}/2 > a + \sqrt{b^2 + 1/2}.$$

The theorem now follows by Lemma 15.4.1.   □

From Lemma 15.4.1, we observe that Theorem 15.4.2 might be improved if the result on the locations of complex roots of $P(G, \lambda)$ could be strengthened. We would like to propose the following conjecture.

Conjecture 15.4.2   *Let G be a graph of order n. For $\lambda \in \mathbb{R}$ with $\lambda \geq n - 1$,*

$$(P(G, \lambda))^2 \geq P(G, \lambda + 1)P(G, \lambda - 1).$$

By Theorem 15.4.2, Conjecture 15.4.2 holds if

$$m \leq (1 + \sqrt{2}/2)n - 2.5 - \sqrt{2}/2.$$

15.5 Mean colour numbers

Let G be a graph with $v(G) = n$. Then

$$P(G, \lambda) = \sum_{k=1}^{n} \alpha(G, k)(\lambda)_k, \qquad (15.12)$$

where $\alpha(G, k)$ is the number of partitions of $V(G)$ into exactly k non-empty independent sets.

For any n-colouring θ of G, let $l(\theta) = |\theta(V(G))|$, i.e., the number of colours used by θ. The *mean colour number* $\mu(G)$ of G, defined by Bartels and Welsh (1995), is the average of $l(\theta)$'s over all n-colourings θ of G. The number of n-colourings θ of G with $l(\theta) = k$ is $\alpha(G, k)(n)_k$. Thus, by the definition of $\mu(G)$, we have

$$\mu(G) = \frac{\sum\limits_{k=1}^{n} k(n)_k \alpha(G, k)}{\sum\limits_{k=1}^{n} (n)_k \alpha(G, k)}. \qquad (15.13)$$

The following expression of $\mu(G)$ in terms of the chromatic polynomial of G was obtained by Bartels and Welsh (1995).

Theorem 15.5.1 *If G is a graph of order n, then*

$$\mu(G) = n\left(1 - \frac{P(G, n-1)}{P(G, n)}\right). \qquad (15.14)$$

Proof. Since

$$k(n)_k = k \cdot n(n-1)\cdots(n-k+1) = n((n)_k - (n-1)_k),$$

we have

$$\mu(G) = \frac{n \sum\limits_{k=1}^{n} ((n)_k - (n-1)_k)\alpha(G,k)}{\sum\limits_{k=1}^{n} (n)_k \alpha(G,k)} = n\left(1 - \frac{P(G,n-1)}{P(G,n)}\right). \quad \square$$

By Theorem 15.5.1, we have $\mu(K_n) = n$ and

$$\mu(O_n) = n(1 - (n-1)^n/n^n). \qquad (15.15)$$

For any graph G of order n, it is clear that $\mu(G) \leq \mu(K_n)$, where equality holds if and only if G is complete. Bartels and Welsh (1995) believed that $\mu(G) \geq \mu(O_n)$.

Conjecture 15.5.1   *For any graph G of order n,*

$$\mu(G) \geq n(1 - (n-1)^n/n^n).$$

By Theorem 15.5.1, the above conjecture is equivalent to the following:

$$P(G,n)/P(G,n-1) \geq n^n/(n-1)^n > e,$$

where $e(= 2.7182818\cdots)$ is the base of natural logarithms. Note that

$$\lim_{n\to\infty} n^n/(n-1)^n = e.$$

Seymour (1997) proved that

$$P(G,n)(P(G,n-1))^{-1} \geq \frac{685}{252}(= 2.7182539\cdots),$$

which is very close to the conjecture.

For graphs G and H of order n, by Theorem 15.5.1, $\mu(G) \geq \mu(H)$ if and only if

$$P(G,n)P(H,n-1) \geq P(G,n-1)P(H,n). \qquad (15.16)$$

For graphs G, H and $\lambda \in \mathbb{R}$, define

$$\tau(G,H,\lambda) = P(G,\lambda)P(H,\lambda-1) - P(G,\lambda-1)P(H,\lambda). \qquad (15.17)$$

Lemma 15.5.1   *For any graphs G and H with $v(G) = v(H) = n$, $\mu(G) \geq \mu(H)$ is equivalent to $\tau(G,H,n) \geq 0$.* $\qquad \square$

Dong (2000) proved the following:

Theorem 15.5.2   *Let G be a connected graph of order n and T be a tree of order n. Then*

$$\tau(G,T,\lambda) = (\lambda-1)(\lambda-2)^{n-1}P(G,\lambda) - \lambda(\lambda-1)^{n-1}P(G,\lambda-1) \geq 0$$

for all real $\lambda \geq n$. Hence $\mu(G) \geq \mu(T)$. $\qquad \square$

By Theorem 15.5.2, it is easy to obtain the following result, which completely proves Bartels and Welsh's conjecture.

Theorem 15.5.3   *Let G be a graph of order n. Then*

$$\tau(G,O_n,\lambda) = (\lambda-1)^n P(G,\lambda) - \lambda^n P(G,\lambda-1) \geq 0$$

for all real $\lambda \geq n$. Hence $\mu(G) \geq \mu(O_n)$.

Proof. Let $H = G + K_1$. Then H is a connected graph of order $n + 1$. By Theorem 15.5.2,

$$(\lambda - 1)(\lambda - 2)^n P(H, \lambda) - \lambda(\lambda - 1)^n P(H, \lambda - 1) \geq 0$$

for all real $\lambda \geq n + 1$. Since $P(H, \lambda) = \lambda P(G, \lambda - 1)$, we have

$$(\lambda - 2)^n P(G, \lambda - 1) - (\lambda - 1)^n P(G, \lambda - 2) \geq 0$$

for all real $\lambda \geq n + 1$; that is,

$$(\lambda - 1)^n P(G, \lambda) - \lambda^n P(G, \lambda - 1) \geq 0$$

for all real $\lambda \geq n$. □

Bartels and Welsh (1995) further proposed the following more general conjecture:

Conjecture 15.5.2 *If H is a spanning subgraph of a graph G, then $\mu(G) \geq \mu(H)$.*

The above conjecture is equivalent to another conjecture that $\mu(G) \geq \mu(G - xy)$ for any graph G and any edge xy in G. But counterexamples have been found by Mosca (1998) as shown below.

Theorem 15.5.4 *Let G be the graph obtained from K_{2m} by adding two new vertices u and v and adding new edges joining u to any m vertices in K_{2m} and joining v to the other m vertices in K_{2m}. If $m \geq 2$, then $\mu(G) > \mu(G + uv)$.* □

The proof of Theorem 15.5.4 is left to the reader (see Exercise 15.5).

Although Conjecture 15.5.2 is not true in general, it holds if H is a tree or an empty graph by Theorems 15.5.2 and 15.5.3. The following two results obtained by Dong (2003) show that it also holds in several other cases.

Theorem 15.5.5 *(i) For any graph G, $\mu(G \cup K_1) > \mu(G)$.*

(ii) If G is chordal and H is a proper spanning subgraph of G, then the inequality $\tau(G, H, \lambda) > 0$ holds for all $\lambda \in \mathbb{R}$ with $\lambda \geq v(G)$.

(iii) If H is a proper subgraph of a chordal graph G, then $\mu(H) < \mu(G)$. □

Note that in Theorem 15.5.5, (iii) follows from (i) and (ii).

A special family Ω of chordal graphs is defined below:

(a) $K_n \in \Omega$ for all n and

(b) $H \in \Omega$ if there is a vertex w of degree 1 in H such that $H - w \in \Omega$.

Theorem 15.5.6 *Let G be a graph and H be a proper subgraph of G such that $H \in \Omega$. Then $\tau(G, H, \lambda) > 0$ for all real $\lambda \geq v(G)$. Hence $\mu(H) < \mu(G)$.* $\square$

Avis, Simone and Nobili (2002) also found upper bounds and lower bounds for $\mu(G)$ in terms of the independence number and clique number of G.

Theorem 15.5.7 *Let G be any graph with independence number α and clique number ω. For any real $\lambda \geq n = v(G)$,*

$$\frac{P(G, \lambda - 1)}{P(G, \lambda)} \leq \frac{\lambda - \omega}{\lambda}\left(\frac{\lambda - 1}{\lambda}\right)^{n-\omega}, \tag{15.18}$$

$$\frac{P(G, \lambda - 1)}{P(G, \lambda)} \geq \frac{\lambda - n + \alpha}{\lambda}\left(\frac{\lambda - n + \alpha - 1}{\lambda - n + \alpha}\right)^{\alpha} \tag{15.19}$$

and thus $n - (n - \omega)\left(\frac{n-1}{n}\right)^{n-\omega} \leq \mu(G) \leq n - \alpha\left(\frac{\alpha-1}{\alpha}\right)^{\alpha}$. $\square$

Avis, Simone and Nobili (2002) also characterized graphs that yield the lower bound and upper bound of $\mu(G)$ in the above result.

The lower bound and upper bound of $\mu(G)$ in Theorem 15.5.7 can actually be derived from Theorem 15.5.6 (see Dong (2003)) and Theorem 15.5.5 (iii) respectively. Indeed, by Theorem 15.5.5(iii),

$$\mu(G) \leq n - (n - k - 1)^{n-k}/(n - k)^{n-k-1},$$

where k is the tree width of G (see Exercise 15.7), and since $k \leq n - \alpha$, the upper bound of $\mu(G)$ in Theorem 15.5.7 follows immediately. For some graphs G, the upper bound of $\mu(G)$ obtained by Theorem 15.5.5 (iii) is in fact better than that obtained by Theorem 15.5.7 (see Exercise 15.8).

We now end this section by listing some known inequalities on chromatic polynomials. Let G be a graph of order n, where $n \geq 2$, x be a vertex in G, e be the base of natural logarithms and $\lambda \in \mathbb{R}$.

(1) (Dong (2000)) If G is connected and x is not a cut-vertex, then for all $\lambda \geq n$,

$$(\lambda - 1)(2\lambda - d(x) - 2)P(G - x, \lambda) \geq (2\lambda - 3)P(G, \lambda). \tag{15.20}$$

(2) (Dong, Teo, Little and Hendy (2001)) If G is not complete, then for all $\lambda \geq n$,

$$\frac{\lambda}{\lambda - (1 - e^{-1})n} < \frac{P(G, \lambda)}{P(G, \lambda - 1)} < \frac{\lambda}{\lambda - n}. \tag{15.21}$$

(3) (Dong, Teo, Little and Hendy (2001)) For all $\lambda \geq n - 1$,

$$\lambda - d(x) \leq \frac{P(G, \lambda)}{P(G - x, \lambda)} \leq \lambda - (1 - e^{-1})d(x). \tag{15.22}$$

(4) (Dong (2003)) For any graphs G and H, if $\lambda > \max\{v(G) - 2, v(H) - 2\}$, then

$$\begin{aligned}
&2(\lambda + 1)P(G, \lambda)P(H, \lambda) \\
\leq\ & (\lambda P(G, \lambda - 1) + P(G, \lambda))\, P(H, \lambda + 1) \\
& + (\lambda P(H, \lambda - 1) + P(H, \lambda))\, P(G, \lambda + 1), \tag{15.23}
\end{aligned}$$

where equality holds if and only if $G \cong H \cong K_n$.

(5) By (15.23), for all $\lambda \geq n - 2$,

$$\begin{aligned}
&(\lambda + 1)(P(G, \lambda))^2 \\
\leq\ & (\lambda P(G, \lambda - 1) + P(G, \lambda))\, P(G, \lambda + 1). \tag{15.24}
\end{aligned}$$

Exercise 15

15.1   Let G be a graph with m edges. Show that

$$(m-3)n(C_3) \geq 2n(K_4).$$

15.2   Let G be a graph of order n with $P(G,\lambda) = \sum\limits_{i=0}^{n-1} a_{n-i}\lambda^{n-i}$. Prove that for any $\lambda \in \mathbb{R}$ with $\lambda \geq n-1$ and any $k \in \mathbb{N}_0$ with $k \leq n-1$,

$$(-1)^k P(G,\lambda) \leq (-1)^k \sum_{i=0}^{k} a_{n-i}\lambda^{n-i}.$$

15.3   Let G be a graph of order n. Prove that for all $\lambda \in \mathbb{N}_0$,

$$P(G,\lambda) \leq \lambda^n,$$

where equality holds for $\lambda \geq 1$ if and only if G is empty.

15.4   Show that for all $\lambda \in \mathbb{N}$ and all graphs $G \in \mathcal{I}$,

$$(P(G,\lambda))^2 \geq P(G,\lambda+1)P(G,\lambda-1).$$

15.5   Let G be the graph obtained from the complete graph K_{2m} by adding two new vertices u and v and adding new edges joining u to any m vertices in K_{2m} and joining v to the other m vertices in K_{2m}. Prove that if $m \geq 2$, then $\mu(G) > \mu(G+uv)$.

15.6   Let G be a graph of order n which is not complete. Show that $P(G,\lambda) > 0$ for all real $\lambda > n-2$.

15.7   Applying Theorem 15.5.5 (iii), show that if G is a graph of order n and tree width k, then

$$\mu(G) \leq n - (n-k-1)^{n-k}/(n-k)^{n-k-1}.$$

15.8   Show that if G is the cycle C_n, then the upper bound of $\mu(G)$ obtained by Theorem 15.5.5(iii) is better than that of $\mu(G)$ obtained by Theorem 15.5.7.

Bibliography

B.D. Acharya (1982), Connected graphs switching equivalent to their iterated line graphs, *Discrete Math.* **41**, 115-122.

A.A. Adam and I. Broere (1993), Chromatic polynomials of graphs in terms of chromatic polynomials of trees, *J. Math. Phys. Sci.* **27**, no. 3, 231-240.

D. Avis, C. De Simone and P. Nobili (2000), Two conjectures on the chromatic polynomials, Lecture Notes, Comput Sci. 1776: 154-162.

D. Avis, C. De Simone and P. Nobili (2002), On the chromatic polynomial of a graph. ISMP 2000, Part 2 (Atlanta, GA), *Math. Program.* **92**, no. 3, Ser. B, 439-452.

X.W. Bao and X.E. Chen (1994), Chromaticity of the graph $\theta(a, b, c, d, e)$ (Chinese, English and Chinese summaries), *J. Xinjiang Univ. Natur. Sci.* **11**, 19-22.

R.A. Bari (1974), Chromatically equivalent graphs, *Graphs and combinatorics* (Proc. Capital Conf., George Washington Univ., Washington, D.C., 1973), pp. 186-200. Lecture Notes in Math, Vol. 406, Springer, Berlin.

R.A. Bari and D.W. Hall (1977), Chromatic polynomials and Whitney's broken circuits, *J. Graph Theory* **1**, no. 3, 269-275.

S. Barnard and J.F. Child, *Higher Algebra,* Macmillan, London, 1955.

S. Barnett (1983), *Polynomials and linear control systems*, Monographs and Textbooks in Pure and Applied Mathematics, 77. Marcel Dekker, Inc., New York.

J.E. Bartels, and D.J.A. Welsh (1995), The Markov chain of colourings, In "Proceedings of the Fourth Conference on Integer Programming and Combinatorial Optimization (IPCO IV)", Lecture Notes in Computer Science, Vol. 920, pp. 373-387, Springer-Verlag, New York/Berlin, 1995.

L .W. Beineke (1971), Derived graphs with derived complements, *Recent Trends in Graph Theory (Proc. Conf., New York, 1970)*, 15-24, Lecture Notes in Mathematics, Vol. 186, Springer, Berlin.

S. Beraha (1975), *Infinite non-trivial families of maps and chromials*, Ph.D. thesis, Johns Hopkins University, 1975.

S. Beraha and J. Kahane (1979), Is the four-color conjecture almost false? *J. Combin. Theory Ser. B* **28**, no. 1, 1-12.

S. Beraha, J. Kahane and R. Reid (1973), B_7 and B_{10} are limit points of chromatic zeros, *Notices Amer. Math. Soc.* **20**, 45.

S. Beraha, J. Kahane and N.J. Weiss (1975), Limits of zeroes of recursively defined polynomials, *Proc. Nat. Acad. Sci. U.S.A.* **72**, no. 11, 4209.

S. Beraha, J. Kahane and N.J. Weiss (1978), Limits of zeros of recursively defined families of polynomials, *Studies in foundations and combinatorics* pp. 213-232, Adv. in Math. Suppl. Stud., 1, Academic Press, New York-London.

S. Beraha, J. Kahane and N.J. Weiss (1980), Limits of chromatic zeros of some families of maps, *J. Combin. Theory Ser. B* **28**, no. 1, 52-65.

H. Bielak (1997), Chromatic uniqueness in a family of 2-connected graphs, *Discrete Math.* **164**, 21-28.

H. Bielak (2001), Roots of chromatic polynomials, *Discrete Math.* **231**, 97-102.

H. Bielak, Chromatic coefficients, *Preprint*.

N.L. Biggs (2002), Chromatic polynomials for twisted bracelets, *Bull. London Math. Soc.* **34**, no. 2, 129-139.

N.L. Biggs, R.M. Damerell, and D.A. Sands (1972), Recursive families of graphs, *J. Combin. Theory Ser. B* **12**, 123-131.

N.L. Biggs, M.H. Klin and P. Reinfeld (2002), Algebraic methods for chromatic polynomials, *Preprint*.

N.L. Biggs, G. H. J. Meredith 1976), Approximations for chromatic polynomials. *J. Combin. Theory Ser. B* **20**, 5-19.

G.D. Birkhoff (1912), A determinant formula for the number of ways of coloring a map, *Annal. Math.* **14**, no. 2, 42-46.

G.D. Birkhoff and D.C. Lewis (1946), Chromatic polynomials, *Trans. Amer. Math. Soc.* **60**, 355-351.

J.A. Bondy and U.S.R. Murty (1976), *Graph theory with applications*, American Elsevier Publishing Co., Inc., New York.

O.V. Borodin and I.G. Dmitriev (1991), Characterization of chromatically rigid polynomials. (Russian) *Sibirsk. Mat. Zh.* **32**, no. 1, 22-27, 219; *translation in Siberian Math. J.* **32**, no. 1, 17-21.

M. Borowiecki and E. Drgas-Burchardt (1993a), Classes of chromatically unique graphs, *Discrete Math.* **111**, 71-75.

M. Borowiecki and E. Drgas-Burchardt (1993b), Classes of chromatically unique or equivalent graphs, *Discrete Math.* **121**, 11-18.

P. Borwein (2002), *Computational Excursions in Analysis and Number Theory*, CMS Books in Mathematics, Springer-Verlag New York, Inc.

K. Braun, M. Kretz, B. Walter and M. Walter (1974), Die chromatischen polynome unterringfreier graphen, *Manuscripa Math.* **14**, 223-234.

F. Brenti (1992), Expansion of chromatic polynomials and log-concavity, *Trans. Amer. Math. Soc.* **332**, 729-755.

F. Brenti, G.F. Royle and D.G. Wagner (1994), Location of zeros of chromatic and related polynomials of graphs, *Canad. J. Math.* **46**, no. 1, 55-80.

J. I. Brown (1998a), Chromatic polynomials and order ideals of monomials, *Discrete Math.* **189**, 43-68.

J. I. Brown (1998b), On the roots of chromatic polynomials, *J. Combin. Theory Ser. B* **72**, 251-256.

J.I. Brown (1999), Subdivisions and chromatic roots. *J. Combin. Theory Ser. B* **76**, 201-204.

J.I. Brown and C.A. Hickman (2002a), On chromatic roots with negative real part. *Ars Combin.* **63**, 211-221.

J.I. Brown and C.A. Hickman (2002b), On chromatic roots of large subdivisions of graphs. *Discrete Math.* **242**, 17-30.

J. I. Brown, C. Hickman, A. D. Sokal and D.G. Wagner (2001), On the chromatic roots of generalized Theta graphs, *J. Combin. Theory Ser. B* **83**, 272-297.

T.H. Brylawski and J.G. Oxley (1992), The Tutte polynomial and its applications, *Matroid Applications*, N. White ed., Cambridge Univ. Press, pp. 123-225.

L. Caccetta and S. Foldes (1997), Symmetric calculation of chromatic polynomials, *Ars Combin.* **4**, 289-292.

S.C. Chang (2001), Chromatic polynomials for lattice strips with cyclic boundary conditions, *Phys. A* **296**, no. 3-4, 495-522.

C.Y. Chao, Z.Y. Guo and N.Z. Li (1992), Some families of chromatically equivalent graphs, *Bull. Malaysian Math. Soc. (2)* **15**, no. 2, 77-82.

C.Y. Chao, Z.Y. Guo and N.Z. Li (1997), On q-graphs, *Discrete Math.* **172**, 9-16.

C.Y. Chao and N.Z. Li (1985), On trees of polygons, *Arch. Math.* **45**, 180-185.

C.Y. Chao, N.Z. Li and S.J. Xu (1986), On *q*-trees, *J. Graph Theory* **10**, no. 1, 129-136.

C.Y. Chao and G. A. Novacky Jr. (1982), On maximally saturated graphs, *Discrete Math.* **41**, no. 2, 139-143.

C.Y. Chao and E.G. Whitehead Jr. (1978), On chromatic equivalence of graphs, *Theory and applications of graphs (Proc. Internat. Conf., Western Mich. Univ., Kalamazoo, Mich., 1976)*, pp. 121-131. Lecture Notes in Math., Vol. 642, Springer, Berlin.

C.Y. Chao and E.G. Whitehead Jr. (1979a), Chromaticity of self-complementary graphs. *Arch. Math. (Basel)* **32**, no. 3, 295-304.

C.Y. Chao and E.G. Whitehead Jr. (1979b), Chromatically unique graphs, *Discrete Math.* **27**, no. 2, 171-177.

C.Y. Chao and L.C. Zhao (1983), Chromatic polynomials of a family of graphs, *Ars Combin.* **15**, 111-129.

Y.M. Chee and G.F. Royle (1990), The chromaticity of a class of K_5-homeomorphs, *Preprint.*

X.B. Chen (1998), Some families of chromatically unique bipartite graphs, *Discrete Math.* **184**, 245-253.

X.E. Chen and X.W. Bao (1994), Chromaticity of the graph $\theta(a, b, c, d, e)$ (Chinese), *J. Xinjiang Univ.* **11**, no. 3, 19-22.

X.E. Chen, X.W. Bao and K.Z. Ouyang (1992), Chromaticity of the graph $\theta(a, b, c, d)$, *J. Shaanxi Normal Univ.* **20**, 75-79.

X.E. Chen and K.Z. Ouyang (1992), Chromaticity of some 2-connected $(n, n+2)$-graphs, *J. Lanzhou University (Natural Sciences)* **28**, 183-184.

X.E. Chen and K.Z. Ouyang (1997a), Chromatic classes of certain 2-connected $(n, n + 2)$-graphs homeomorphic to K_4, *Discrete Math.* **172**, 17-22.

X.E. Chen and K.Z. Ouyang (1997b), Chromatic classes of certain 2-connected $(n, n + 2)$-graphs II, *Discrete Math.* **172**, 31-38.

G.L. Chia (1986), A note on chromatic uniqueness of graphs, *J. Graph Theory* **10**, 541-543.

G.L. Chia (1988a), Some remarks on the chromatic uniqueness of graphs, *Ars Combin.* **26A**, 65-72.

G.L. Chia (1988b), The Petersen graph is uniquely determined by its chromatic polynomial, *Research Report No. 31/88, Department of Mathematics, Univ. Malaya.*

G.L. Chia (1990), The chromaticity of wheels with a missing spoke, *Discrete Math.* **82**, 209-212.

G.L. Chia (1995a), On the join of graphs and chromatic uniqueness, *J. Graph Theory* **19**, 251-261.

G.L. Chia (1995b), On chromatic uniqueness of graphs with connectivity two, *Malaysian J. Science* **16B**, 61-65.

G.L. Chia (1996a), The chromaticity of wheels with a missing spoke, II, *Discrete Math.* **148**, 305-310.

G.L. Chia (1996b), On the chromatic equivalence class of a family of graphs, *Discrete Math.* **162**, 285-289.

G.L. Chia (1997a), Some problems on chromatic polynomials, *Discrete Math.* **172**, 39-44.

G.L. Chia (1997b), A bibliography on chromatic polynomials, *Discrete Math.* **172**, 175-191.

G.L. Chia (1998), On the chromatic equivalence class of graphs, *Discrete Math.* **178**, 15-23.

G.L. Chia, B.H. Goh and K.M. Koh (1988), The chromaticity of some families of complete tripartite graphs, *Scientia, Ser.* **2**, 27-37.

G.L. Chia and C.K. Ho (2001), On the chromatic uniqueness of edge-gluing of complete bipartite graphs and cycles, *Ars Combin.* **60**, 193-199.

G.L. Chia and C.K. Ho (2003), On the chromatic uniqueness of edge-gluing of complete tripartite graphs and cycles, *Bull. Malaysian Math. Sc. Soc. (Second Series)* **26**, 87 - 92.

G.L. Chia and C.K. Ho (2004), A result on chromatic uniqueness of edge-gluing of graphs, *preprint.*

G.L. Chia and C.K. Ho (2005), Chromatic equivalence classes of complete tripartite graphs, *preprint.*

V. Chvátal (1970), A note on coefficients of chromatic polynomials, *J. Combin. Theory* **9**, 95-96.

L.Y. Cui, L.C. Zhao and X.J. Ma (2000), Adjoint uniqueness of three families of connected graphs, *J. Northeastern Univ. (Natural. Sc.)* **21**, no. 4, 218-221.

O.M. Dántona, C. Mereghetti and F. Zamparini (2001), The 224 non-chordal graphs on less than 10 vertices whose chromatic polynomials have no complex roots, *Discrete Math.* **226**, 387-396.

M. Dhurandhar (1984), Characterization of quadratic and cubic σ-polynomials, *J. Combin. Theory Ser. B* **37**, 210-220.

G.A. Dirac (1952), A property of 4-chromatic graphs and some results on critical graphs, *J. London Math. Soc.* **27**, 85-92.

G.A. Dirac (1961), On rigid circuit graphs, *Abh. Math. Sem. Univ. Hamburg* **25**, 71-76.

I.G. Dmitriev (1980), Weakly cyclic graphs with integral chromatic number, (in Russian), *Metody Diskret. Analiz.* **34**, 3-7.

I.G. Dmitriev (1982), Characterization of a class of k-trees, (Russian) *Metody Diskret. Analiz.* No. 38, 9-18.

K. Dohmen (1993a), Chromatische Polynome von Graphen und Hypergraphen, *Dissertation, Universität Düsseldorf.*

K. Dohmen (1993b), Lower bounds and upper bounds for chromatic polynomials. *J. Graph Theory* **17**, no. 1, 75-80.

K. Dohmen (1996), On graph invariants satisfying the deletion-contraction formula, *J. Graph Theory* **21**, no. 3, 311-316.

K. Dohmen (1998), Bounds to the chromatic polynomial of a graph, *Results Math.* **33**, no. 1-2, 87-88.

K. Dohmen (1999), On partial sums of chromatic polynomials. *Ars Combin.* **52**, 125-127.

K. Dohmen, A. Pönitz and P. Tittmann (2002), A new two-variable generalization of the chromatic polynomial, *Preprint.*

F.M. Dong (1990), Uniqueness of the chromatic polynomial of a generalized wheel graph (Chinese), *J. Math. Res. Exposition* **10**, no. 3, 447-454.

F.M. Dong (1993), On chromatic uniqueness of two infinite families of graphs, *J. Graph Theory* **17**, 387-392.

F.M. Dong (1997), *Structures and chromaticity of graphs*, Ph. D. Thesis, Dept. of Math., National University of Singapore.

F.M. Dong (2000), Proof of a chromatic polynomial conjecture, *J. Combin. Theory Ser. B* **78**, no. 1, 35-44.

F.M. Dong (2003), Bounds for mean colour numbers of graphs, *J. Combin. Theory Ser. B* **87**, 348-365.

F.M. Dong (2004), The largest non-integer zero of chromatic polynomials of graphs with fixed order, *Discrete Math.* **282**, 103-112.

F.M. Dong and K.M. Koh (1997a), On graphs in which any pair of colour classes but one induces a tree, *Discrete Math.* **169**, 39-54.

F.M. Dong and K.M. Koh (1997b), On the structure and chromaticity of graphs in which any two colour classes induce a tree, *Discrete Math.* **176**, 97-113.

F.M. Dong and K.M. Koh (1998), Non-chordal graphs having integral-root chromatic polynomials, *Bull. Combin. Appl.* **22**, 67-77.

F.M. Dong and K.M. Koh (1999a), Structures and chromaticity of some extremal 3-colourable graphs, *Discrete Math.* **203**, 71-82.

F.M. Dong and K.M. Koh (1999b), The acyclic colouring, triangle number and chromatic polynomial of a graph, *Algebras and combinatorics* (Hong Kong, 1997), 217-236, Springer, Singapore.

F.M. Dong and K.M. Koh (2004a), On upper bounds for real roots of chromatic polynomials, *Discrete Math.* **282**, 95-101.

F.M. Dong and K.M. Koh (2004b), Two results on real zeros of chromatic polynomials, *Combin. Probab. Comput.* **13**, 809-813.

F.M. Dong and K.M. Koh (2004c), A note on a zero-free interval for chromatic polynomials of bipartite graphs, *preprint.*

F.M. Dong and K.M. Koh (2004d), Domination numbers and zeros of chromatic polynomials, *preprint.*

F.M. Dong and K.M. Koh (2005a), On plane graphs whose chromatic polynomials are zero-free in $(1, 2)$, *preprint.*

F.M. Dong, K.M. Koh and C.A. Soh (2004), Divisibility of certain coefficients of the chromatic polynomials, *Discrete Math.* **275**, 311-317.

F.M. Dong, K.M. Koh and K.L. Teo (2001), Structures and chromaticity of extremal 3-colourable sparse graphs, *Graphs and Combin.* **17**, no. 4, 611-635.

F.M. Dong, K.M. Koh, K.L. Teo, C.H.C. Little and M.D. Hendy (2000a), An attempt to classify bipartite graphs by chromatic polynomials, *Discrete Math.* **222**, 73-88.

F.M. Dong, K.M. Koh, K.L. Teo, C.H.C. Little and M.D. Hendy (2000b), Chromatically unique bipartite graphs with low 3-independent partition numbers, *Discrete Math.* **224**, 107-124.

F.M. Dong, K.M. Koh, K.L. Teo, C.H.C. Little and M.D. Hendy (2001), Sharp bounds for the number of 3-independent partitions and the chromaticity of bipartite graphs, *J. Graph Theory* **37**, no. 1, 48-77.

F.M. Dong and Y.P. Liu (1996), On the chromatic uniqueness of the graph $W(n, n - 2, k)$, *Graphs and Combin.* **12**, no. 3, 221-230.

F.M. Dong and Y.P. Liu (1998), All wheels with two missing consecutive spokes are chromatically unique, *Discrete Math.* **184**, 71-85.

F.M. Dong, Y.P. Liu and K.M. Koh (1997), The chromaticity of odd wheels with a missing spoke, *New Zealand J. Math.* **26**, no. 1, 31-44.

F.M. Dong, K.L. Teo and K.M. Koh (2002), A note on the chromaticity of some 2-connected $(n, n + 3)$-graphs, *Discrete Math.* **243**, 217-221.

F.M. Dong, K.L. Teo, K.M. Koh and M.D. Hendy (2002), Non-chordal graphs having integral-root chromatic polynomials II, *Discrete Math.* **245**, 247-253.

F.M. Dong, K.L. Teo, C.H.C. Little and M.D. Hendy (2001), Some inequalities on chromatic polynomials, *New Zealand J. Math.* **30**, no. 2, 111-118.

F.M. Dong, K.L. Teo, C.H.C. Little and M.D. Hendy (2002a), Two invariants for adjointly equivalent graphs. *Australas. J. of Combin.* **25**, 133-143.

F.M. Dong, K.L. Teo, C.H.C. Little and M.D. Hendy (2002b), Zeros of adjoint polynomials of paths and cycles. *Australas. J. Combin.* **25**, 167-174.

F.M. Dong, K.L. Teo, C.H.C. Little and M.D. Hendy (2002c), Chromaticity of some families of dense graphs, *Discrete Math.* **258**, 303-321.

F.M. Dong, K.L. Teo, C.H.C. Little, M.D. Hendy and K.M. Koh (2004), Chromatically unique multibridge graphs, *Electronic J. of Combin. Theory* **11**, #R12.

Q.Y. Du (1992), Chromatic polynomials of triangulated graphs (Chinese, English summary), *Nei Mongol Daxue Xuebao Ziran Kexue* **23**, 148-151.

Q.Y. Du (1993a), On σ-polynomials and a class of chromatically unique graphs, *Discrete Math.* **115**, 153-165.

Q.Y. Du (1993b), A partial ordering of the σ-polynomials (Chinese, English and Chinese summaries), *Neimenggu Daxue Xuebao Ziran Kexue* **24**, 563-566.

Q.Y. Du (1994), New upper bounds for the coefficients of σ-polynomials (Chinese, English and Chinese summaries), *Neimenggu Daxue Xuebao Ziran Kexue* **25**, 14-16.

Q.Y. Du (1995), On the parameter $\pi(G)$ of graph G and graph classification, *Acta Sc. Naturalium*, Univ. Neimonggol, **26**, no. 3, 258-262.

Q.Y. Du (1996a), Chromaticity of the complements of paths and cycles, *Discrete Math.* **162**, 109-125.

Q.Y. Du (1996b), On σ-equivalence and χ-equivalence of graphs, *J. Graph Theory* **21**, 211-217.

Q.Y. Du (1996c), The σ-polynomials of unlabeled graphs (Chinese), *Neimenggu Daxue Xuebao Ziran Kexue* **27**, no. 5, 601-605.

R.J. Duffin (1965), Topology of series-parallel networks, *J. Math. Anal. Appl.* **10**, 303-318.

B. Eisenberg (1971), Generalized lower bounds for the absolute values of the coefficients of chromatic polynomials, *Lecture Notes in Math.* **186** (Springer), 85-94.

B. Eisenberg (1972), Characterization of a tree by means of coefficients of the chromatic polynomial, *Trans. New York Acad. Sci.* **34**, 146-153.

Y. Fang (1998), The chromatically equivalence classes on the K_r-gluing of wheels W_{n+1} and W_{m+1} and generalized trees (Chinese), *J. Shanghai Teachers Univ. (Natural Sc.)* **27**, no. 3, 28-34.

Y. Fang (2000), The chromatically equivalence classes on the K_r-gluing of two complete graphs K_n and K_{r+2} (Chinese), *J. Shanghai Teachers Univ. (Natural Sc.)* **29**, no. 2, 24-29.

Y. Fang, Y.B. Shi and M.Y. Tang (2002), Classes of chromatically equivalent graphs on the gluing of two wheels, *Preprint*.

G.E. Farr (1993), A correlation inequality involving stable set and chromatic polynomials, *J. Comb. Theo., Series B* **58**, 14-21.

E.J. Farrell (1979), An introduction to matching polynomials, *J. Comb. Theo., Series B* **27**, 75-86.

E.J. Farrell (1980a), Chromatic roots—some observations and conjectures, *Discrete Math.* **29**, no. 2, 161-167.

E.J. Farrell (1980b), On chromatic coefficients, *Discrete Math.* **29**, no. 3, 257-264.

E.J. Farrell (1993), The impact of F-polynomials in graph theory, *Ann. Discrete Math.* **55**, 173-178.

E.J. Farrell (2000), On the derivative of the chromatic polynomial, *Bull. Inst. Combin. Appl.* **29**, 33-38.

E.J. Farrell and E.G. Whitehead Jr. (1990), On matching and chromatic properties of circulants, *J. Combin. Math. Combin. Comput.* **8**, 79-88.

E.J. Farrell and E.G. Whitehead Jr. (1991), Matching, rook and chromatic polynomials and chromatically vector equivalent graphs, *J. Combin. Math. Combin. Comput.* **9**, 107-118

E.J. Farrell and E.G. Whitehead Jr. (1992), Connections between the matching and chromatic polynomials, *Internat. J. Math. Math. Sci.* **15**, no. 4, 757-766.

S. Frank and D. Shier (1986), The chromatic polynomial revisited, Proceedings of the seventeenth Southeastern international conference on combinatorics, graph theory, and computing (Boca Raton, Fla., 1986). *Congr. Numer.* **55** , 57-68.

R.W. Frucht (1985), A new method of computing chromatic polynomials of graphs, *Analysis, geometry, and probability (Valparaso, 1981)*, 69-77, Lecture Notes in Pure and Appl. Math., 96, Dekker, New York.

R.W. Frucht and R.E. Giudici (1983), Some chromatically unique graphs with seven points, *Ars Combin.* **16A**, 161-172

R.W. Frucht and R.E. Giudici (1985), A note on the matching numbers of triangle-free graphs, *J. Graph Theory* **9**, 455-458

D.Fulkerson and O. Gross (1965), Incidence matrices and interval graphs, *Pacific J. Math.* **15**, 835-855.

F.R. Gantmacher (1960), *Matrix Theory, vol. II*, Chelsea, New York.

D. Gernert (1984), Recent results on chromatic polynomials, *Contributions to operations research and mathematical economics*, Vol. I, 307-314, *Methods Oper. Res.*, 51, Athenum/Hain/Hanstein, Knigstein.

D. Gernert (1985), A survey of partial proofs for Read's conjecture and some recent results, *IX symposium on operations research. Part I. Sections 1-4 (Osnabrück, 1984), 233-238,* Methods Oper. Res., 49, Athenäum/Hain/ Hanstein, Königstein.

R.E. Giudici (1982), Formas recursivas para los polinomios cromaticos de los prismas y sus asociados, *Report No. 76, Dpto. de Mat. y Ciencia de la Comp. Univ. Simón Bolivar.*

R.E. Giudici (1985), Some new families of chromatically unique graphs, *Analysis, Geometry, and Probability, Lecture Notes in Pure and Appl. Math.* **96** Dekker, New York, 147-158.

R.E. Giudici (1988), A note on chromatic equivalence of graphs, *Scientia*, Valparaiso, Chile.

R.E. Giudici and Lima de Sá (1990), Chromatic uniqueness of certain bipartite graphs, *Congr. Numer.* **76**, 69-75.

R.E. Giudici and M.A. López (1985), Chromatic uniqueness of $\overline{sK_n}$, *Report No. 85-03, Dpto. de Mat. y Ciencia de la Comp. Univ. Simón Bolivar.*

R.E. Giudici, M.A. López and P.M. Salzberg (1986), Chromatic uniqueness for some bipartite graphs $K_{m,n}$, (Spanish, English summary), *Acta Cient. Venezolana* **37**, 484-494.

R.E. Giudici and C. Margaglio (1988), Chromatically equivalent graphs, *Ars Combin.* **25B**, 221-229.

R.E. Giudici and C. Margaglio (1991), Chromaticity of supercycles of four cells, *Proc. Sixth Caribbean Conf. on Combin. and Comput.* **6**, 185-198.

R.E. Giudici and M.Y. Melián (1988a), Chromatic uniqueness of 3-face graphs, *Report No. 88-05, Dpto. de Mat y Ciencia de la Comp., Univ. Simón Bolivar.*

R.E. Giudici and M.Y. Melián (1988b), Chromatic uniqueness of the supercycle $C(b, a, a, a)$, *Report No. 3-88, Escuela de Matemáticas, Universidad Metropolitana.*

R.E. Giudici and R.M. Vinke (1980), A table of chromatic polynomials, *J. Combin. Inform. and Syst. Sci.* **5**, 323-350.

C.D. Godsil and I. Gutman (1981), On the theory of matching polynomial, *J. Graph Theory* **5**, 137-144

Z. Gracia and P.M. Salzberg (1985), Chromatic classification of $K_p - G_6$. *Ars Combin.* **20**, B, 107-111.

Z.Y. Guo (1997), On T-chromatic uniqueness of graphs, *Discrete Math.* **172**, 45-51.

Z.Y. Guo and Y.J. Li (1989), Chromatic uniqueness of complement of the cycle union, *J. Wuhan Urban Construction Inst.* **6** (No.1), 1-9

Z.Y. Guo and E.G. Whitehead Jr (1997), Chromaticity of a family of K_4 homeomorphs, *Discrete Math.* **172**, 53-58.

P. Gută (1997), The chromaticity of a generalized wheel graph, *Australas. J. of Combin.* **15**, 295-298

G. Haggard and T.R. Mathies (1999), The computation of chromatic polynomials, *Discrete Math.* **199**, 227-231.

G. Haggard and T.R. Mathies (2001), Using thresholds to compute chromatic polynomials, *Ars Combin.* **58**, 85-95.

B.T. Han (1986), Chromaticity of q-trees (Chinese), *Kexue Tongbao* **31** (15), 1200.

B.T. Han (1988), Chromaticity of q_k-trees (Chinese), *Acta Math. Appl. Sinica* **11**, no. 4, 457-467.

S. Hernández and F. Luca (2005), Integer roots chromatic polynomials of non-chordal graphs and the Prouhet-Tarry-Escott problem, *preprint*.

S.G. Hoggar (1974), Chromatic polynomials and logarithmic concavity, *J. Combin. Theory Ser. B* **16**, 226-239.

N. Homobono and C. Peyrat (1989), Graphs such that every two edges are contained in a shortest cycle, *Discrete Math.* **76**, 37-44.

Y. Hong (1984), A note on the coefficients of chromatic polynomials (Chinese), *J. East China Norm. Univ. Natur. Sci. Ed.*, no. 4, 33-35.

S.B. Hu (1998), Chromatic uniqueness of a family of K_4-homeomorphs, *Pure and Appl. Math.* **14**, no. 1, 14-16.

B. Jackson (1993), A zero-free interval for chromatic polynomials of graphs, *Combin. Probab. Comput.* **2**, 325-336.

B. Jackson and A.D. Sokal (2003), Maxmaxflow and Counting Subgraphs, draft.

K.R. James and W. Riha (1975), "Algorithm 24: Algorithm for deriving the chromatic polynomial of a graph", *Computing* **14**, 195-203.

R. Jiang (1999), The chromatic uniqueness of the complements of graphs that are $G = (p, p+1)$ and $R(G) = -2$, *J. Qinghai Normal Univ. (Natural Sc.)*, no. 3, 8-13.

R. Jiang and S.Z. Wang (2001), The chromatic uniqueness of the complements of several classes of graphs that are $G = (p, p+1)$ and $R(G) = -2$, *J. Qinghai Normal Univ. (Natural Sc.)*, no. 4, 5-9.

Z.C. Jin (2001), Chromatic uniqueness of the complement of $\left(\bigcup_{i=1}^{s} a_i D_{3m_i} \right) \cup \left(\bigcup_{j=1}^{t} b_j D_{3n_j+1} \right)$, *J. Qinghai Normal Univ. (Natural Sc.)*, no. 2, 4-8.

K.M. Koh and B.H. Goh (1990), Two classes of chromatically unique graphs, *Discrete Math.* **82**, 13-24.

K.M. Koh and C.P. Teo (1990a), The chromatic uniqueness of graphs related to broken wheels, Research Report No. 413, Department of Math, National University of Singapore.

K.M. Koh and C.P. Teo (1990b), Some results on chromatically unique graphs, *Proc. of Asian Math. Conf.*, 258-262.

K.M. Koh and C.P. Teo (1991a), The chromatic uniqueness of certain broken wheels, *Discrete Math.* **96**, 65-69.

K.M. Koh and C.P. Teo (1991b), Chromatic equivalence of a graph and its complement, *Bull. Inst. Combin. Applications* **3**, 81-82.

K.M. Koh and C.P. Teo (1996), Chromaticity of series-parallel graphs, *Discrete Math.* **154**, 289-295

K.M. Koh and K.L. Teo (1990), The search for chromatically unique graphs, *Graphs and Combin.* **6**, no. 3, 259-285.

K.M. Koh and K.L. Teo (1994), Chromatic classes of 2-connected $(n, n + 3)$-graphs with at least two triangles, *Discrete Math.* **127**, 243-258.

K.M. Koh and K.L. Teo (1997), The search for chromatically unique graphs, II, *Discrete Math.* **172**, 59-78.

R.R. Korfhage (1978), σ-polynomials and graphs coloring, *J. Combin. Theory Ser. B* **24**, 137-153.

R.R. Korfhage (1988), A note on quadratic s-polynomials, *Discrete Math.* **69**, 195-196

A. Kostochka (1982), A lower bound for the Hadwiger number of a graph as a function of the average degree of its vertices, *Discret. Analyz. Novosibirsk* **38**, 37-58.

R.C. Laskar and W.R. Hare (1975), The chromatic polynomial of a complete r-partite graph, *Amer. Math. Monthly* **82**, no. 7, 752-754.

F. Lazebnik (1990), New upper bounds for the greatest number of proper colorings of a (ν, e)-graph, *J. Graph Theory* **14**, no. 1, 25-29.

X.Q. Lan (1991), On the chromaticity of graphs (Chinese), *J. Inner Mongolia Teachers Univ. (Natural Sc.)*, no. 2, 23-28.

N.Z. Li (1985), A criterion for σ-polynomial of a graph, *J. Shanghai Second Polytech university*, no.1, 85-90

N.Z. Li (1987), On the new development of theory of chromatic polynomial of graph, *J. Shanghai Second Polytech. Univ.*, no. 1, 1-9.

N.Z. Li (1988), On coefficients of σ-polynomials of graphs, *J. Shanghai Sec. Polytechnic Univ.* **1**, 1-5.

N.Z. Li (1992a), The chromatic polynomials of disconnected unlabeled graphs (Chinese, English summary), *J. Shanghai Sec. Polytechnic Univ.* **9**, 14-21.

N.Z. Li (1992b), On graphs having σ-polynomials of the same degree, *Discrete Math.* **110**, 185-196.

N.Z. Li (1997), The list of chromatically unique graphs of order seven and eight, *Discrete Math.* **172**, 193-221.

N.Z. Li, X.W. Bao and R.Y. Liu (1997), Chromatic uniqueness of the complements of certain forests, *Discrete Math.* **172**, 79-84.

N.Z. Li and Z.Z. Feng (2002), Counterexamples against some families of chromatically unique graphs, *Discrete Math.* **243**, 257-258.

N.Z. Li and R.Y. Liu (1990), The chromaticity of the complete t-partite graph $K(1, p_2, \cdots, p_t)$, *J. Xinjiang Univ. Natur. Sci.* **7**, 95-96.

N.Z. Li, H.S.H. Sun and I.X. Wen (1995), Properties of chromatic polynomials of unlabeled graphs, *J. Combin. Math. Combin. Comput.* **17**, 21-32.

N.Z. Li and E.G. Whitehead Jr. (1989), Classification of graphs having cubic σ-polynomials, *Graph Theory and Its Applications: East and West, Ann. New York Acad. Sci.* **576**, 328-335.

N.Z. Li and E.G. Whitehead Jr. (1992), The chromatic uniqueness of W_{10}, *Discrete Math.* **104**, 197-199.

N.Z. Li and E.G. Whitehead Jr. (1993), The chromaticity of certain graphs with five triangles, *Discrete Math.* **122**, 365-372.

N. Z. Li, E.G. Whitehead Jr. and S.J. Xu (1987), Classification of chromatically unique graphs having quadratic σ-polynomials, *J. Graph Theory* **11**, 169-176.

W.M. Li (1987), Almost every K_4-homeomorph is chromatically unique, *Ars Combin.* **23**, 13-35.

W.M. Li (1991), Some new results on chromatic uniqueness of K_4 homeomorphs (Chinese, English summary), *Math. Appl.* **4**, 43-47.

W.X. Li and F. Tian (1978), Some problems concerning the chromatic polynomials of graphs (Chinese, English summary), *Acta Math. Sinica* **21**, 223-230.

X.F. Li and X.S. Wei (2001), The chromatic uniqueness of a family of 5-bridge graphs (Chinese), *J. Qinghai Normal Univ.*, no. 2, 12-17.

Y.X. Lin and F.J. Zhang (1989), Two short proofs on chromatic polynomials, *Ars Combin.* **27**, 221-222.

B.L. Liu, Z.H. Zhou and G.Q. Tan (1993), The Akiyama-Harary problem for chromatic polynomials (Chinese, Chinese summary), *Acta Math. Sci. (Chinese)* **13**, 252-255.

R.Y. Liu (1987), A new method to find chromatic polynomial of graph and its applications, *Kexue Tongbao* **32**, 1508-1509.

R.Y. Liu (1990), Adjoint polynomial of graphs (Chinese, English summary), *J. Qinghai Normal University*, 1-9.

R.Y. Liu (1992a), Chromatic uniqueness of $K_n - E(kP_s \cup rP_t)$ (Chinese, English summary), *J. System Sci. Math. Sci.* **12**, 207-214.

R.Y. Liu (1992b), Several results on adjoint polynomials of graphs (Chinese, English), *J. Qinghai Normal University*, 1-6.

R.Y. Liu (1993), Chromatic uniqueness of complementary graph of P_{q-1}, *Pure and Applied Math. Supplement* (2) **9**, 86-87.

R.Y. Liu (1994), Chromatic uniqueness of complements of union of irreducible cycles (Chinese, English and Chinese summaries), *Math. Appl.* **7**, 200-205.

R.Y. Liu (1996), Chromatic uniqueness of a class of complementary graphs of trees (Chinese), *Math. Appl.* **9**, suppl., 170-173.

R.Y. Liu (1997), Adjoint polynomials and chromatically unique graphs, *Discrete Math.* **172**, 85-92.

R.Y. Liu and X.W. Bao (1993), Chromatic uniqueness of the complements of 2-regular graphs, *Pure and Applied Math. Supplement* (2) **9**, 69-71.

R.Y. Liu and J.F. Wang (1992), On chromatic uniqueness of complement of union of cycles and paths (Chinese, English summary), *Theoretical Computer Science* **1**, 112-126.

R.Y. Liu and L.C. Zhao (1997), A new method for proving chromatic uniqueness of graphs, *Discrete Math.* **171**, 169-177.

R.Y. Liu, H.X. Zhao and C.F. Ye (2002), A complete solution to a conjecture on chromatic uniqueness of complete tripartite graphs, *Preprint*.

B. Loerinc (1978), Chromatic uniqueness of the generalized θ-graphs, *Discrete Math.* **23**, 313-316.

B.M. Loerinc (1980), Computing chromatic polynomials for special families of graphs, *Courant Computer Science Report #19*, New York University.

B. Loerinc and E.G. Whitehead Jr. (1981), Chromatic polynomials for regular graphs and modified wheels, *J. Combin. Theory Ser. B* **31**, no. 1, 54-61.

L. Lovász (1993), *Combinatorial problems and exercises*, Second edition, North-Holland Publishing Co., Amsterdam.

Lundow and Markström (2002), *Research Report No 3*, Math. Inst., Umeå Uni., Sweden.

W. Mader (1967), Homomorphieeigenschaften und mittlere Kantendichte von Graphen. *Math. Ann.* **174**, 265-268.

M. Marden (1966), Geometry of polynomials, Second edition, Mathematical Surveys, No. 3 *American Mathematical Society, Providence*, R.I. xiii+243 pp.

G.H.J. Meredith (1972), Coefficients of chromatic polynomials, *J. Combin. Theory Ser. B* **13**, 14-17.

M. Mosca (1998), Removing edges can increase the average number of colors in the colorings of a graph, *Combin. Probab. Comput.* **7**, 211-216.

A. Nijenhuis and H.S. Wilf (1978), Combinatorial algorithms, New York.

B. Omoomi and Y.H. Peng (2001), Chromatic equivalence classes of certain cycles with edges, *Discrete Math.* **232**, 175-183.

B. Omoomi and Y.H. Peng (2003a), Chromatic equivalence classes of certain generalized polygon trees, II, *Ars Combin.* **68**, 115-124.

B. Omoomi and Y.H. Peng (2003b), Chromatic equivalence classes of certain generalized polygon trees, III, *Discrete Math.* **271**, 223-234.

Y.H. Peng (1991), On the chromatic uniqueness of certain bipartite graphs, *Discrete Math.* **94**, no. 2, 129-140.

Y.H. Peng (1992a), On the chromatic coefficients of a bipartite graph, *Ars Combin.* **34**, 107-117.

Y.H. Peng (1992b), Three families of chromatically unique graphs, *Serdica* **18**, no. 1-2, 10-16.

Y.H. Peng (1992c), New infinite families of chromatically unique graphs, *Sains Malaysiana.* **21**, no.4, 15-25.

Y.H. Peng (1993), On the chromatic uniqueness of certain trees of polygons, *J. Australas. Math. Soc. Ser. A* **55**, no. 3, 403-410.

Y.H. Peng (1994), On the chromatic coefficients of a graph and chromatic uniqueness of certain n-partition graphs, in: *Combinatorics, Graph Theory, Algorithms and Applications* (Beijing, 1993) (World Scientific. River Edge, NJ), 307-316.

Y.H. Peng (1995), Another family of χ-unique graphs, *Graphs and Combin.* **11**, 285-291.

Y.H. Peng (1998), Two new classes of chromatically unique K_4 homeomorphs, *Proceedings of the Second Asian Mathematical Conference 1995* (Nakhon Ratchasima), 283-286, World Sci. Publishing, River Edge, NJ.

Y.H.Peng and G.C.Lau (2004), Chromatic Classes of 2-connected $(n, n + 4)$-graphs with at least four triangles, *Discrete Math.* **278**, 209-218.

Y.H. Peng, C.H.C. Little, K.L. Teo and H. Wang (1997), Chromatic equivalence classes of certain genralized polygon trees, *Discrete Math.* **172**, 103-114.

Y.L. Peng (2003), Some new results on chromatic uniqueness of K_4-homeomorphs, *Preprint.*

Y.L. Peng and R.Y. Liu (2002), Chromaticity of a family of K_4-homeomorphs, *Discrete Math.* **258**, 161-177.

M. Petkovšek (1984), Comparability line graphs, *Graph theory* (Novi Sad, 1983), 231-244, Univ. Novi Sad, Novi Sad.

P. Pitteloud (2003), Estimates of coefficients of chromatic polynomials and numbers of cliques of (c, n, m)-graphs, *J. Graph Theory* **42**, 81-94.

R.B. Potts (1952), Some generalized order-disorder transformations, *Math. Proc. Cambridge Philos. Soc.* **48** (1952), 106-109.

A. Procacci, B. Scoppola and V. Gerasimov (2002), Potts model on infinite graphs and the limit of chromatic polynomials. *Preprint.*

R.C. Read (1967), Machine computation of the chromatic polynomials of graphs, with application to a cataloque of 7-node graphs, *University of West Indies, Scientific Report UWI/CC8.*

R.C. Read (1968), An introduction to chromatic polynomials, *J. Combin. Theory* **4**, 52-71.

R.C. Read (1975), Reviewer's remarks, MR50: 6906.

R.C. Read (1978), Some applications of computers in graph theory, in *Selected Topics in Graph Theory,* Academic Press, 417-444.

R.C. Read (1979), Algorithms in graph theory, in *Applications of Graph Theory,* Academic Press, 381-417.

R.C. Read (1981), A large family of chromatic polynomials, *Proc. Caribbean Conf. on Combin. and Computing,* University of the West Indies, Barbados, 23-41.

R.C. Read (1986), Broken wheels are SLC, *Ars Combin.* **21A**, 123-128.

R.C. Read (1987a), An improved method for computing the chromatic polynomials of sparse graphs, *Research Report CORR 87-20, C & O Dept. Univ. of Waterloo.*

R.C. Read (1987b), Connectivity and chromatic uniqueness, *Ars Combin.* **23**, 209-218.

R.C. Read (1987c), On the chromatic properties of graphs up to 10 vertices, *Congr. Numer.* **59**, 243-255.

R.C. Read (1988a), A note on the chromatic uniqueness of W_{10}, *Discrete Math.* **69**, 317.

R.C. Read (1988b), Recent advances in chromatic polynomial theory, *Proc. Fifth Caribbean Conf. on Combin. and Comput. Barbados.* .

R.C. Read (1991), The computation of chromatic polynomials, *Preprint.*

R.C. Read and G.F. Royle (1991), Chromatic roots of families of graphs, *Graph Theory, Combinatorics, and Applications*, 1009-1029.

R.C. Read and W.T. Tutte (1988), Chromatic polynomials, in *Selected Topics in Graph Theory 3*, Academic Press, 15-42.

R.C. Read and E.G. Whitehead Jr. (1999), Chromatic polynomials of homeomorphism classes of graphs, *Discrete Math.* **204**, 337-356.

H.Z. Ren (2002a), The chromaticity of two families of K_4-homeomorphs, *Preprint.*

H.Z. Ren (2002b), On the chromaticity of K_4-homeomorphs, *Discrete Math.* **252**, 247-257.

H.Z. Ren and R.Y. Liu (2001), On chromatic uniqueness of K_4-homeomorphs (Chinese), *J. Math. Study* **34**, no. 1, 94-99.

H.Z. Ren and S.M. Zhang (2001), Chromatic uniqueness of a family of K_4-homeomorphs, *J. Qinghai Normal Univ. (Natural Sc.)*, no. 2, 9-11.

J. Riordan (1958), *An introduction to combinatorial analysis*, Wiley Publications in Mathematical Statistics, John Wiley & Sons, Inc., New York; Chapman & Hall, Ltd., London.

J. Rodriguez and A. Satyanarayana (1997), Chromatic polynomials with least coefficients, *Discrete Math.* **172**, 115-119.

H. A. Roslan (2004), *Chromaticity of certain bipartite graphs*, Ph.D. thesis, Universiti Putra Malaysia.

A. Sakaloglu and A. Satyanarayana (1995), Graphs with least number of colorings, *J. Graph Theory* **19**, no. 4, 523-533.

A. Sakaloglu and A. Satyanarayana (1997), Planar graphs with least chromatic coefficients, *Discrete Math.* **172**, 121-130.

P.M. Salzberg (1984), Chromatic classification of the graphs $K_p - Z$ for $|edges(Z)| \leq 5$, unpublished notes.

P.M. Salzberg, M.A. López and R.E. Giudici (1986), On the chromatic uniqueness of bipartite graphs, *Discrete Math.* **58**, 285-294.

P. Seymour (1997), Two chromatic polynomial conjectures, *J. Combin. Theory Ser. B* **70**, 184-196.

H. Shahmohamad and E.G. Whitehead Jr. (2002), Homeomorphs and amallamorphs of the Petersen graph, *Discrete Math.* **250**, 281-289.

R. Shrock (2001), Chromatic polynomials and their zeros and asymptotic limits for families of graphs, 17th British Combinatorial Conference (Canterbury, 1999), *Discrete Math.* **231**, 421-446.

R. Shrock and S.H. Tsai (1997), Families of graphs with chromatic zeros lying on circles, *Physical Review* **56**, no. 2, 1342-1345.

R. Shrock and S.H. Tsai (1998), Ground state entropy of Potts antiferromagnets: cases with noncompact W boundaries having points at $1/q = 0$, *J. Phys. A: Math. Gen.* **31**, 9641-9665.

R. Shrock and S.H. Tsai (1999), Ground state entropy of Potts antiferromagnets on cyclic polygon chain graphs, *J. Phys. A* **32**, no. 27, 5053-5070.

A. D. Sokal (2000), Chromatic polynomials, potts models and all that, *Physica A* **279**, 324-332.

A. D. Sokal (2001), Bounds on the complex zeros of (Di)chromatic polynomials and potts-model partition functions, *Combin. Probab. Comput.* **10**, 41-77.

A. D. Sokal (2004), Chromatic roots are dense in the whole complex plane, *Combin. Probab. Comput.* **13**, no. 2, 221-261.

R.P. Stanley (1973), Acyclic orientations of graphs, *Discrete Math.* **5**, 171-178.

R.P. Stanley (1998), Graph colorings and related symmetric functions: ideas and applications: a description of results, interesting applications, & notable open problems, Selected papers in honor of Adriano Garsia (Taormina, 1994), *Discrete Math.* **193**, 267-286.

D.M. Strickland (1997), Building fences around the chromatic coefficients, *J. Graph Theory* **26**, 123-128.

J.R. Swenson (1973), The chromatic polynomial of a complete bipartite graph, *Amer. Math. Monthly* **80**, 797-798.

M.Y. Tang (1996), The chromaticity of generalized q-wheels $W(5+q)$ and $W(7+q)$ (Chinese), *J. Shanghai Teachers Univ. (Natural Sc.)* **25**, no. 1, 31-32.

M.Y. Tang (1999), The chromaticity of generalized trees (Chinese), *J. Shanghai Teachers Univ. (Natural Sciences)* **28**, no.3, 21-25.

C.P. Teo and K.M. Koh (1990), The chromaticity of complete bipartite graphs with at most one edge deleted, *J. Graph Theory* **14**, 89-99.

C.P. Teo and K.M. Koh (1992), The number of shortest cycles and the chromatic uniqueness of a graph, *J. Graph Theory* **16**, 7-15.

C.P. Teo and K.M. Koh (1994), On chromatic uniqueness of uniform subdivisions of graphs, *Discrete Math.* **128**, 327-335.

K.L. Teo and K.M. Koh (1991), Chromatic classes of certain 2-connected $(n, n+2)$-graphs, *Ars Combin.* **32**, 65-76.

V. Thier (1983), "Graphen and Polynome", Diploma thesis, TU, München.

A. Thomason (1984), An extremal function for complete subgraphs, *Math. Proc. Camb. Phil. Soc.* **95**, 261-165.

C. Thomassen (1997), The zero-free intervals for chromatic polynomials of graphs, *Combin. Probab. Comput.* **6**, 497-506.

C. Thomassen (2000), Chromatic roots and hamiltonian paths, *J. Combin. Theory Ser. B* **80**, 218-224.

C. Thomassen (2001), Chromatic graph theory, Challenges for the 21st century (Singapore, 2000), 183-195, World Sci. Publishing, River Edge, NJ.

I. Tomescu (1972), Le nombre minimal de colorations d'un graphe, *C.R. Acad. Sci. Paris Sér. A-B* **274**, A539-A542.

I. Tomescu (1985), *Problems in combinatorics and graph theory*, Translated from the Romanian by Robert A. Melter. Wiley-Interscience Series in Discrete Mathematics. A Wiley-Interscience Publication. John Wiley & Sons, Ltd., Chichester.

I. Tomescu (1987), On 3-colorings of bipartite p-threshold graphs, *J. Graph Theory* **11**, 327-338.

I. Tomescu (1989), Some extremal results concerning the number of graph and hypergraph colorings, *Proc. Combinatorics and Graph Theory, Banach Center Publ.* **25**, 187-194.

I. Tomescu (1990), Maximal chromatic polynomials of connected planar graphs, *J. Graph Theory* **14**, 101-110.

I. Tomescu (1994a), On the sum of all distances in chromatic blocks, *J. Graph Theory* **18**, 83-102.

I. Tomescu (1994b), Maximum chromatic polynomials of 2-connected graphs, *J. Graph Theory* **18**, 329-336.

I. Tomescu (1997), Maximum chromatic polynomial of 3-chromatic blocks, *Discrete Math.* **172**, 131-139.

W.T. Tutte (1954), A contribution to the theory of chromatic polynomials, *Canad. J. Math.* **6**, 80-91.

W.T. Tutte (1970), On chromatic polynomials and the golden ratio, *J. Combin. Theory* **9**, 289-296.

W.T. Tutte (1973), Chromatic sums for rooted planar triangulations: the cases $\lambda = 1$ and $\lambda = 2$, *Canad. J. Math.* **25**, 426-447.

W.T. Tutte (1984), *Graph Theory*, Encyclopedia of Mathematics and its Applications, Addison-Wesley, Reading.

P. Vaderlind (1988), Chromaticity of triangulated graphs, *J. Graph Theory* **12**, 245-248.

O.V. Vorodin and I.G. Dmitriev (1991), A characterization of chromatically rigid polynomials, *Sibirskii Math. Zhurnal* **32**, no. 1, 22-27.

D.G. Wagner (2000), Zeros of reliability polynomials and f-vectors of matroids, *Combin. Probab. Comput.* **9**, no. 2, 167-190.

D.R. Wagner (1937), Über eine Erweiterung eines Satzes von Kuratowski, *Deusche Math.* **2**, 280-285.

C.D. Wakelin (1993), The chromatic polynomial relative to the complete graph basis, *Manuscript.*

C.D. Wakelin and D.R. Woodall (1992), Chromatic polynomials, polygon trees, and outerplanar graphs, *J. Graph Theory* **16**, 459-466.

T.R. Walsh (1997), Worst-case analysis of Read's chromatic polynomial algorithm, *Ars Combin.* **46** , 145-151.

L.G. Wang (1997), Chromatic uniqueness of three classes of graphs $K_m - E(G)$, *J. Qinghai Normal University (Natural Sc)*, no. 3, 6-12

L.G. Wang and R.Y. Liu (2001), Chromatic uniqueness for the complement of the union of irreducible t-shape trees (Chinese), *Pure Appl. Math.* **17**, no. 2, 126-132, 137.

T. Wanner (1989), On the chromaticity of certain subgraphs of a q-tree, *J. Graph Theory* **13**, 597-605.

M.E. Watkins (1970), Connectivity of transitive graphs, *J. Combin. Theory* **8**, 23-29.

W. Weickert (1983), Untersuchungen zum chromatischen Polynom anhand spezieller Basisdarstellungen, Diploma Thesis, Techo0 Univ. Munchen.

D.J.A. Welsh (1993), *Complexity: Knots, Colourings, and Counting*, Vol. 186 of *London Mathematical Society Lecture Notes*, Cambridge Univ. Press, Cambridge/New York.

E.G. Whitehead Jr. (1975), Chromatic polynomials for chorded cycles, *Proc. Sixth SEA Conf. on Combinatorics, Graph Theory and Computing, Utilitas Math.*, 619-625.

E.G. Whitehead Jr. (1978), Stirling number identities from chromatic polynomials, *J. Combin. Theory Ser. A* **24**, 314-317.

E.G. Whitehead Jr. (1985), Chromaticity of two-trees, *J. Graph Theory* **9**, 279-284.

E.G. Whitehead Jr. (1988), Chromatic polynomials of generalized trees, *Discrete Math.* **72**, 391-393.

E.G. Whitehead Jr. (1989), Chromatic polynomials and the structure of graphs, *Graph Theory and Its Applications: East and West, Ann. New York Acda. Sci.* **576**, 630-632.

E.G. Whitehead Jr. (1995), An improved algorithm for computing matching polynomials, *Graph theory, combinatorics and algorithms*, Vol. 1, 2 (Kalamazoo, MI, 1992), 1243-1247, Wiley-Intersci. Publ., Wiley, New York.

E.G. Whitehead Jr. (1996), Chromatic equivalent K_4-homeomorphs, *Combinatorics, graph theory and algorithms vol. I, II* (Kalamazoo, MI), 867-872.

E.G. Whitehead Jr. and L.C. Zhao (1984a), Chromatic uniqueness and equivalence of K_4 homeomorphs, *J. Graph Theory* **8**, 355-364.

E.G. Whitehead Jr. and L.C. Zhao (1984b), Cutpoints and the chromatic polynomial, *J. Graph Theory* **8**, 371-377.

H. Whitney (1932a), The coloring of graphs, *Ann. Math.* **(12)** 33, 688-718.

H. Whitney (1932b), A logical expansion in mathematics, *Bull. Amer. Math. Soc.* **38**, 572-579.

D.R. Woodall (1977), Zeros of chromatic polynomials, in *Combinatorial Survey, Proc. Sixth British Combin. Conf.* (ed. P.J. Cameron), Academic Press, 199-223.

D.R. Woodall (1992a), An inequality for chromatic polynomials, *Discrete Math.* **101**, 327-331.

D.R. Woodall (1992b), A zero-free interval for chromatic polynomials, *Discrete Math.* **101**, 333-341.

D.R. Woodall (1997), The largest real zero of the chromatic polynomial, *Discrete Math.* **172**, 141-153.

D.R. Woodall (2002), Tutte polynomial expansions for 2-separable graphs, *Discrete Math.* **247**, 201-213.

J. Xu and H. Li (1992), Chromatic polynomials of connection n-cycle graphs, *J. Northwest Univ.* **22**, 147-152.

J. Xu and H. Li (1994), A new method of calculating chromatic polynomials: a recursive algorithm using vertex contraction (Chinese, English and Chinese summaries), *J. Northwest Univ.* **24**, 103-105.

J. Xu and Z.H. Liu (1995), The chromatic polynomial between graph & its complement— about Akiyama and Harary's open problem, *Graphs and Combin.* **11**, no. 4, 337-345.

S.J. Xu (1983), Complete-graph-basis and the chromaticity of graphs with more edges, *J. Shanghai Teachers College* **3**, 18-23.

S.J. Xu (1984), The chromaticity of regular graphs and bigraphs with more edges, *J. Shanghai Teachers College* **1**, 10-18.

S.J. Xu (1987), Some notes on chromatic uniqueness of graphs (Chinese, English summary), *J. Shanghai Teach. Univ., Nat. Sci. Ed.* **2**, 10-12.

S.J. Xu (1988), On σ-polynomials, *Discrete Math.* **69**, 189-194.

S. J. Xu (1991a), The chromatic uniqueness of complete bipartite graphs, *Discrete Math.* **94**, 153-159.

S. J. Xu (1991b), A lemma in studying chromaticity, *Ars Combin.* **32**, 315-318.

S.J. Xu (1992), Corrigendum, *Discrete Math.* **104**, 217.

S.J. Xu (1993), Chromaticity of a family of K_4-homeomorphs, *Discrete Math.* **117**, 293-297.

S.J. Xu (1994), Classes of chromatically equivalent graphs and polygon trees, *Discrete Math.* **133**, 267-278.

S.J. Xu (1997), Chromaticity of chordal graphs. *Graphs and Combin.* **13**, no. 3, 287-294.

S.J. Xu and N.Z. Li (1984), The chromaticity of wheels, *Discrete Math.* **51**, 207-212.

S.J. Xu, J.J. Liu and Y.H. Peng (1994), The chromaticity of s-bridge graphs and related graphs, *Discrete Math.* **135**, 349-358.

S.J. Xu and J. Z. Zhang (1995), An inverse problem of the weighted shortest path problem, *Japan J. Indust. Appl. Math.* **12**, no. 1, 47-59.

C.F. Ye (2001a), The chromatic uniqueness of 5-bridge graphs, *J. Qinghai Normal Univ.*, no. 3, 1-5.

C.F. Ye (2001b), The chromatic uniqueness of 6-bridge graphs (Chinese), *J. Math. Study.* **34**, no. 4, 399-421.

C.F. Ye (2001c), Graphs characterized by $P(G, \lambda) = \sum \frac{n}{k} \begin{bmatrix} k \\ n-k \end{bmatrix} (\lambda)_{k+l}$ (Chinese), *Pure Appl. Math.* **17**, no. 3, 246-251.

C.F. Ye (2002), The chromatic uniqueness of s-bridge graphs (Chinese), *J. Xinjiang Univ. Natur. Sci.* **19**, no. 3, 261-265.

C.F. Ye and N.Z. Li (2002), Graphs with chromatic polynomial $\sum_{l \leq m_0} \binom{l}{m_0 - l}(\lambda)_l$, *Discrete Math.* **259**, 369-381.

C.F. Ye and R.Y. Liu (1994), Chromatic uniqueness of a new family of graphs (Chinese), *Pure Appl. Math.* **10**, Special Issue, 46-53.

Z.X. Yin (2001), Chromatic uniqueness of a family of K_4-homeomorphs (Chinese), *Northeast. Math. J.* **17**, no. 2, 138-142.

Y.H. Zang (1999), Judgement on chromatic uniqueness of a family of K_4-homeomorphs (Chinese), *J. Ningbo Univ.* **12**, no. 2, 20-23.

Y.H. Zang and B.C. Yuan (1998), Characteristics of a class chromatic equivalence graph (Chinese), *J. Northeast Normal Univ.*, no. 3, 46-48.

B.R. Zhang (1997), Chromatic uniqueness of $K_n - E(k_0 P_3 \bigcup_{i=1}^{r} k_i P_{q_i - 1})$, *Pure and Appl. Math.* **13**, no. 1, 61-67.

S.M. Zhang (2000), The chromatic uniqueness of a kind of K_4-homeomorphs (Chinese), *J. Qinghai Normal Univ. (Natural Sc.)*, no. 1, 12-16.

X.Y. Zhang (1997), Chromatic uniqueness of $K_n - E(kP_5 \cup rK_3)$, *J. Northeast Normal Univ.*, no. 2, 8-14.

H.X. Zhao (1997), Adjoint uniqueness of graphs with $R(G) = -1$, *J. Qinghai Normal Univ. (Natural Sc.)*, no. 4, 6-10.

H.X. Zhao, B.F. Huo and R.Y. Liu (2000), Chromaticity of the complements of paths, *J. Math. Study* **33**, no. 4, 345-353.

H.X. Zhao, X.L. Li, and R.Y. Liu (2003), A solution for some problems and conjectures on adjoint equivalent graphs, *Preprint.*

H.X. Zhao, X.L.Li and R.Y. Liu (2004), Chromaticity of the complements of some sparse graphs, *Preprint.*

H.X. Zhao, X.L. Li, R.Y. Liu and L.S. Wang (2002a), On the chromaticity of 4-partite graphs, *Preprint.*

H.X. Zhao, X.L. Li, R.Y. Liu and L.S. Wang (2002b), On the algebraic properties of the adjoint polynomials and chromaticity of two classes of graphs, *Preprint.*

H.X. Zhao, X.L. Li, R.Y. Liu and C.F. Ye (2004), The chromaticity of certain complete multipartite graphs, *Graphs and Combin.* **20**, 423-434.

H.X. Zhao, X.L, Li, S.G. Zhang and R.Y. Liu (2004), On the minimum real roots of the σ-polynomial and chromatic uniqueness of graphs, *Discrete Math.* **281**, 277-294.

H.X. Zhao and R.Y. Liu (2001a), On the character of the matching polynomial and its application to circuit characterization of graphs, *J. Combin. Math. Combin. Comput.* **37**, 75-86.

H.X. Zhao and R.Y. Liu (2001b), The necessary and sufficient condition of the irreducible paths, *J. Northeast Normal Univ.* **33** (2001), no. 2, 18-21.

H.X. Zhao, R.Y. Liu, X.L. Li and S.G. Zhang (2002), Chromatic uniqueness of the complement of L-shape graphs, *preprint.*

H.X. Zhao, R.Y. Liu and C.F. Ye (2003), On the chromatic uniqueness of tripartite graphs, *preprint.*

H.X. Zhao, R.Y. Liu and S.G. Zhang (2004), Classifications of complete 5-partite graphs and chromaticity of 5-partite graphs with $5n$ vertices, *Appl. Math. J. Chinese Univ. Ser. B* **19**, no. 1, 116-124.

L.C. Zhao (1994), The chromaticity of $K_n - E(C_n)$ and $K_n - E(\sigma_n)$, Unpublished Notes.

H.W. Zou (1998a), The chromaticity of the complete tripartite graph $K(n-4, n, n)$ (Chinese), *J. Shanghai Teachers Univ. (Natural Sc.)* **27**, no. 1, 37-43.

H.W. Zou (1998b), The chromaticity of the complete tripartite graph $K(2, 4, 6)$ (Chinese), *J. Shanghai Teachers Univ. (Natural Sc.)* **27**, no. 4, 14-17.

H.W. Zou (2000a), Classes of chromatically normal graphs of bipartite graphs $K(m, n) - A$ (Chinese), *Jiangxi Science* **18**, no. 2, 63-67.

H.W. Zou (2000b), On the chromatic uniqueness of complete tripartite graphs $K(n_1, n_2, n_3)$, *J. Sys. Sc. and Math. Sc.* **20**, no. 2, 181-186.

H.W. Zou (2002), Chromatic uniqueness of certain bipartite graphs $K(m, n) - A(|A| \geq 2)$ (Chinese), *Tongji Daxue Xuebao Ziran Kexue Ban* **30**, no. 8, 1014-1018.

H.W. Zou (2004), The chromatic uniqueness of certain complete t-partite graphs, *Discrete Math.* **275**, 375-383.

H.W. Zou and Y.B. Shi (1999), On the chromaticity of the complete tripartite graph $K(n-k, n, n)$ (Chinese), *J. Shanghai Teachers Univ. (Natural Sc.)* **28**, no. 1, 15-22.

H.W. Zou and Y.B. Shi (2000), On the chromaticity of the complete tripartite graph $K(n, n, n+k)$ (Chinese), *J. Shanghai Teachers Univ. (Natural Sc.)* **29**, no. 3, 29-35.

A.A. Zykov (1949), On some properties of linear complexes, (Russian) *Mat. Sbornik N.S.* **24** (66), 163-188.

Index